2nd International Conference on Manufacturing Technologies (ICMT 2018)

IOP Conference Series: Materials Science and Engineering
Volume 398

Orlando, Florida, USA
19 – 21 January 2018

ISBN: 978-1-5108-7011-6
ISSN: 1757-8981

Printed from e-media with permission by:

Curran Associates, Inc.
57 Morehouse Lane
Red Hook, NY 12571

Some format issues inherent in the e-media version may also appear in this print version.

Printed by Curran Associates, Inc. (2018)

For permission requests, please contact the Institute of Physics
at the address below.

Institute of Physics
Dirac House, Temple Back
Bristol BS1 6BE UK

Phone: 44 1 17 929 7481
Fax: 44 1 17 920 0979

techtracking@iop.org

Additional copies of this publication are available from:

Curran Associates, Inc.
57 Morehouse Lane
Red Hook, NY 12571 USA
Phone: 845-758-0400
Fax: 845-758-2633
Email: curran@proceedings.com
Web: www.proceedings.com

Table of contents

Volume 398

2nd International Conference on Manufacturing Technologies (ICMT 2018)

19–21 January 2018, Orlando, FL, USA

Accepted papers received: 25 July 2018
Published online: 20 August 2018

Structural Design and Mechanics

Vehicle Engineering

Manufacturing Systems and Industrial Development

Modern Mechanical Engineering and Applications

Material structure design and preparation

IOP Conf. Series: Materials Science and Engineering **398** (2018) 011001 doi:10.1088/1757-899X/398/1/011001

PREFACE

Dear authors from ICMT2018;

On behalf of all conference committees, I sincerely appreciate your participation and support for ICMT2018.

The 2nd International Conference on Manufacturing Technologies was successfully held in Florida Polytechnic University, FL, USA during January 15-17, 2018. Our conference venue, Florida Polytechnic University, is Florida's accredited public university with an exclusive focus on the core STEM disciplines of engineering, technology and applied science. ICMT2018 reflected the technology needs manufacturing technologies science. This proceeding contains the reviewed papers presented at the ICDME 2018 and covers trending and important issues related to manufacturing technologies and other main ICMT Topics.

ICMT2018 conference papers consisted of different types of presentations including keynote, oral, and poster presentations offered by researchers, engineers, and graduate students. ICMT2018 received 97 papers from different counties, only 41 papers were accepted to present at the conference. As a platform for technical exchange, each author has 15 minutes to introduce his or her research. In addition to the presentation, each author is encouraged to connect and exchange ideas with other scholars. The conference is also an opportunity to explore research on manufacturing technologies presented by different participants. With the fast globalization of technology and manufacturing, technical issues will be discussed and disseminated in a global forum in which researchers and engineers with international experiences are invited to share their expertise and views.

The support from all authors and committee members will surely lead to the success of ICMT2018. We hope that the participants of ICMT2018 will have fruitful discussions among each other and collaborate on future projects on manufacturing technologies.

Professor. A. R.Al-Ali

American University of Sharjah, UAE

January 16, 2018

IOP Conf. Series: Materials Science and Engineering **398** (2018) 011002 doi:10.1088/1757-899X/398/1/011002

Conference Chairs

Prof. Ramesh K. Agarwal, Washington University in St. Louis, USA
Prof. Dr. Ridha Ben Mrad, University of Toronto, Canada
Prof. A. R.Al-Ali, American University of Sharjah, UAE

Conference Co-Chairs

Prof. Houssain Kettani, Florida Polytechnic University, USA

Technical Program Committee Chairs

Prof. Yong Suk Yang, Pusan National University, Korea
Prof. Chih-Shing (Stan) Wei, The Cooper Union for the Advancement of Science and Art, USA

International Technical Committees

Prof. Guo Xinli, School of Materials Science and Engineering, Southeast University, Nanjing, China
Prof. TANG Guoyi, Tsinghua University, China
Prof. Md Mustafizur Rahman, Universiti Malaysia Pahang, Malaysia
Prof. Faiz Mohammad, Aligarh Muslim University, India
Prof. K, Lawrence DeVries, The University of Utah, USA
Assoc. Prof. Gang Ke, Guangzhou University, China
Assoc. Prof. Recai KUS, Selcuk University, Turkey
Asst. Prof. Dr. Zhao Ping, University of Minnesota, USA
Dr. Qingshan Mu, University Of Texas Health Science Center At Houston, USA
Dr. Xiaoyi Xu, Stanford university, school of medicine, USA
Dr. Dongsoo Kim, Korea Institute of Materials Science
Zhauyt Algazy, Kazakh National Research Technical University, Kazakhstan
M.P Jahan, Miami University, USA

IOP Conf. Series: Materials Science and Engineering **398** (2018) 011003 doi:10.1088/1757-899X/398/1/011003

Peer review statement

All papers published in this volume of *IOP Conference Series: Materials Science and Engineering* have been peer reviewed through processes administered by the proceedings Editors. Reviews were conducted by expert referees to the professional and scientific standards expected of a proceedings journal published by IOP Publishing.

IOP Conf. Series: Materials Science and Engineering **398** (2018) 012001 doi:10.1088/1757-899X/398/1/012001

Real-time Visualization of Lubricant Films on Head Slider Using an Ellipsometric Microscope

Q Liu and F Yan

C-MEIC & CICAEET, School of Information and Control, Nanjing University of Information Science & Technology, Nanjing, China
E-mail: sherryliuqq@163.com

Abstract. As the recording density increased in hard disk drives (HDDs), the head disk interface (HDI) spacing has been decreased to 2-3 nm during its operation. The head slider may pick up the lubricant at this contact or near-contact HDI. Therefore, a real-time visualization of lubricant thickness distribution on a slider surface is essential to study the physical mechanisms for reducing the head disk interface spacing. A vertical-objective-based ellipsometric microscope (VEM) was presented as the effective method for real-time visualization of nm-thick lubricant films on disks with high lateral resolution. However, extinction coefficient of head slider is low, which means the intensity of reflected light may be reduced. In this report, the feasibility of VEM used for head observation is discussed, and the images of polar perfluoropolyether (PFPE) lubricant Zdol4000 applied on head slider are obtained by the current setup. It indicates that this method can be used in real-time visualization of nm-thick lubricant films on head slider.

1. Introduction

Recently, the study on physical mechanisms of the nm-thick liquid lubricant films plays an important role in the performance of HDDs due to decrease of head-disk interface spacing to 2 to 3nm with the increasing of recording density. The contact or near-contact HDI is needed to increase the recording density. However, lubricant pick-up by a head slider will occur. Therefore, a real-time observation of lubricant thickness distribution on a slider surface is essential to study the physical mechanisms or phenomena for reducing the head disk interface spacing. High lateral resolution is required for this observation because the contact area is small, of the order of 1μm [1-5]. Ellipsometry is widely used for the investigation for thin films [6-8]. Ellipsometric microscopes (EMs), which are based on ellipsometry, can provide thickness distribution as a dark-bright contrast image by using an imaging device as a photo detector. In conventional EMs, a parallel light source is set oblique to the sample with well-defined polarization, as well as the objective lens. Narrow field of view can be involved by this oblique setup when using high lateral resolution lenses. Therefore, a vertical-objective-based EM (VEM) has been improved to solve this problem. In the VEM, the illumination light comes from the lubricant film side and the off-axis light is focused onto the back focal plane of the objective lens to obtain the oblique illumination. This VEM can provide both high lateral resolution and thickness resolution on disks; however it has not been applied to the head slider observation since the low extinction coefficient of head slider may cause extremely low reflectivity. In this report, the feasibility and the improved method of VEM used for head observation is discussed.

2. Setup of VEM

Based on ellipsometry, in EMs, a sample is illuminated by parallel, oblique light with well-defined polarization. The measured quantity is the change of polarization due to reflection of light at the sample. The polarized conditions are different due to different film thicknesses, so on the captured image, the light intensity changes depends on the film thickness. Figure 1 shows the schematic of our VEM. As the oblique illumination is necessary to get ellipsometric contrast in EMs, in our VEM, the objective lenses are set vertically and the off-axis light is focused onto the back focal plane of the objective lens. A commercial microscope was modified and an objective lens at NA = 0.95 and WD = 0.3 mm was used. A LED was used as the light source, with wavelength of λ = 546nm. The light reflected from the sample was imaged through the objective and imaging lenses onto a highly sensitive electron multiplying CCD camera. This type of VEM the has the thickness resolution of about 0.4 nm, which has a large field of view of 160×160 μm^2, and can be used for real-time observation[9-10]. However, it has not been proved whether this type of VEM can be used for head observation.

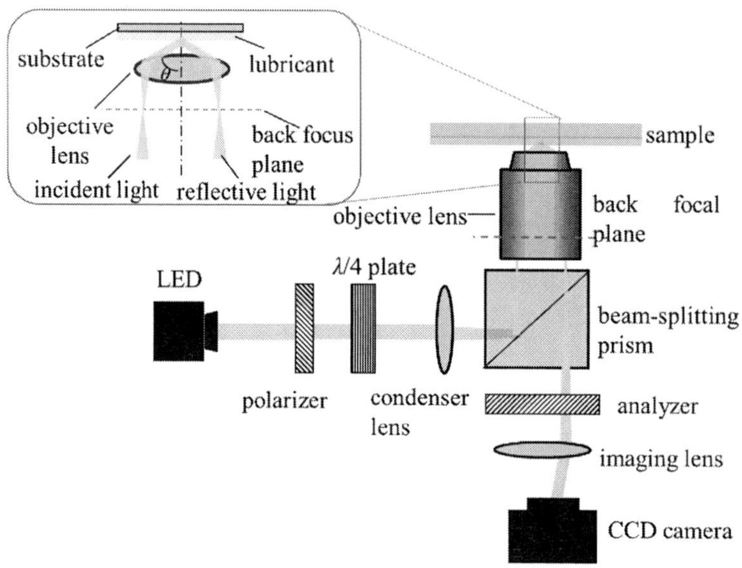

Figure 1. Schematic of vertical-objective-based ellipsometric microscope.

3. Method

A schematic structure of the head with the lubricant film on the slider surface is shown in figure 2. The incident light travels through the lubricant film on the slider surface and is then reflected by the slider surface. Different film thicknesses results in different reflected light intensity which can be detected by the system.

As mentioned before, the measured quantity is the change of polarization due to reflection of light at the sample. The ellipsometric angles ψ and Δ, which are used to describe the change in polarization, are given by

$$\tan\psi \cdot \exp(\mathrm{i}\Delta) = \frac{R_p}{R_s} = \frac{|R_p|}{|R_s|} \cdot \exp[\mathrm{i}(\phi_p - \phi_w)], \quad (1)$$

where R_p and R_s are the complex reflection coefficients of the p- and s-polarized lights, which can be calculated by 2×2 matrix method[11]. 2×2 matrix method is well used in the film thickness measurement, especially for the multilayer structures. When light incidents on the multilayer structure, the change of the reflected light can be decomposed into the effects of the propagation through the interface between different layers and the inside of a layer. Using different 2×2 matrices to describe these effects, the complex reflection coefficients are obtained by the matrix operations.

Moreover, the intensity signal of ellipsometric microscope I is generally given by[12]

IOP Conf. Series: Materials Science and Engineering **398** (2018) 012001 doi:10.1088/1757-899X/398/1/012001

$$I = \frac{R_s^2 I_0}{2} \times \left[r^2 \cos^2 A + 2r \sin A \cos A \sin(2P + \Delta) + \sin^2 A \right], \tag{2}$$

where r is the amplitude ratio for p- and s-polarized lights, I_0 is the intensity of incident light, and P and A are the angles of the polarizer and analyzer.

In our simulation, the layer model of the head slider consisted of air, lubricant film and head slider as a substrate, the refractive indices were 1, 1.3, and 1.98+0.377i, respectively. Moreover, a lubricated magnetic disk for HDDs was used for comparison. The layer model consisted of air, lubricant film and magnetic disk as a substrate, the refractive indices were 1, 1.3, and 2.35+3.93i, respectively. For both simulations, the thicknesses of lubricant and substrate were 1-10 nm and ∞, respectively. The light source with the wavelength λ of 546 nm and the incident angle θ of 60deg was used as well.

4. Results and discussion

4.1. Simulation results

In figure 2, the simulated relationships between ellipsometric intensity I and film thickness h for head and disk are shown, which are in the extinction condition, which means the reflected ellipsometric intensity I = 0, when the film thickness h = 0. It should be noted that the reflected ellipsometric intensity I has been normalized by the intensity of incident light I_0 = 1 in all the simulations. From this simulation, we know both the situations have the same tendency when film thickness comes from 0 to 10 nm; however the amplitude of ellipsometric intensity I of head is about 1/4 of the magnetic disk for the same h. As the thickness resolution is determined by the signal-to-noise ratio, and the signal-to-noise ratio of the observed image is proportional to the square root of the intensity of the illumination light, which means the thickness resolution of head observation is about 2 times of the one of disk observation, theoretically. For example, if the thickness resolution is about 0.4 nm for the disk observation, it is about 0.8 nm for head observation. As the lubricant film is just a few nanometers thick, the thickness resolution of 0.8 nm is low and should be improved.

For 2×2 matrix method, the complex reflection coefficients R_p and R_s are not only the functions of refractive indices n of each layer, but also the incident angle θ and wavelength λ, so we did the simulations for different incident angles and wavelengths.

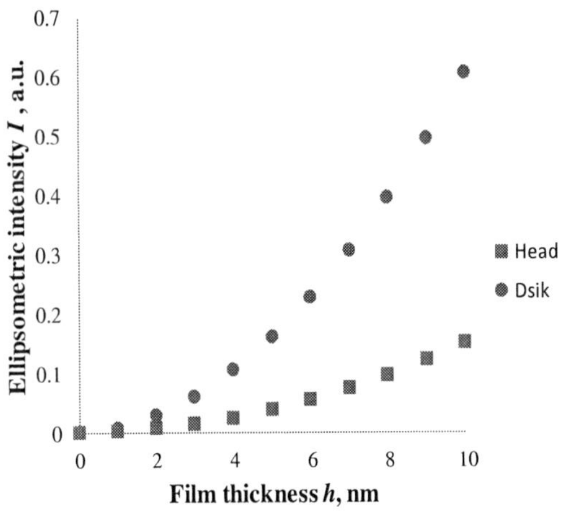

Figure 2. Simulated relationships between ellipsometric intensity I and film thickness h for head and disk (wavelength λ = 546 nm and the incident angle θ= 60deg).

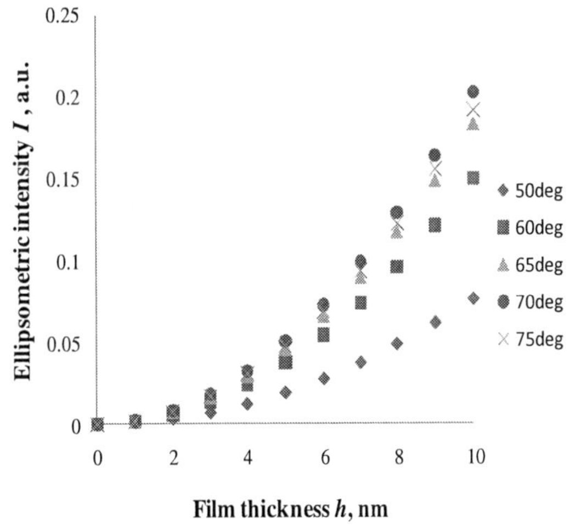

Figure 3. Simulated relationships between ellipsometric intensity I and film thickness h of different incident angles θ for head.

IOP Conf. Series: Materials Science and Engineering **398** (2018) 012001 doi:10.1088/1757-899X/398/1/012001

Figure 3 shows the results of simulation of the relationships between the ellipsometric intensity I and film thickness h for head at different incident angles with wavelengh $\lambda = 546$ nm. This calculation indicates that the ellipsometric intensity varies depending on the incident angle for the each film thickness. It reaches the maximum when the incident angle $\theta = 70$deg.

The simulation of the relationships between the ellipsometric intensity I and film thickness h with different wavelength has been done as well. Commonly used wavelengths of visible light were calculated, and the result at the incident angle $\theta = 70$deg is shown figure 4. For example, the film thickness $h = 3$nm, the difference of ellipsometric intensity I for different wavelengths is shown in figure 5. It is obviously concluded that shorter wavelength leads to larger intensity of the reflected light. The thickness resolution of the VEM is inversely proportional to the square root of the intensity of the illumination light as mentioned before, therefore the thickness resolution can be improved by decreasing the wavelength of the incident light, which means more information can been obtained in the observation. Thus, in our simulation, by choosing wavelength of 385nm and incident angle of 70deg for head observation can get the optimal measurement effect.

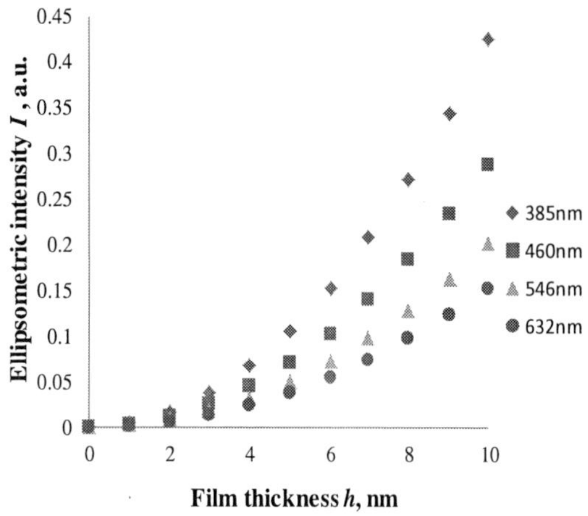

Figure 4. Simulated relationships between ellipsometric intensity I and film thickness h of different wavelength λ for head.

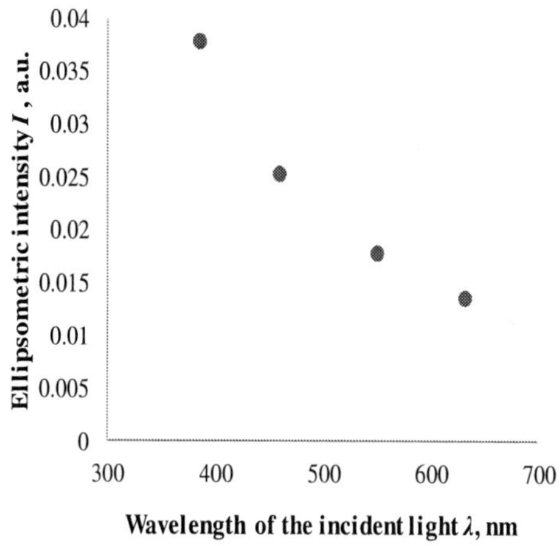

Figure 5. Simulated relationships between ellipsometric intensity I and wavelength of the incident light λ for head (film thickness $h = 3$nm).

Comparing the simulated optimum of head with disk, as shown in figure 6, the amplitude of intensity I of head is improved and becomes about 2/3 of magnetic disk observation for the same film thickness h. That means if the disk observation has the thickness resolution of 0.4nm, it is about 0.5nm for head observation. This means that the improved VEM is feasible if signal-to-noise ratio of I is improved. And if incident light intensity I is improved by 1.5 times, it can get the same thickness resolution as disk observation.

In the previous work, the linear relationship between the ellipsometric intensity I and film thickness h has been obtained by both simulation and experiment method for disk observation [13]. In disk observation, the ellipsometric intensity I is approximately proportional to the film thickness h, when the angle of analyzer A has been set in the extinction condition and the angle of polarizer P has been adjusted. This means that the intensity can be directly converted into film thickness by the VEM, which enables real-time observation. Therefore, in this study the same simulation has been done for head observation. The wavelength of 385nm and incident angle of 70deg has been used for this simulation. The result is shown in figure 7. When the angle of analyzer A has been set at 8.7deg, as well as the angle of polarizer $P = 65$deg, the linear relationship between ellipsometric intensity I and film thickness h for head observation has been obtained. However, the intensity is inversely proportional to the thickness comparing with disk observation.

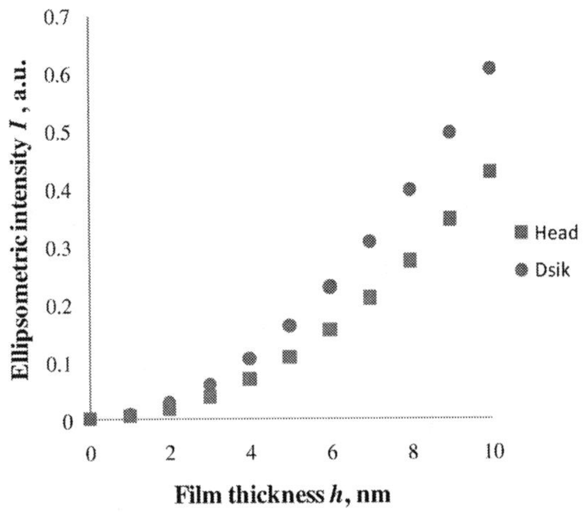

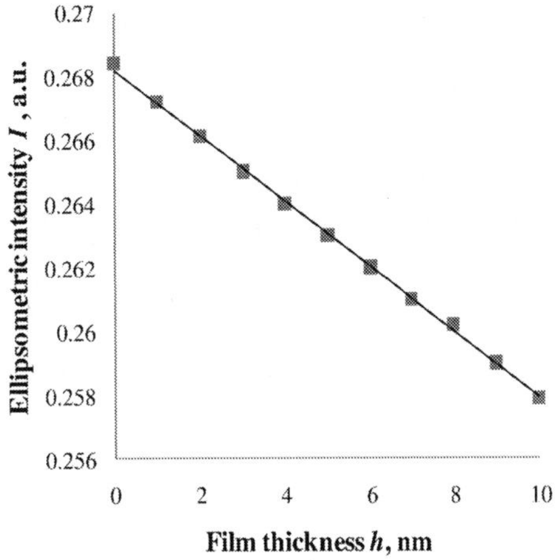

Figure 6. Simulated relationships between ellipsometric intensity I and film thickness h for head (wavelength λ = 385nm and the incident angle θ= 70deg) and disk (wavelength λ = 546nm and the incident angle θ= 60deg).

Figure 7. Linear relationship between intensity I and film thickness h for head(wavelength λ = 385nm and incident angle θ = 70deg).

4.2. Real-time visualization of head slider

Figure 8 shows the dewetting images of the 6.7nm-thick films on head slider observed in real time taken by the VEM. The observation was started at about 5min 30sec after the film applied to the slider. The lubricant in this experiment was Zdol4000 which was polar lubricant. Therefore, the lubricant on the disk was unstable and spontaneously dewetted the slider. The thickness of the lubricant was measured by AFM. The figure shows that dewetting started from the right to the left on the slider, and the part of slider with lubricant films was darker than the bared part, which was consistent with the simulation result. The temporal change of the nm-thick lubricant film on the head slider can by observed by VEM.

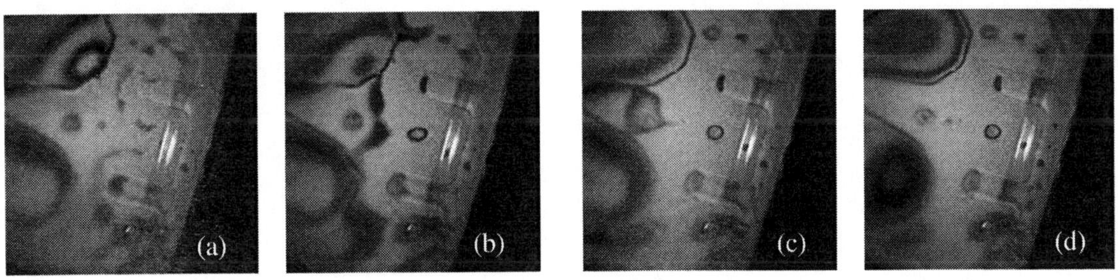

Figure 8. Dewetting images of lubricant film on head slider taken by VEM, (a)-(d) are the images obtained at 5min30s, 5min50s, 6min10s and 6min30s after film applying, respectively (h=6.7nm).

5. Summary

The VEM is expected to be a useful method for real-time observation of the lubrication phenomena and physical mechanisms that uses thin liquid lubricants. However, it has low thickness resolution for head slider observation due to the low extinction coefficient of head slider. An improved VEM is presented, of which the thickness resolution can be increased by choosing the suitable wavelength and incident angle of the incident light and the feasibility of this type of VEM has been proved. The linear relationship between the ellipsometric intensity and film thickness for head observation has been obtained. However, the intensity is inversely proportional to the thickness, which is confirmed by the visualization of lubricant films on head slider by VEM.

6. References

[1] Marchon B, Guo X, Moser A, Spool A, Kroeker R and Crimi F 2009 *J Appl. Phys.* **105** 074313
[2] Mate C, Marchon B, Murthy A and Kim S 2010 *Tribol. Lett.* **37** 581
[3] Tani H, Kubota M, Tsujiguchi Y and Tagawa N 2011 *Microsyst. Technol.* **17** 1175
[4] Vangipuram S Canchi and Bogy D 2010 *IEEE Trans. Magn.* **46** 764
[5] Vangipuram S Canchi and Bogy D 2012 *Microsyst. Technol.* **18** 1517
[6] Henon S and Meunier J 1991 *Rev. Sci. Instrum.* **62** 936
[7] Gang J, Roger J and Hans A 1996 *Rev. Sci. Instrum.* **67** 2930
[8] Linke F and Merkel R 2005 *Rev. Sci. Instrum.* **76** 063701
[9] Liu Q, Fukuzawa K, Kajihara Y, Zhang H and Itoh S 2011 *Tribology Online* **6** 251
[10] Fukuzawa K, Q Liu, Tarukado T, Kajihara Y, Watanabe R, Itoh S and Zhang H 2013 *IEEE Trans. Magn.* **49** 2530
[11] Azzam R M A and Bashara N M 1987 *Ellipsometry and Polarized Light* (Amsterdam: North-Holland) p 332
[12] Fukuzawa K, Noda T and Mitsuya Y 2003 *IEEE Trans. Magn.* **39** 898
[13] Fukuzawa K, Yoshida T, Itoh S and Zhang H 2008 *Langmuir* **24** 11645

Acknowledgments

This project has been supported by National Natural Science Foundation of China (No. 61701244), and the Scientific Research Foundation of Nanjing University of Information Science & Technology (No. 2013r106).

IOP Conf. Series: Materials Science and Engineering **398** (2018) 012002 doi:10.1088/1757-899X/398/1/012002

Research on the Prediction Methods of Air Material Consumption

YU Jin-hao[1], LIU Shen-yang[2], LIU Yan[3], LUO Guang-xu[2]

[1]Department of Air Material Management, Air Force Logistics College, Xuzhou 221000, China
[2]I Department of Aviation Four Stations, Air Force Logistics College, Xuzhou 221000, China
[3]Department of Basic, Air Force Logistics College, Xuzhou 221000, China

Abstract. Through making an analysis of the maintenance methods of air material in some complex equipment, this paper does research on the total air material consumption rule and applies the reliability theory and probability theory to establish air material consumption prediction models. Applicability of the methods is given by way of a numerical example. The methods provide a theoretical basis for calculating reserves of air material scientifically and have the vital important guiding significance.

1. Introduction

In recent years, with the application of high-tech and information technology, equipment of formed units has become more and more complex with more and more species. Different maintenance strategies also tend to be applied for different components of equipment, making it difficult to grasp the law of air material consumption and leading to heavier workload of air material consumption forecasting. To meet the needs of equipment maintenance, it is necessary to store a certain variety and quantity of air material in advance. If the storage capacity of air material is too small, the equipment's successful completion of the training mission cannot be guaranteed; if the storage capacity of each air material is too much, it will cause overstock which affects economic benefit of the components. To ensure that air material stored in an organization is of reasonable quantity and good quality and can timely and reliably guarantee the equipment maintenance needs, a scientific and valid method of equipment air material consumption forecasting must be given.

Many scholars have made scientific researches on air material consumption prediction. Yang Shimei proposes a combination forecasting method of air material based on least squares support vector machines and information entropy in order to achieve precision support of aviation equipment, which solves the problems of the existing methods in difficulty in accurately predicting the aviation material air material under conditions of small samples[1]. What's more, improved models for air material consumption prediction have been developed to avoid the limitations of traditional prediction methods, utilize all the information , and improve the precision of air material consumption prediction[2].

However, through analyzing the scientific researches on air material consumption prediction, we could find that few papers have been published as of now regarding the total air material consumption rule, which is formulated by combining air material consumption rules under both corrective maintenance and condition-based maintenance.

A certain amount of condition-based replaced units should be stocked according to equipment maintenance requirement. If the stock is too small, a completed equipment maintenance task can hardly be guaranteed; if the stock is too big, the caused backlog will affect the economic benefits. Therefore, a good knowledge about the consumption rule of condition-based replaced units as well as a determination of the proper stock of condition-based replaced units are the key subjects in this paper. Based on the abstraction of above problems, a general solution is given to such kind of problems.
This paper presents methods to master the air material consumption law under condition-based maintenance. Then we could predict the number of air material consumption under condition-based maintenance in a period of time and determine a reasonable number of stored air material.
The predictive precision of models applied to predict the air material consumption under condition-based maintenance can be enhanced effectively by using the information of the reliability of the units and the maintenance strategy.

2. The rule of air material consumption

Suppose Suppose the failure density function for the No. i condition-based replacement unit is $f_{T_i}(t)$ and the failure density function for the total service life of k numbers of condition-based replacement units is $f_{\sum_{i=1}^{k} T_i}(t)$ [3]. he service life of a unit may obey the exponential distribution, the gamma distribution, the normal distribution, or the weibull distribution [4]. Figure (1) shows the graph of the exponential distribution, Figure (2) shows the graph of the gamma distribution, Figure (3) shows the graph of the normal distribution, and Figure (4) shows the graph of the weibull distribution [5].

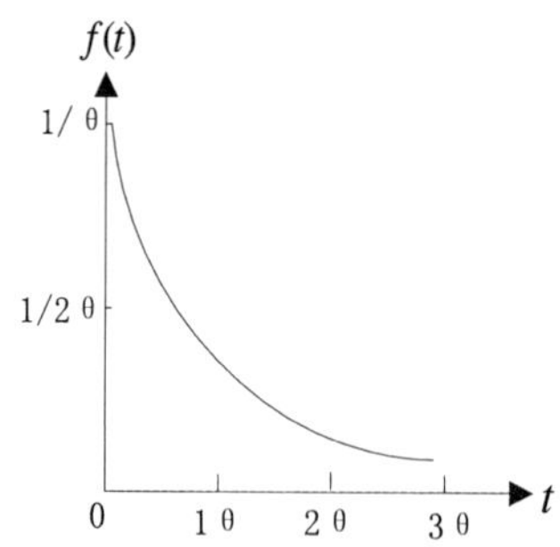

Figure 1. The Graph of the Exponential Distribution

When the service life of the unit obeys the exponential distribution, the fault probability function is

$$f(t) = \lambda e^{-\lambda t} = \frac{1}{\theta} e^{-\frac{1}{\theta} t} \tag{1}$$

In Equation (1), λ is the failure rate, and θ is the average lifespan.

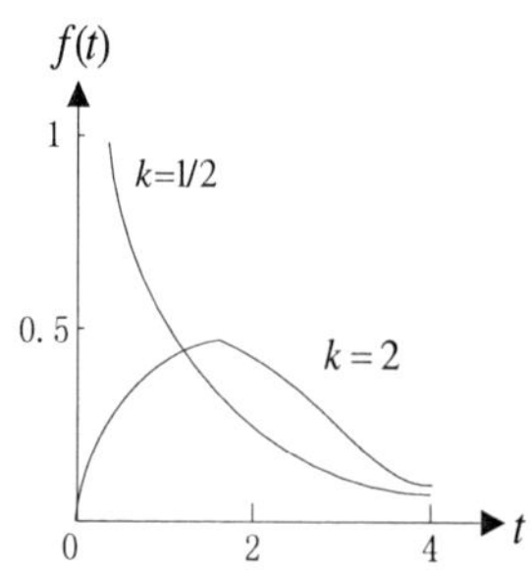

Figure 2. The Graph of the Gamma Distribution

When the service life of the unit obeys the gamma distribution, the fault probability function is

$$f(t)=\frac{\lambda^k}{\Gamma(k)}t^{k-1}\mathrm{e}^{-\lambda t} \tag{2}$$

In Equation (2), k is the shape parameter, and λ is the scale parameter.

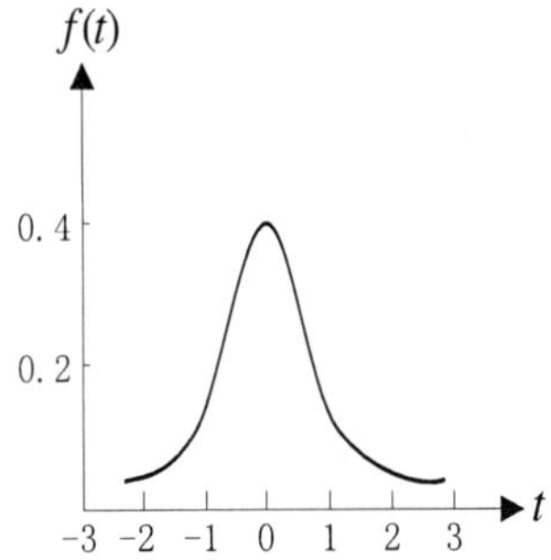

Figure 3. The Graph of the Normal Distribution

When the service life of the unit obeys the normal distribution, the fault probability function is

$$f(t)=\frac{1}{\sqrt{2\pi}\sigma}\mathrm{e}^{-\frac{(t-u)^2}{2\sigma^2}} \tag{3}$$

In Equation (3), u is the average lifespan, and σ^2 is the variance of lifespan.

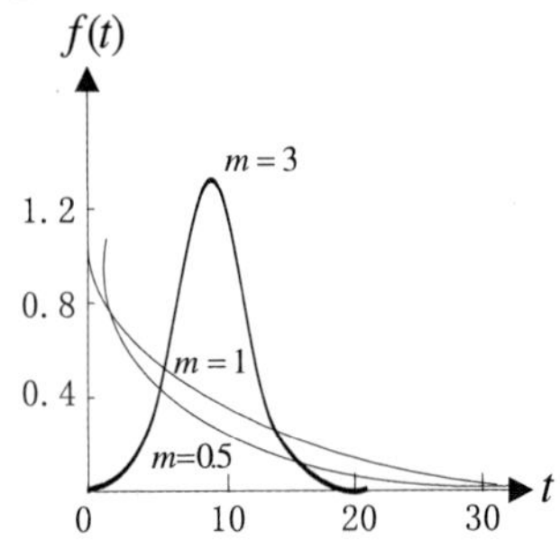

Figure 4. The Graph of the Weibull Distribution

When the service life of the unit obeys the weibull distribution, the fault probability function is

$$f(t)=\frac{m}{\eta}\left(\frac{t-\gamma}{\eta}\right)^{m-1}\mathrm{e}^{-\left(\frac{t-\gamma}{\eta}\right)^m} \tag{4}$$

In Equation (4), m is the shape parameter, η is the scale parameter, and γ is the location parameter. Through a comprehensive reference on air material consumption rule of corrective maintenance within $[0,T)$ and the air material consumption rule of condition-based maintenance at T time, it can obtain the total air material consumption rule within $[0,T]$.

Suppose the air material consumption in corrective maintenance is y_1 and the total air material consumption is y_2.

The probability for air material consumption $y_2=k$ ($k\geq 2$) within $[0,T]$ is

$$\begin{aligned}&P(y_2=k)\\&=P\left(\sum_{i=1}^{k-1}T_i<T-T_0,y_1=k-1\right)+P\left(\sum_{i=1}^{k}T_i>T-T_0,y_1=k\right)\\&=P\left(\sum_{i=1}^{k-1}T_i<T-T_0\middle|y_1=k-1\right)P(y_1=k-1)\end{aligned}$$

IOP Conf. Series: Materials Science and Engineering **398** (2018) 012002 doi:10.1088/1757-899X/398/1/012002

$$
\begin{aligned}
&+P\left(\sum_{i=1}^{k} T_i > T-T_0 \middle| y_1 = k\right) P(y_1 = k) \\
&= \int_0^{T-T_0} \int_{T-\sum_{i=1}^{k-1} t_i}^{+\infty} f_{\sum_{i=1}^{k-1} T_i}\left(\sum_{i=1}^{k-1} t_i\right) f_{T_k}(t_k) \mathrm{d}\left(\sum_{i=1}^{k-1} t_i\right) \mathrm{d}t_k + P\left(\sum_{i=1}^{k} T_i < T\right) \\
&- \int_0^{T-T_0} \int_{T-\sum_{i=1}^{k} t_i}^{+\infty} f_{\sum_{i=1}^{k} T_i}\left(\sum_{i=1}^{k} t_i\right) f_{T_{k+1}}(t_{k+1}) \mathrm{d}\left(\sum_{i=1}^{k} t_i\right) \mathrm{d}t_{k+1} - P\left(\sum_{i=1}^{k+1} T_i < T\right)
\end{aligned} \tag{5}
$$

The probability for air material consumption $y_2 = 1$ within time $[0, T]$ is

$$
\begin{aligned}
&P(y_2 = 1) \\
&= P(T_1 > T, y_1 = 0) + P(T_1 > T - T_0, y_1 = 1) \\
&= P(T_1 > T) + P(T_1 > T - T_0 | y_1 = 1) P(y_1 = 1) \\
&= \int_T^{+\infty} f_{T_1}(t_1) \mathrm{d}t_1 + P(T_1 < T) - P\left(\sum_{i=1}^{2} T_i < T\right) \\
&- \int_0^{T-T_0} \int_{T-t_1}^{+\infty} f_{T_1}(t_1) f_{T_2}(t_2) \mathrm{d}t_1 \mathrm{d}t_2
\end{aligned} \tag{6}
$$

The probability for air material consumption $y_2 = 0$ within time $[0, T]$ is 0.

Combining the average value of air material consumption caused by corrective maintenance within time $[0, T)$ and the average value of air material consumption at T time, the average value of air material consumption within time $[0, T]$ can be obtained:

$$
\begin{aligned}
\bar{y}_2 &= \sum_{k=1}^{+\infty} k \left[P\left(\sum_{i=1}^{k} T_i < T\right) - P\left(\sum_{i=1}^{k+1} T_i < T\right) \right] \\
&+ \int_T^{+\infty} f_{T_1}(t_1) \mathrm{d}t_1 + \sum_{k=1}^{+\infty} \left[\int_0^{T-T_0} \int_{T-\sum_{i=1}^{k} t_i}^{+\infty} f_{\sum_{i=1}^{k} T_i}\left(\sum_{i=1}^{k} t_i\right) f_{T_{k+1}}(t_{k+1}) \mathrm{d}\left(\sum_{i=1}^{k} t_i\right) \mathrm{d}t_{k+1} \right]
\end{aligned} \tag{7}
$$

After learning the consumption rule of condition-based replacement units, a proper reserve plan of condition-based replacement units can be formulated. Firstly based on equation (7), we can calculate the average stock amount of a certain condition-based replacement unit and then according to the equipment quantity and the number of units in one piece of equipment, the optimal stock amount of condition-based replacement units for all the equipment can be determined.

3. Applications

Suppose a certain organization possesses one pieces of equipment, and each piece of equipment is maintained by the combination of periodic maintenance and corrective maintenance. During the normal operation, the service life of a certain unit obeys the gamma distribution, of which the shape parameter is $\alpha = 6$, the scale parameter is $\beta = 0.01/\mathrm{h}$. Predict the spare part consumption amount of all equipment within $[0, 5000\mathrm{h}]$ and determine the stock amount of air material.

The probability for air material consumption $y_2 = k$ ($k = 2,3,4\cdots$) within $[0, 5000\mathrm{h}]$ is

$$
P(y_2 = k) = \int_0^{5000} \frac{0.01^{6(k-1)}}{\Gamma[6(k-1)]} t^{6(k-1)-1} \mathrm{e}^{-0.01t} \mathrm{d}t - \int_0^{5000} \frac{0.01^{6k}}{\Gamma(6k)} t^{6k-1} \mathrm{e}^{-0.01t} \mathrm{d}t
$$

The probability for air material consumption $y_2 = 1$ within $[0, 5000\mathrm{h}]$ is

$$
P(y_2 = 1) = 1 - \int_0^{5000} \frac{0.01^6}{\Gamma(6)} t^{6-1} \mathrm{e}^{-0.01t} \mathrm{d}t
$$

The probability for air material consumption $y_2 = 0$ within $[0, 5000\mathrm{h}]$ is 0.

The distribution of air material consumption is shown in Figure (5).

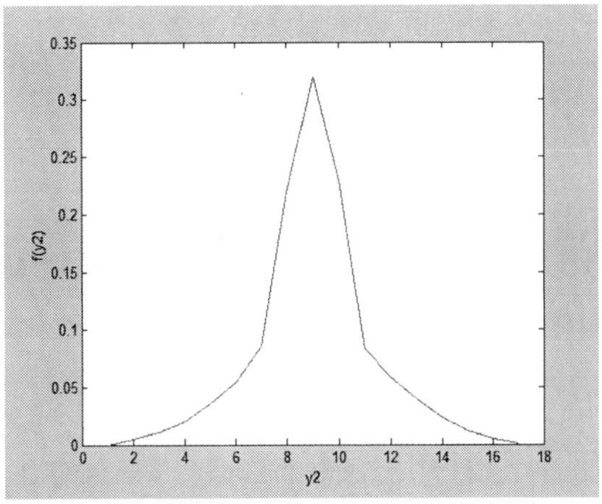

Figure 5. Distribution of Air Material Consumption

Using MATLAB programming software, we can obtain

$$\bar{y}_2 = \sum_{k=1}^{+\infty} \int_0^{5000} \frac{0.01^{6k}}{\Gamma(6k)} t^{6k-1} e^{-0.01t} dt + 1 = 9.10$$

Conclusions

In this paper, through making a analysis of condition-based replacement units of certain complex equipment in a certain organization as well the simplification of related problems, using the reliability theory and probability theory, the air material consumption models under condition-based maintenance can be derived, of which the practicability is verified by an example. The methods proposed in this paper can help equipment management personnel to grasp the air material consumption rule under condition-based maintenance, and to accurately predict air material consumption amount, which provide theoretical basis for proper stock amount of air material.

Acknowledgements

This work is supported by supported by the national natural science foundation of China (No. 71401173).

References

[1] Yang Shimei, Guo Jiansheng and Dong Xinglu 2012 " Combination Prediction Method for Aerial Spare Parts Based on LSSVM and Entropy". *J. Fire Control & Command Control* **37** 154-157

[2] Zhao Jianzhong, Xu Tingxue, Liu Yong and Yin Yantao 2012 "Consumption Forecasting of Missile Spare Parts Based on Rough Set, Entropy Weight and Improved SVM". *J. Acta Armamentarii* **33** 1258-1265

[3] He Shuyuan 2013 Stochastic Processes. *J. Beijing, Peking University Press* **2** 11-12

[4] Liu Jiakun and Wang Gongshu 2012 Applied Stochastic Processes. Beijing. *J. Science Press* **1** 34-39

[5] Sheng Zhou, Xie Shiqian and Pan Chengyi 2013 Probability Theory and Mathematical Statistics. *J. Beijing. Higher Education Press* **2** 11-12

IOP Conf. Series: Materials Science and Engineering **398** (2018) 012003 doi:10.1088/1757-899X/398/1/012003

A Novel Method for Mechanical Constitutive Parameters Estimation of Human Left Ventricle

Z Q Huang [1] **and Y Q Li** [1,2]

[1] Department of Applied Mathematics, Northwestern Polytechnical University,Xi'an,710129, China

[2] NPU-UoG International Cooperative Lab for Computation & Application in Cardiology Northwestern Polytechnical University Xi'an, 710129, China

E-mail: liyiqiang@mail.nwpu.edu.cn

Abstract. In this paper, a new framework for estimation of myocardium constitutive parameters is established. A more realistic, cardiac magnetic resonance image based realistic human left ventricular finite element analysis model is introduced for analyzing the deformation of left ventricle during diastole firstly, the material behavior is described by the anisotropic nonlinear Holzapfel-Ogden constitutive model; then a novel hybrid simplex and quantum behavioral particle swarm optimization algorithm which is proposed to estimate the constitutive model parameters of myocardium as the inverse problem of left ventricle deformation. Numerical results show that finite element analysis results and the estimated parameters are in good agreement with the experimental data reported in literature, which demonstrate that parameter estimation framework is valid. Comparing with current optimization algorithm, the presented hybrid optimal algorithm can estimate the mechanical parameters more efficiently.

1. Introduction

In worldwide, heart disease seriously affects the quality of human beings' life, early diagnosis and treatment can significantly reduce the incidence and fatality rate of heart disease [1]. Mechanical modelling and simulation of cardiac functionality is the main method for people to understand the mechanism of heart disease. Recent studies have shown that organ damage of human heart will cause the changes of myocardial constitutive parameters [2], so estimations of myocardial constitutive parameters have a great meaning for disease diagnosis.

The development of modern medical imaging technology, such as Cardiac Magnetic Resonance (CMR) [1,2], can provide high-quality, high-resolution medical images that accurately describe the anatomical structure and functionality of human heart, it become an important auxiliary method for the diagnosis of heart disease. Based on these real-time medical image data, how to extract the effective measure of heart failure, such as the changes of myocardial parameters, is an important part of the diagnosis of heart disease [2]. Myocardial material is generally considered as a fibre reinforced hyper-elastic material, which has a strong nonlinear stress response. Many constitutive models of myocardium, such as Fung-type, Neo-Hookean, Saint Venant-Kirchhoff [3], and Holzapfel-Ogden constitutive model [4], were developed to simulate the mechanical behaviour of myocardial. Holzapfel-Ogden constitutive model can consider structure of myocardial fibers and lamellar, it can describe mechanical behaviour of myocardial more accurately. Meng et al [2] used data obtained by medical image analysis and Kalman filter method to estimate the two-dimensional linear elastic constitutive myocardial parameters; Guchhait et al. Used Saint Venant-Kirchhoff constitutive model to estimate the mechanical parameters of a two-dimensional model, Aggarwal et al. used the Fung-type constitutive model [5], and estimated the mechanical parameters of the heart valve. These studies are based on simplified mechanical constitutive models or geometric models. Unlike these studies, a three-dimensional finite element model of human left ventricular which is constructed by medical images analysis is used to estimate the mechanical constitutive parameters in this paper.

Given the mechanical parameters of myocardial tissue, studies of the mechanical behaviour of the myocardium under certain mechanical conditions are recognized as a forward problem. While mechanical states of the myocardium are known, parameter identification problem of material is an inverse problem, which is based on the minimization of misfit functional between the observation data and the responses of the computational model under certain material parameters. In order to obtain the parameter estimation, an optimization problem must be solved. The traditional optimization algorithm includes gradient-based optimization algorithm and gradient-free optimization algorithm [6]. Gradient based methods, such as Newton algorithm, LM algorithm, are effective tools to solve optimization problems, these methods usually convergent fast, but these methods does not convergent globally, and the convergence of these methods depends on the selection of initial value. Another drawback of these methods is that calculation of gradient matrix costs huge time, to calculating the gradient, finite difference methods is usually used, it require the evaluate the misfit function many times which require huge amount calculation, especially for the problems whose constitutive model have many material parameters. Gradient-free optimization methods, such as the simplex method [6], is another kind of optimization methods, these methods does not need to the gradient matrix, but they does not convergent globally. In recent years, intelligent based optimization algorithms, such as simulated annealing algorithm, particle swarm algorithm, quantum particle swarm algorithm, genetic algorithm [7,8] are developed. Unlike the traditional algorithms, this kind of optimization algorithms has global convergence, but the speeds of such algorithms are relatively slow [7,8]. In recent years, people are focusing on the studies of hybrid algorithm [4], which can combine the advantages of both intelligent based algorithms and traditional algorithms. In this paper, a novel hybrid simplex and quantum particle swarm algorithm is developed for estimate the parameters efficiently.

The main work of paper is organized as follows: Firstly, a framework of myocardial parameter estimation is established by CMR medical image based finite element analysis model of left ventricular and optimal parameter estimation algorithm. In this framework, the mechanical behaviour of human left ventricular is described by Holzapfel-Ogden constitutive model. Then, a novel hybrid simplex and quantum particle swarm algorithm is introduced to solve the parameter estimation problem. In the end, a numerical example is given to demonstrate the effectiveness and efficiency of the proposed algorithm.

2. Framework of constitutive parameter estimation

2.1. Left ventricular diastolic finite element simulation model

In this section, CMR medical image based finite element analysis model of mechanical behaviours of left ventricle of human heart is described firstly as the forward problem, then the inverse problem for estimating the material parameters is introduced.

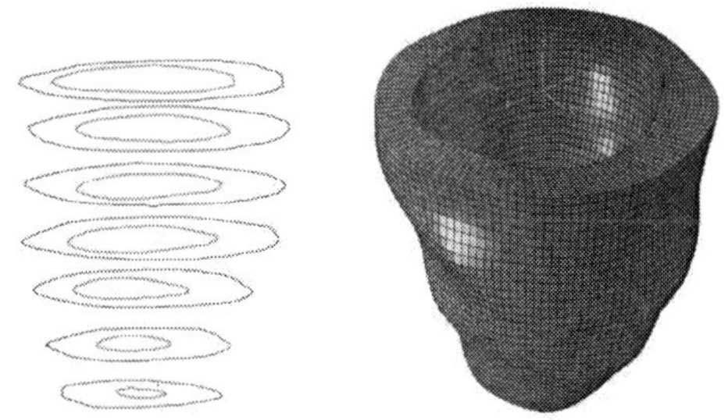

Figure 1. Image segmentation by level set method and the FE mesh of LV [1].

We first establish the geometry of the left ventricle of human heart. Based on early diastolic stage CMR image of human heart, an improved level set image segmentation method is used to construct the inner and outer walls of left ventricle, by using this series of segmentation results, the three-dimensional geometry of left ventricular is reconstructed [1], and then the finite element mesh is generated, as shown in Figure. 1.

Assuming that the deformation of left ventricle during diastole is a quasi-static process, the myocardium is considered as an anisotropic incompressible material. Let $\Omega \subset R^3$ be an open and bounded domain with coordinates X and boundary $\partial\Omega$, occupied by an incompressible hyperplastic body, such as the whole left ventricle. The deformation of left ventricle can be analysed by finite element discretize the virtual work principle, which can be written as, find displacement **u** and hydrostatic pressure p, such as

$$\Phi(\mathbf{u}, p; \delta\mathbf{u}, \delta p) = \int_{\Omega}\left(\frac{\partial\varphi(\mathbf{C},\boldsymbol{\theta})}{\partial\mathbf{F}} + pJ\mathbf{F}^{-T}\right) : \delta\mathbf{u} + p(J-1)\delta p d\Omega - \int_{\partial\Omega_{\text{in}}} p_b J\mathbf{F}^{-T}\cdot\delta\mathbf{u}d\Omega = 0, \tag{1}$$

over the space of admissible displacements and pressures satisfying any given Dirichlet boundary conditions. In equation (1) , J = det **F**, where $\mathbf{F} = \nabla\boldsymbol{x} = Grad\,\boldsymbol{u} + \boldsymbol{I}$ is deformation gradient tensor, $\boldsymbol{I}$ is the identity tensor in R^3 , $C = J^{-\frac{2}{3}}\mathbf{F}^{-\mathrm{T}}\mathbf{F}$ is the volume-preserving right Cauchy–Green deformation tensor, φ is the strain energy density function , $\boldsymbol{\theta} = \left[a, b, a_f, b_f, a_s, b_f, a_{\text{fs}}, b_{\text{fs}}\right]$ is the vector of constitutive parameters of cardiac material, p_b is the pressure on the inner wall of left ventricle.

The incompressible Holzapfel-Ogden constitutive relationship of myocardial tissue is described by the following strain energy density function

$$\varphi(C,\theta) = \frac{a}{2b}e^{b}(I_1 - 3) + \sum_{i=f,s}\frac{a_i}{2b_i}\left(e^{(b_i(I_{4i}-1)^2)-1}\right) + \frac{a_{\text{fs}}}{2b_{\text{fs}}}\left(e^{(b_{\text{fs}}(I_{8fs})^2))} - 1\right), \tag{2}$$

where $I_1 = \text{tr}(C), I_{4f} = f_0\cdot(Cf_0),\ I_{4s} = s_0\cdot(Cs_0), I_{8\text{fs}} = f_0\cdot(Cs_0)$ are the first, fibre, sheet, and fibre sheet invariants of **C** respectively, f_0, s_0 are fibre and sheet directions. After linearizing equation (1) by a finite element discretization, then Newton-Raphson method is used to solve the discretized nonlinear equation. In our numerical experiments, the open sourced package FEniCS [9] is used to implement the finite element analysis procedure.

2.2. *Parameter estimation by inverse approach*

As the constitutive parameters of the myocardial material cannot be measured directly, the estimation of material parameters are an inverse approach of quasi-static nonlinear hyper elasticity deformation by using a given set of partially measured data, at one or more loading situations. A common approach to solve this inverse problem is to minimizing the men squared error between measured and computed responses under constrained or unconstrained optimization framework. Such as the following constraint optimization problem:

$$\begin{aligned}
&\boldsymbol{\theta} = \arg\min_{\theta} f(w(\mathbf{u}, p; \boldsymbol{\theta}), \overline{w}) = \| w(\mathbf{u}, p; \boldsymbol{\theta}) - \overline{w}\|^2,\\
&s.t.\begin{cases}\Phi(\mathbf{u}, p; \delta\mathbf{u}, \delta p; \boldsymbol{\theta}) = 0; \forall\delta\mathbf{u}\in \mathrm{U}, \delta p \in P,\\ \qquad\boldsymbol{\theta}_L \le \boldsymbol{\theta} \le \boldsymbol{\theta}_H,\end{cases}
\end{aligned} \tag{3}$$

where f is the error functional between measured data w and computed responses, the equality constrain Φ is the virtual work principle described in (1), inequality of constrain represents $\boldsymbol{\theta}$ is bounded by a lower bound $\boldsymbol{\theta}_L$ and a higher bound $\boldsymbol{\theta}_H$. For notational convenience, the reduced formulation of the misfit functional is used, which is only depends on the material parameter $\boldsymbol{\theta}$. In particular, we introduce the following reduced optimization problem to estimate the parameter $\boldsymbol{\theta}$,

$$\boldsymbol{\theta} = \arg\min_{\boldsymbol{\theta}} \hat{f}(w(\mathbf{u}(\boldsymbol{\theta}), p(\boldsymbol{\theta}); \boldsymbol{\theta}), \overline{w}) . \tag{4}$$

3. Simplex-QPSO algorithm for parameter estimation

Many optimal algorithms can be used to solving the inverse parameters estimate problem (4), in this work, a novel hybrid simplex and Quantum-Behaved particle swarm algorithm is developed for estimating the parameters. Take parameters $\boldsymbol{\theta}$, as the coordinates of particles, if we choose m particle whose coordinates are denoted as $\theta^L, L = 1,2,\cdots,m$, the best position of m particles is $P^L(k)$, the global best position is denoted as $P^g(k)$,then algorithm 1 is introduced for updating the positions of m particles.

Algorithm 1. (Position updating by quantum-behaved swarm algorithms)

Step 1. Initializing the position of particles;

Step 2. Calculating the average optimal positions of m particles;

$$C = \frac{1}{m}\sum_{L=1}^{m} P^L(k), \tag{5}$$

Step 3. Calculating the fitness of M particles, compare with fitness of last iteration, if fitness $f(\theta^{i+1}) < f(P_i)$, then update the current best position of this particle;

Step 4. Calculating the current global best position of current M particles, so as to find

$$P_g(k+1) = \arg\min\{f(P^L)\}.$$

Comparing with the global best position of last iterations, if current global position is better, the global best position is updated.

Step 5. Updating the new positions of particle by using following equations

$$\boldsymbol{p}_d^L(k) = \phi_d(k)\boldsymbol{P}_d^L(k) + (1-\phi_d(k))\boldsymbol{P}_d^g(k), \tag{6}$$

$$\boldsymbol{\theta}_d^L(k) = \boldsymbol{p}_d^L \pm \alpha(k)\left|\boldsymbol{C}_d(k) - \boldsymbol{\theta}_d^L(k)\right| \times \ln(1/u_d^L(t)) \tag{7}$$

where, ϕ_d, u_d^L are uniform distributed random number on (0,1), the subscript d is denote the d th component of a vector.

Step 6. If the given convergence condition is satisfied, stop this algorithm, otherwise return to step 2.

Algorithm 2: (The hybrid simplex quantum-behaved particle swarm algorithm)

Step 1. Initializing the solver: setting the bounds of parameters, optimal target function and prescribed error limits ε, the positions of initial $3m+1$ particles and current iteration number $n = 1$;

Step 2. Calculating value of function $f_{obj}(\boldsymbol{\theta})$ with at $3m+1$ particles, and sort it by ascending ordering,

$$f_{obj}(\boldsymbol{\theta}_1) \le f_{obj}(\boldsymbol{\theta}_2) \le \cdots \le f_{obj}(\boldsymbol{\theta}_{3m+1}).$$

Step 3. Using simplex algorithm to do iteration with first $m+1$ optimal particle, update the positions of those particles;

Step 4. Using algorithm 1 to update rest 2m particles' positions;

Step 5. If the given convergence condition is satisfied, terminate this algorithm, otherwise let $n = n+1$ and return to step 2.

4. Numerical examples

In this section, a numerical example is presented for validating the proposed algorithm, which is used to estimate constitutive parameters of human left Ventricle. The constitutive parameters of the myocardial material which is used in Ref [10], is taken as the true value of our numerical example, as shown in Table II. The geometry and finite element mesh of human left ventricular is generated based on the CMR medical images and level set method, which is described in this article. The ventricular inner wall pressure is taken as 8 mmHg. By using the finite element formulation of deformations of left ventricular in section 2, a convergence result is obtained. The results of the finite element analysis are used as the observed data for parameter estimation to verify the correctness of the proposed algorithm to estimate material parameters.

To estimation of myocardial parameters, the objective functional is constructed by using the measured strain data and the ventricular volume data, which can be written as following,

$$f_{\mathrm{obj}}(\varepsilon,V)=\sum_{k}\sum_{i,j}\left|1-\varepsilon_{\mathrm{ij}}^{\ k}/\bar{\varepsilon}_{ij}^{\ k}\right|+\left|1-V/\bar{V}\right|, \tag{8}$$

where $\bar{\varepsilon}_{ij}^{\ k}$ are the measured strains at the kth observation point measurements, $\bar{V}$ is the measured ventricular volume. To obtain the estimations parameters estimates the optimization problem (4) is solved by the proposed optimal algorithm. The Levenberg-Marquardt (LM) algorithm [4] is used to solve problem (4) as comparison of the proposed algorithm. The parameter estimation results are shown in Table 1, the parameters a,b,a_f,b_{fs} obtained by the LM algorithm are less than the true value, the parameters b_f,a_s,a_{fs} are bigger than the true value, the estimation errors of b_s are quite large. Comparing with the results of LM algorithm, the results of proposed algorithm has smaller error with parameter $a,b,a_f,b_f,a_s,a_{fs},b_{fs}$, but the estimation of b_s also has a little difference with the true value. Figure 2 shows the pressure volume curves which is obtained by using the two sets of estimated parameter. It is shown that the two sets of estimated parameter are valid estimates of the left ventricular mechanical parameters.

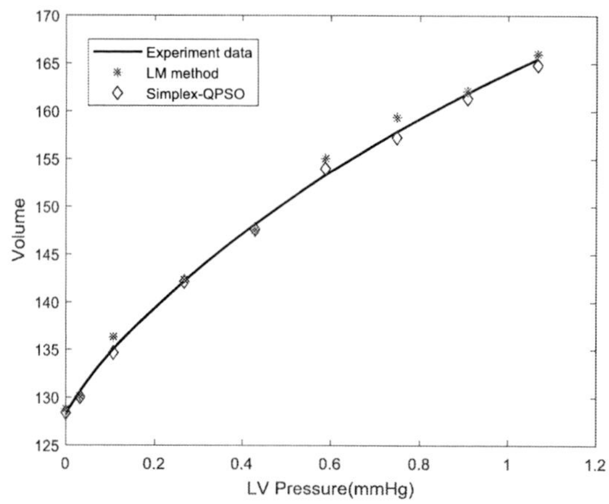

Figure 2. Ventricular pressure volume curve.

The computational efficiency of proposed algorithm is shown in Table 2. Comparing with LM algorithm, both algorithms require a lot of finite element computation, but the proposed algorithm requires less computation time than LM algorithm.

Table 1. Resaults of Parameter Estimation

	a	b	a_f	b_f	a_s	b_s	a_{fs}	b_{fs}
True value[10]	0.236	10.810	20.037	14.154	3.725	5.165	0.411	11.300
LM algorithm	0.238	10.677	18.980	18.459	4.083	3.317	0.433	10.290
HS-QPSO	0.235	10.829	20.015	14.217	3.875	1.900	0.410	11.330

Table 2. Efficiency Comparison of Two Algorithms

	Iterations	Time	Error limit
LM algorithm	773	18h	1e-6

	Iterations	Time	Error limit
HS-QPSO	482	12h	1e-6

5. Conclusions

In this study, a novel framework is established for estimation of myocardium constitutive parameters; this framework contains a finite element analysis model of realistic human left ventricular and a new hybrid simplex and quantum behaviour particle swarm optimization algorithm which is proposed to estimate the constitutive model parameters of myocardium. The realistic geometry of human left ventricular is constructed by using cardiac magnetic resonance image (CMR) and level set method, then mechanical behaviour of human left ventricular during diastolic stage is analysed by finite element method, we consider the left ventricle as an anisotropic nonlinear material, which is described by Holzapfel-Ogden constitutive model. Using this simulation model and the observation data, estimation of mechanical parameters of left ventricular myocardial materials is taken as an inverse problem of finite element analysis. A novel hybrid quantum particle swarm and simplex algorithm is proposed for estimation of mechanical parameters; this approach directly uses the observed data in medical images, and uses the hybrid quantum behaviour particle swam with global convergence. The numerical analysis shows that the method can effectively estimate the constitutive parameters of myocardial materials; this approach provides an effective method for the diagnosis of heart disease.

6. Acknowledgment

This work is funded by the National Natural Science Foundation of China(11471261, 11101333), the Natural Science Foundation of Shaanxi (2017JM1005) and NPU Foundation for Fundamental Research, and also supported by Center for high performance computing of Northwestern Polytechnical University.

7. References

[1] Cai L, Gao H, Luo X Y and Nie Y F 2015 Multi-scale modelling of the human left ventricle *Scientia Sinica Physica, Mechanica and Astronomica* **45**

[2] Meng J, Liu H-F, Yue M-X and Hu H-J 2012 Biomechanical model guided dual estimation of myocardial motion and material parameters *Journal of Zhejiang University. Engineering Science* **46** 912–917

[3] Guchhait S and Banerjee B 2015 Constitutive error based material parameter estimation procedure for hyperelastic material *Computer Methods in Applied Mechanics and Engineering* **297** 455–475

[4] Holzapfel G A and Ogden R W 2009 Constitutive modelling of passive myocardium: a structurally based framework for material characterization *Philosophical Transactions of the Royal Society of London A: Mathematical, Physical and Engineering Sciences* **367** 3445–3475

[5] Aggarwal A and Sacks M S 2014 Parameter estimation of heart valve leaflet hyperelastic mechanical behavior using an inverse modeling approach *Bioengineering Conference (NEBEC), 2014 40th Annual Northeast* (IEEE) pp 1–2

[6] Werner J 2013 *Optimization theory and applications* (Springer-Verlag)

[7] Shi X H, Liang Y C, Lee H P, Lu C and Wang L M 2005 An improved GA and a novel PSO-GA-based hybrid algorithm *Information Processing Letters* **93** 255–261

[8] Fan S-K S and Zahara E 2007 A hybrid simplex search and particle swarm optimization for unconstrained optimization *European Journal of Operational Research* **181** 527–548

[9] Alnaes M, Blechta J, Hake J, Johansson A, Kehlet B, Logg A, Richardson C, Ring J, Rognes M E and Wells G N 2015 The FEniCS project version 1.5 *Archive of Numerical Software* **3** 9–23

[10] Wang H M, Luo X Y, Gao H, Ogden R W, Griffith B E, Berry C and Wang T J 2014 A modified Holzapfel-Ogden law for a residually stressed finite strain model of the human left ventricle in diastole *Biomechanics and modeling in mechanobiology* **13** 99–113

IOP Conf. Series: Materials Science and Engineering **398** (2018) 012004 doi:10.1088/1757-899X/398/1/012004

Investigation on Sound Absorbing Performance of the Polyester Fiber for Noise Reduction in Large-scale Equipment

Panfeng Bai[1], Xiaocui Yang[1], Xinmin Shen[1, 2, *], Xiaonan Zhang[1], Jingwei Zhu[1], Qin Yin[1], Zhizhong Li[1], Chao Wang[1], and Lei Xu[1]

[1]College of Field Engineering, Army Engineering University, No. 1 Haifu Street, Nanjing, Jiangsu, P. R. China
[2]Department of Industrial and Systems Engineering, The Hong Kong Polytechnic University, Hung Hom, Kowloon, Hong Kong SAR, P. R. China

Email: shenxmjfjlgdx2014@163.com

Abstract. Polyester fiber was attempted to reduce the noise in large-scale equipment. Sound absorbing performance of the polyester fiber with different parameter was investigated based on the standing wave method. Experimental result showed that the sound absorbing coefficients were ameliorated along with the increases of thickness of the polyester fiber. Meanwhile, for the same thickness, the corresponding sound absorbing coefficient of polyester fiber with density of 0.3g/cm^3 was better that of 0.2 g/cm^3. Furthermore, it could be observed that sound absorbing coefficients were almost increasing linearly along with increase of frequency when thickness of the polyester fiber was low. It would be found that there existed peak of the sound absorbing coefficients when the thickness was large. Especially when the thickness reached 24mm, the sound absorbing coefficients in the frequency range of 2000Hz ~ 6000Hz exceeded 90%. The excellent sound absorbing performances of the polyester fiber would be propitious to reduce noise in large-scale equipment.

1. Introduction

Noise reduction is a significant problem in the large-scale equipment, which affects its viability and working efficiency, especially for the submarine and surface warship [1]. For the noise reduction in the cabin, polyester fiber is considered as a promising sound absorbing material [2, 3], because it has the excellent sound absorbing coefficient and can be installed on the inner wall. However, with the different parameters, such as different density or different thickness, sound absorbing performance of the polyester fiber is varied [4]. Meanwhile, sound absorbing coefficient of polyester fiber corresponds to the sound frequency [5]. Therefore, the investigation on sound absorbing performances of the polyester fibers with different parameters is necessary and significative.

Two kinds of polyester fibers with densities of 0.2g/cm^3 and 0.3g/cm^3 were investigated in this study. For each kind of polyester fiber, the investigated thicknesses of the samples were 8mm, 16mm, and 24mm, respectively. Meanwhile, the sound absorbing coefficients of each sample corresponding to the different sound frequencies were achieved according to the standing wave method [6]. Through comparing sound absorbing performances of the polyester fibers with different parameters, the optimal option for the noise reduction in a certain frequency range was achieved.

2. Experimental apparatus

Sound absorbing coefficients of the polyester fibers were detected by the AWA6128A detector, as shown in Figure 1. In order to meet the requirement of the detector, dimensions of the samples were set to ϕ 96mm for the detection in the low sound frequency range and ϕ 30mm for the detection in the high sound frequency range, which were realized by the laser cutting technique [7].

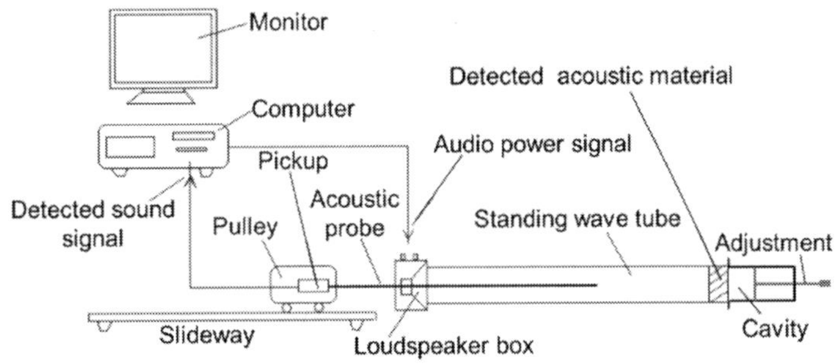

Figure 1. Schematic diagram of the experimental apparatus

The obtained data of the detector were peak sound level L_{max} and the corresponding valley sound level L_{min} of the detected sample for the certain frequency, and the corresponding sound absorbing coefficient α can be calculated by Eq. 1.

$$\alpha = \left(4\times 10^{(L_{\max}-L_{\min})/20}\right) \Big/ \left(1+10^{(L_{\max}-L_{\min})/20}\right)^2 \quad (1)$$

3. Results and discussions

The summarized data of sound absorbing coefficients for the polyester fibers with different parameters were shown in Table. 1. The detected low frequency of each sample included 100Hz, 200Hz, 300Hz, 400Hz, 500Hz, 600Hz, 700Hz, 800Hz, 950Hz, 1100Hz, 1300Hz, 1500Hz, 1700Hz, and 1900Hz, and the detected high frequency included 2000Hz, 2300Hz, 2600Hz, 2900Hz, 3200Hz, 3500Hz, 3800Hz, 4100Hz, 4400Hz, 4700Hz, 5000Hz, 5300Hz, 5600Hz, and 6000Hz, which were conducted according to instruction of the detector. It could be easily observed that for different densities and thicknesses, distribution of sound absorbing coefficients corresponding to varied sound frequency was distinct.

Table 1. Summarized Data of Sound Absorbing Coefficients.

Density	$0.2g/cm^3$			$0.3g/cm^3$		
Thickness / **Frequency**	8mm	16mm	24mm	8mm	16mm	24mm
100Hz	3.2	3.8	4.5	3	3.6	4.3
200Hz	2.7	4.5	7	2.4	3.9	7.3
300Hz	4	6.5	11.8	3.1	7.5	12.2
400Hz	4.8	9.3	17.7	4.7	11.3	23
500Hz	7.7	12.6	25.5	5.8	15.4	32.4
600Hz	8.3	16.2	30.6	12.4	23.2	40.8
700Hz	9.2	19.4	39	7.8	25.2	50.5
800Hz	9.9	24.2	47	10.3	30.5	59.3
950Hz	12.8	29	56	12.4	37.9	68.8
1100Hz	8.3	30.5	61.9	9.1	47.8	76.3
1300Hz	18.8	43.3	75.7	22.5	59.7	84.5
1500Hz	19.5	51.9	83.2	22	67	89.7
1700Hz	22.5	58.7	89.9	25.8	73.6	93.3
1900Hz	24.5	66.2	94.4	30.2	82	94.4
2000Hz	28.7	74.3	98.7	40.5	82.1	98.7

2300Hz	33.6	80	99.7	41.3	90.1	96.9
2600Hz	38.2	87.3	98.8	46.9	93.9	96.3
2900Hz	42.9	92.3	98	55.1	98.2	93.5
3200Hz	48.5	94.7	94.9	57.3	99.1	92.2
3500Hz	54.2	97.9	93.2	65.7	100	91.3
3800Hz	57.6	97.6	90.5	68.4	99.7	90.2
4100Hz	65.5	97.8	89.4	73.7	99.5	92.2
4400Hz	65.7	97.5	89.6	77.9	97.9	90
4700Hz	69.7	95.7	88.3	81.7	97.5	92.6
5000Hz	72.3	95.4	89.6	86.2	95.3	93
5300Hz	75.8	92.6	90.4	87.2	94.7	94.5
5600Hz	79.7	93.6	93.6	90.6	93.4	95.8
6000Hz	81.5	90.6	94.9	94.6	91.6	96.3

In order to intuitionally exhibit the evolution principle of the acoustic performances along with the increase of thicknesses, the sound absorbing coefficients for the three samples with the density of 0.2g/cm^3 and those with the density of 0.2g/cm^3 were shown in Figure. 2. It could be observed that along with the increase of thicknesses of the sample, sound absorbing coefficient for each frequency was entirely improved. However, along with the increase of frequency, sound absorbing coefficient was not increasing by degrees.

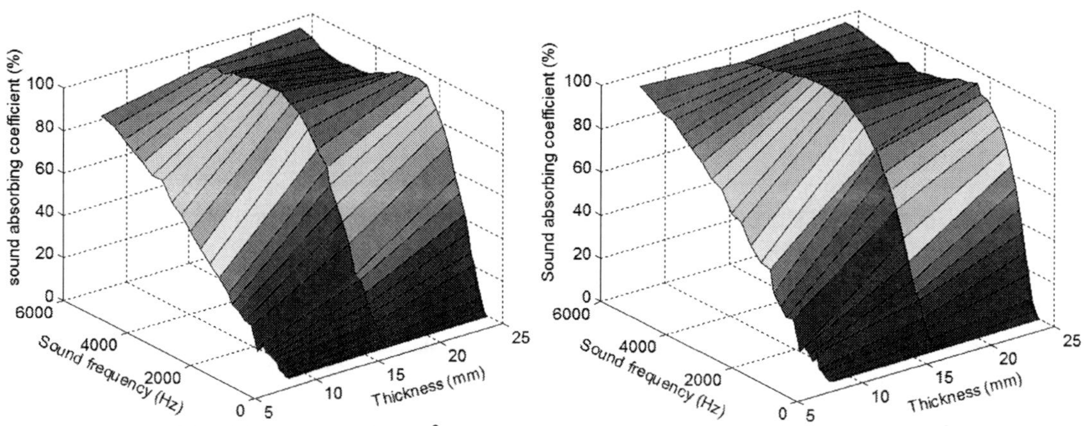

(a) with density 0.2g/cm^3 (b) with density 0.3g/cm^3

Figure 2. Sound absorbing coefficients for samples

For the purpose of the further quantitative analysis of the sound absorbing coefficient, the whole data was summarized in Figure 3. It could be found that thicknesses of the polyester fibers had a great influence to sound absorbing coefficients. Along with the increase of thickness, the sound absorbing coefficients in the low frequency range was raised, and that in the high frequency range was steadier, which was suitable for the samples with two densities. It had been considered that the sound wave with a high frequency was primarily absorbed by the surface of the sound absorbent, and that with a low frequency was mainly absorbed by the inner structures of the polyester fiber or porous material [8]. Meanwhile, it could be observed that the sound absorbing coefficient of the polyester fibers with density of 0.3 g/cm^3 was better than that of the polyester fibers with density of 0.2 g/cm^3.

It could be observed that the sound absorbing coefficients of polyester fiber were almost increasing linearly along with increase of frequency when thickness of the polyester fiber was 8mm, no matter the density was 0.2g/cm^3 or 0.3g/cm^3. It would be found that there existed peak value of the sound absorbing coefficients when thickness of the polyester fiber was 16mm or 24mm. Especially when the thickness reached 24mm, sound absorbing coefficients of the polyester fiber in frequency range of 2000Hz ~ 6000Hz exceeded 90%.

For the six series of experimental data, the relationships between sound absorbing coefficients and sound frequencies were investigated by data fitting according to the triple sine function, as shown in Eq. 2, and the related parameters which corresponded to the thickness 8mm with density $0.2g/cm^3$, thickness 16mm with density $0.2g/cm^3$, thickness 24mm with density $0.2g/cm^3$, thickness 8mm with density $0.3g/cm^3$, thickness 16mm with density $0.3g/cm^3$, and thickness 24mm with density $0.3g/cm^3$ were summarized in Table 2.

Comparisons of the data fitting results and the original experimental data were shown in Fig. 5. It could be observed that these fits were fine. The total departure of the regressive average value $R-square$ (R^2) was introduced to quantitative evaluate prediction accuracy of data

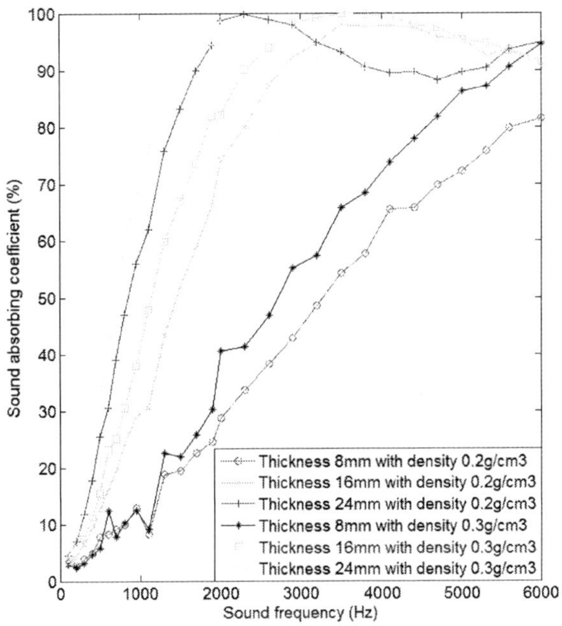

Figure 3. Summarized sound absorbing coefficient

fitting [9], and it could be obtained by Eq. 3. Here α_{ei}, α_{pi}, and N were the actual sound absorbing coefficient obtained by the experiment, the sound absorbing coefficient achieved by data fitting, and the experiment time, respectively. In this study, the detected frequency points were 27.

$$\alpha(\omega)=a_1 \sin(b_1\omega+c_1)+a_2 \sin(b_2\omega+c_2)+a_3 \sin(b_3\omega+c_3) \quad (2)$$

$$R^2=1-\frac{\sum_{i=1}^{N}(\alpha_{ei}-\alpha_{pi})^2}{\sum_{i=1}^{N}\alpha_{ei}^2} \quad (3)$$

Table 2. The Related Parameters in the Data Fitting.

Density	$0.2g/cm^3$			$0.3g/cm^3$		
Thickness / **Parameters**	8mm	16mm	24mm	8mm	16mm	24mm
a_1	128.7	168.5	103.3	10010	106.6	106.3
b_1 **(10^{-4})**	4.558	3.578	3.364	4.592	3.731	3.768
c_1	-0.368	1.241	0.1774	0.0976	0.0138	0.1132

a_2	92.53	153.7	270.1	9956	8.217	70.46
b_2 **(10^{-4})**	6.927	5.513	14.97	4.604	13.17	16.17
c_2	2.246	4.31	-2.096	3.24	1.765	-1.552
a_3	25.18	32.06	257.1	2.153	17.13	54.94
b_3 **(10^{-4})**	9.588	10.4	15.1	17.03	13.39	17.38
c_3	4.546	5.71	7.249	3.186	5.038	7.413

According to Eq. 3, the calculated R^2 of the six models and the six groups of experimental data was 0.9972, 0.9987, 0.9979, 0.9946, 0.9974, and 0.9967, respectively, which corresponded to the thickness 8mm with density 0.2g/cm^3, thickness 16mm with density 0.2g/cm^3, thickness 24mm with density 0.2g/cm^3, thickness 8mm with density 0.3g/cm^3, thickness 16mm with density 0.3g/cm^3, and thickness 24mm with density 0.3g/cm^3. It was considered that the prediction was reliable when the evaluation index R^2 is larger than 0.99. The larger R^2 is, the more reliable the prediction is. It could be found that all R^2 were larger than 0.99, and the average value reached 0.9971.

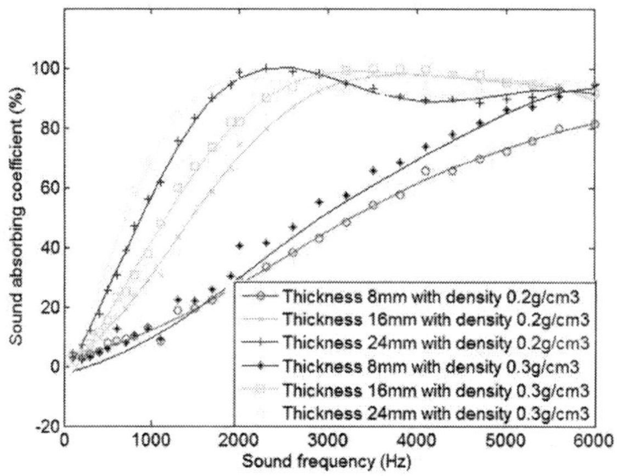

Figure 4. Comparison of theoretical and experimental result.

4. Theoretical analysis

Structure of the inner polyester fiber was a complicated topological structure, which indicated that the absorption of sound in the polyester fiber was a complex dynamic system. The absorption mechanism of sound in the polyester fiber included the following three ways [10]. Firstly, when the sound penetrated into the polyester fiber, air in the interspace among the fibers would be excited vibration and the air could rub wall of the fibers, which resulted in reduction of sound energy by the viscosity resistance. Secondly, there existed temperature gradients among the different particles in the polyester fiber, which would consume some sound energy by thermal conduction [11]. Thirdly, the vibration of fibers also would lead to dissipation of the sound energy. Therefore, the sound absorbing performance of the polyester fiber was influenced by the thickness, density, dimension of the cavity, surface treatment, and other parameters of the polyester fiber.

According to the diffusion characteristics of the sound in the round tubule, the relationship between sound absorbing coefficient α and these parameters of the porosity ϕ, pore tortuosity α_∞, angle frequency ω, glutinous permeability q_0, glutinous characteristic length Λ, thermal permeability q_0', and thermal characteristic length Λ' were shown in Eq. 4 and Eq. 5. The media density and volume bulk modulus could meet the relationships in Eq. 6 and Eq. 7.

$$\alpha(\omega)=\frac{v\phi}{j\omega q_0}\left\{\sqrt{\left[1+\left(\frac{2\alpha_\infty q_0}{\varnothing\Lambda}\right)^2\frac{j\omega}{v}\right]}\right\}+\alpha_\infty \tag{4}$$

$$\alpha'(\omega)=\frac{v'\phi}{j\omega q_0'}\left\{\sqrt{\left[1+\left(\frac{2q_0'}{\emptyset\Lambda'}\right)^2\frac{j\omega}{v'}\right]}\right\}+1 \tag{5}$$

$$\rho=\rho_0\alpha(\omega) \tag{6}$$

$$K=\frac{\gamma P_0}{\gamma-(\gamma-1)\alpha'(\omega)} \tag{7}$$

It could be found that the theoretical expression of the sound absorbing coefficient was extraordinary complicated, which limited its application in predicting sound absorbing coefficient [12]. What's more, establishments of these parameters in Eq. 4 were time-consuming and inefficient [13], which increased the difficulties in practice. Furthermore, it had been reported that accuracy of the model was high in the middle frequency and low in the high or low frequency [14], which meant that novel method was urgently required. In order to promote the application of polyester fiber in the fields of acoustics, development of the practical prediction model would be research focus in future [15, 16].

5. Conclusions

Investigation on the sound absorbing performance of the polyester fiber for noise reduction in large-scale equipment was conducted in this study. Experimental results indicated that the polyester fiber was an excellent acoustic absorption, which could be propitious to reduce the noise in large-scale equipment and improve its acoustic stealth performance.

6. Acknowledgments

This work was supported by a grant from National Natural Science Foundation of China (51505498), a grant from Natural Science Foundation of Jiangsu Province (BK20150714), and a grand from National Key Research and Development Program of China (2016YFC0802900). Xinmin Shen was grateful for support from the Hong Kong Scholars Program (No. XJ2017025).

7. References

[1] Krishna S R, Krishna A R, Ramji K 2011 Reduction of motor fan noise using CFD and CAA simulations *Applied Acoustics* **72(12)** 982-992

[2] Yang X C, Peng K, Shen X M, Zhang X N, Bai P F, Xu P J 2017 Geometrical and Dimensional Optimization of Sound Absorbing Porous Copper with Cavity *Materials & Design*, **131** 297-306

[3] Kucuk M, Korkmaz Y 2015 Sound absorption properties of bilayered nonwoven composites *Fibers and Polymers* **43(4)** 627-640

[4] Jiang S, Xu Y Y, Zhang H P, White C B, Yan X 2012 Seven-hole hollow polyester fibers as reinforcement in sound absorption chlorinated polyethylene composites *Applied Acoustics* **73(3)** 243-247

[5] Naoki K, Takayasu U 2008 Evaluation of acoustical and non-acoustical properties of sound absorbing materials made of polyester fibres of various cross-sectional shapes *Applied Acoustics* **69(7)** 575-582

[6] Emanov A F, Krasnikov A A 2016 Use of the standing wave method to study seismically insulated buildings *Seismic Instruments* **56(4)** 323-349

[7] Lopez A, Assunção E, Pires I, Quintino L 2017 Secondary emissions during fiber laser cutting of nuclear material *Seismic Instruments* **315** 69-76

[8] Koga S I 2006 Sound absorbing material made of polyester nonwovens *The Journal of the Acoustical Society of America* **120(5)** 3147

[9] Manning J, Panneton R 2013 Acoustical model for Shoddy-based fiber sound absorbers *Textile Research Journal* **83(13)** 1356-1370

[10] Yilbas B S, Shaukat M M, Ashraf F 2017 Laser cutting of various materials: Kerf width size analysis and life cycle assessment of cutting process *Optics and Laser Technology* **93** 67-73

[11] Nefske D J 2008 An Equivalent-Acoustic Finite Element Method for Modeling Sound

Absorbing Materials *The Journal of the Acoustical Society of America* **123(5)** 3570

[12] Yang X C, Shen X M, Xu P J, Zhang X N, Bai P F, Peng K, Yin Q, Wang D 2018 Comparison of sound absorbing performances of copper foam and iron foam with the same parameters *IOP Conference Series: Materials Science and Engineering* **292(1)** 012084

[13] Wang J J Leistner P, Li X D 2012 Prediction of sound absorption of a periodic groove structure with rectangular profile *Applied Acoustics* **73** 960-968

[14] Liu X J, Liu J L, Su X Z 2012 Simulation Model for the Absorption Coefficients of Double Layered Nonwovens *FIBRES & TEXTILES in Eastern Europe* **4(93)** 102-107

[15] Yang S, Yu W D, Pan N 2010 Investigation of the sound-absorbing behavior of fiber assemblies *Textile Research Journal* **81(7)** 673-682

[16] Zhu J L, Sun J, Tang H P, Wang J Z, Ao Q B, Bao T F, Song W D 2016 Gradient-structural optimization of metal fiber porous materials for sound absorption *Powder Technology* **301** 1235-1241

Surface Temperature Measurement with Unknown Emissivity Using a Two-color Pyrometer Placed with a Reflector

Xunjian Che[1], Zhi Xie[1]

[1] Department of Information Science and Technology, Northeastern University, Shenyang, China

E-mail: 67139712@qq.com

Abstract. An instrument for measuring surface temperature above 773 K has been developed. A gold-plated reflector is placed ahead of a two-color pyrometer. When the reflector is positioned facing the measured target, the received radiation is increased, and the difference between the effective emissivities of the two received wavelengths is shortened. As a result, the measurement error of the two-color pyrometer is reduced compared with the one without a reflector. The validation experiment has been conducted. The measurement uncertainty is evaluated, and the standard uncertainty is 4.0 K at 973 K.

1. Introduction

Precise measurement of surface temperature is crucial for control of industrial process such as surface temperature measurement of billet of steel, aluminium, glass, copper and so on, where numerous optical pyrometers and infrared cameras are widely used [1-2]. However, these non-contact methods rely on accurate determination of emissivity, and also influenced by background radiation. Emissivity is hard to be determined online because it varies with temperature, wavelength, and physical state of the surface.

Two-color is free of emissivity as long as the emissivities of the two receiving wavelength are equal, we call it gray body. However, the gray body does not exist most of case, a slight discrepancy between the two emissivities results in measurement error.

A gold plated reflector is used as an emissivity enhancer for it integrates the intrinsic and reflected radiation to produce pseudo-blackbody conditions [3-4]. However, the enhanced emissivity (effective emissivity) will never be 1 *in situ* condition and hard to be estimated, because it is influenced by various factors such as the reflectivity and shape of reflector, temperature and emissivity of the measured surface, especially the distance to the measured surface, the readout of a reflector-headed pyrometer fluctuates widely with the distance varying [5]. As the measurement error cannot be eliminated completely, one still have to set an empirical emissivity value before use, so it is not an optimal choice for reference pyrometer.

In this paper, we combines the advantages of the two-color method and the gold plated reflector. The two-color pyrometer is equipped with a reflector ahead. The discrepancy of the effective emissivities can be greatly reduced by the reflector. As a result, the measurement error is decreased exponentially compared with the one without a reflector.

2. Measurement principle

Surface temperature measurement by optical pyrometry is largely affected by the emissivity of the target. For monochromatic optical pyrometry, the relationship of spectral radiance temperature T_s (K), true temperature T_0 (K), and emissivity $\varepsilon(\lambda)$ is expressed [6]:

$$\frac{1}{T_s} = \frac{1}{T_0} - \frac{\lambda}{c_2} \ln \varepsilon(\lambda) \tag{1}$$

where $c_2 = 1.438\times10^{-9}$ m K is the Planck's secondary radiation constant, λ is the received wavelength, (m). We can see from (1) that unless $\varepsilon(\lambda) = 1$ (blackbody), the spectral radiance temperature T_s measured by a monochromic pyrometer is always less than the true temperature T_0. So precise measurement needs prior knowledge of $\varepsilon(\lambda)$. However, $\varepsilon(\lambda)$ is hard to be determined because it varies with temperature, wavelength, and surface condition.

Two-color (ratio) pyrometer has been the most frequently used technique without any knowledge of the emissivity value if the emissivities $\varepsilon(\lambda_1)$ and $\varepsilon(\lambda_2)$ at the two wavelengths λ_1 and λ_2 (m) are equal (graybody). The ratio temperature T_r (K) measured by a two-color pyrometer is expressed as follows [7]:

$$\frac{1}{T_r} = \frac{1}{T_0} + \frac{1}{c_2\left(1/\lambda_2 - 1/\lambda_1\right)} \ln \frac{\varepsilon(\lambda_1)}{\varepsilon(\lambda_2)} \tag{2}$$

From (2) we can see that if $\varepsilon(\lambda_1) = \varepsilon(\lambda_2)$, then $T_r = T_0$. However, gray body seldom exists *in situ* condition. Slight difference between $\varepsilon(\lambda_1)$ and $\varepsilon(\lambda_2)$ would result in significant measurement error. For example, assuming $\varepsilon(\lambda_1) = 0.6$ for the measured target at 973 K, and the difference is $\Delta\varepsilon = \varepsilon(\lambda_2) - \varepsilon(\lambda_1) = 0.05$, the measurement error is $T_r - T_0 = -14.5$ K ($\lambda_1 = 0.96$ μm and $\lambda_2 = 1.46$ μm).
Here we propose an improved two-color pyrometer to reduce the measurement error by decreasing the difference $\Delta\varepsilon$ between the two emissivities. The measurement principle is illustrated in Figure 1. We borrow the idea from the Land Surface Pyrometer (Land SP) [1], and place a gold-plated hemispherical reflector ahead of a pyrometer. The detector receives more radiation than that without the reflector: not only the intrinsic radiation emitted from the target itself, but also the infinite reflected radiation between the target and the reflector. So the effective emissivity $\varepsilon_a(\lambda)$ is enhanced, the difference $\Delta\varepsilon$ is shortened, and the measurement error of a two-color pyrometer is reduced.

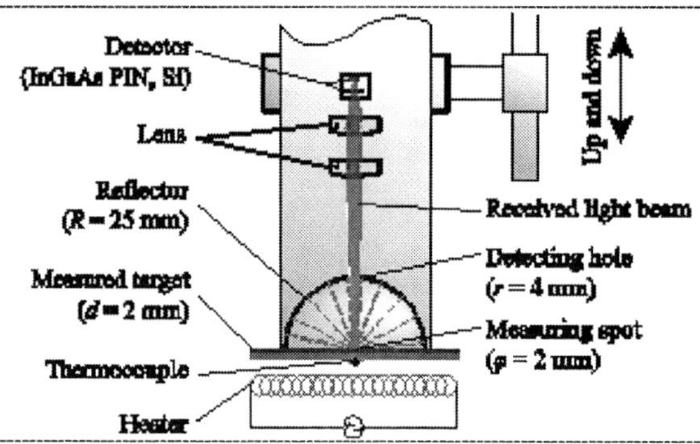

Figure 1. Measurement principle.

3. Instrument design

We use a two-color detector manufactured by Hamamatsu Photonics Co., Ltd. (model K3413-09). It is a hybrid detector containing an infrared-transmitting Si photodiode mounted over an InGaAs PIN photodiode, along with the same optical axis. This structure delivers a wide spectral response range: 0.3–1.1 μm for Si and 0.9–1.8 μm for InGaAs PIN. The effective wavelengths are 0.96 μm and 1.46 μm, respectively. The built-in thermoelectric cooler makes the detector maintain a constant temperature during measurement, allowing low system uncertainty with an improved S/N ratio.
When taking a measurement, we put the reflector facing the target and make sure that the focal point of the hemispherical reflector is right on the surface of the target. At this circumstance, we say the

reflector is focused on the target. To make sure focused, we put the hemisphere reflector contact with the flat target when measuring. Assuming specular reflectance for the focused reflector, the light beam emitted from the focal point is never reflected back to non-focal points, so no heat exchange happens between the focal and non-focal points. The detector receives radiation from the focal point only: either from the focal point itself (intrinsic radiation), or the reflected radiance between the focal point and the reflector (reflected radiation). The measuring spot is designed right at the focal point on the target. The size of the measuring spot is neglected compared with that of the reflector. The focused reflector brings an advantage that the pyrometer measures the temperature of the measuring spot only, so the measurement is not affected by the non-isothermal distributed surface. Assuming Lambertian surface, we have that the normal emissivity is equal to the hemisphere emissivity, and the effective emissivity $\varepsilon_a(\lambda)$ at the focal point is expressed as follows[1]:

$$\varepsilon_a = \frac{\varepsilon}{1-(1-\varepsilon)\rho(1-r^2/R^2)}, \tag{3}$$

where $\rho(\lambda)$ is the reflectance of the reflector, r and R are the radius of the detecting hole and the hemispherical reflector, respectively, (m). When considering the fraction of specular reflection, the expression of $\varepsilon_a(\lambda)$ is [8]:

$$\varepsilon_a = \frac{\varepsilon-\varepsilon(1-\varepsilon)\rho(1-r^2/R^2)(1-\eta)}{1-(1-\varepsilon)\rho(1-r^2/R^2)}, \tag{4}$$

where η is the diffusion coefficient, $\eta = 0$ represents absolutely specular reflection, while $\eta = 1$ represents absolutely diffuse reflection. The $\varepsilon_a(\lambda)$ goes down along with the decrease of the reflectivity $\rho(\lambda)$ and the diffusion coefficient η. Extremely, when $\eta = 0$, we have $\varepsilon_a(\lambda) = \varepsilon(\lambda)$, which means if the measuring spot is absolutely specular reflection, the reflector is invalid. In another words, the focused hemispherical reflector enhances the effective emissivity attributing to the property of diffuse reflectance of the measuring spot.

The difference $\Delta\varepsilon$ would be greatly shortened by the reflector. Followed by the example at the beginning, assuming $\rho(\lambda_1) = \rho(\lambda_2) = 0.9$, the effective emissivities would be enhanced to $\varepsilon_a(\lambda_1) = 0.924$ and $\varepsilon_a(\lambda_2) = 0.938$ for Lambertian surface, and the difference $\Delta\varepsilon$ decreases from 0.05 to 0.014. As a consequence, the measurement error is reduced from -14.5 K to -2.7 K.

The use of reflector inevitably disturbs the surface temperature of the target [9]. The boundary condition of the target alters when the reflector is approaching, and the surface temperature will rise accordingly. For preventing this problem, we make a fast and repeatable measurement with the help of machinery [10]. The pyrometer could make vertical motion on a linear guide driven by a servo motor (Panasonic Motor, model MSMD082G1V) with a resolution of 0.0125 mm. The pyrometer was lifted 400 mm away from the target. When taking a measurement, the pyrometer was driven down and stopped precisely at the position where the reflector is focused, then the signals were captured and the ratio temperature T_r was derived. It took only 0.7 s for a measurement. At last the pyrometer was lifted up.

4. Validation experiment

For accuracy validation, we compared the proposed method with a K-type thermocouple. We glued the thermocouple onto the backside of the target with heat-resistance insulation adhesive. The temperature drop in the thickness direction cannot be ignored when the target is exposed to a cold surrounding, so the surface temperature is lower than the output of the thermocouple. We determined the surface temperature T0 (K) by solving the heat flux function [11]:

$$T_0^4 + BT_0 + A = 0, \tag{5}$$

with coefficients:

$$A = -(T_{Tc}k/d + hT_E)/(\varepsilon_n\sigma) - T_E^4, \tag{6}$$

$$B = (k/d + h)/(\varepsilon_n\sigma), \tag{7}$$

where T_{Tc} (K) is the temperature at the bottom of the target measured by the thermocouple; k is the heat conductivity coefficient, (W/(m.K)); d = 2×10^{-3} m is the thickness of the target; h is the convective heat-transfer coefficient, (W/(m^2 K)); T_E = 303 K is the ambient temperature; σ = 5.67×10^{8} (W m^{-2} K^{-4}) is the Stefan–Boltzmann constant; $\varepsilon_\cap$ is the total hemispherical emissivity, and we assume $\varepsilon_\cap$ = 1 for heat flux calculation.

Table 1. Measurement results of various materials.

	Ti alloy(K)	Stainless steel(K)	Brass(K)	Copper(K)
T_{Tc}	1038.4	1010.9	1088.5	1033.2
T_0	1031.1	1008.4	1087.1	1032.8
$T_{sa}(\lambda_1) - T_0$	**-1.5**	-2.4	-5.5	-1.5
$T_{sa}(\lambda_2) - T_0$	-5.6	-2.7	-8.0	-2.5
$T_{ra} - T_0$	6.5	**-1.8**	**-0.7**	**0.4**
$\varepsilon_a(\lambda_1)$	0.979	0.965	0.932	0.979
$\varepsilon_a(\lambda_2)$	0.949	0.974	0.935	0.977

We measured the target with different materials, and gathered the measurement results in Table I. T_0 is the surface temperature calculated by (5) according to the temperature T_{Tc} measured by the thermocouple. $T_{sa}(\lambda_1)$, $T_{sa}(\lambda_2)$, and T_{ra} are the spectral radiance temperature at the two wavelengths and the ratio temperature enhanced by the reflector. $\varepsilon_a(\lambda_1)$ and $\varepsilon_a(\lambda_2)$ are the effective emissivities calculated from (1) according to $T_{sa}(\lambda_1)$, $T_{sa}(\lambda_2)$, and T_0. In the calculation of T_0, the heat conductivity coefficient k for Ti alloy, stainless steel, brass, and copper are 18, 48, 117, and 366 W/(m K). The convection can be neglected compared with the radiation at elevated temperature above 773 K, so h = 0. As shown in Table 1, the temperature drop of copper is only 0.4 K attributing to high heat conductivity, while the Ti alloy is 7.3 K for low heat conductivity. The ratio temperature T_{ra} produces the lowest measurement error except for the Ti alloy. For Ti alloy, $T_{sa}(\lambda_1)$ yields the lowest error, because the shortened difference $\Delta\varepsilon$ = 0.03 for a two-color pyrometer is relatively large compared with the $\varepsilon_a(\lambda_1)$ = 0.979 for a monochromic pyrometer in terms of measurement error.

5. Uncertainty evaluation

The measurement uncertainty of the proposed method T_{ra} was evaluated. For comparison, we also investigated three other methods:

1) Two-color pyrometer without the reflector T_r;
2) Monochromic pyrometer $T_{sa}(\lambda_1)$ and $T_{sa}(\lambda_2)$ enhanced by the reflector;
3) Monochromic pyrometer $T_s(\lambda_1)$ and $T_s(\lambda_2)$ without the reflector.

As our methods are non-linear, a Monte Carlo method (MCM) was used for evaluating the uncertainty according to Supplement 1 of the GUM [12]. The uncertainty components and their respective distributions are given in Table 2. In the table, a Gaussian distribution (G) is described in terms of expectation and standard deviation, while a rectangular distribution (R) in terms of expectation and semi-width. For more generalized case, the measured emissivity $\varepsilon(\lambda_1)$ was treated as outside uncertainties, ranging from 0.2 to 1 with rectangular distribution [13]. The difference $\Delta\varepsilon = \varepsilon(\lambda_2) - \varepsilon(\lambda_1)$ was with Gaussian distribution, the standard deviation was 0.03. When the reflector was unfocused, and the surface of target was non-isothermal, there existed heat exchanges between focal and non-focal points, thus the effective emissivity would be decreased. Due to the complex path of reflectance light beam when non-focused, we utilized MCM to calculate the effective emissivity [5]. Assume that the surface temperature was linear decrease or increase in the radius direction r, and the maximum difference of surface temperature was dT. We omit the calculating process and give one example: for dT = -30 K, the drop value of $\varepsilon_a(\lambda)$ is about 0.006 when the focal point is 1 mm below the surface. We assumed Gaussian distribution for the distance between the focal point and the target surface, and rectangular distribution for the maximum temperature difference dT. We assumed the reflectivity at the two wavelengths were equal: $\rho(\lambda_1) = \rho(\lambda_2) = \rho$, and ρ was with Gaussian distribution. We also considered non-diffusion reflectance for the measured surface: $\eta < 1$ in (4).

Table 2. Measurement results of various materials.

Uncertainty Component (Distribution)	Expectation	Standard deviation or Semi-width	Type
$\varepsilon(\lambda_1)$ (R)	0.6	0.4	B
$\varepsilon(\lambda_2)$ - $\varepsilon(\lambda_1)$ (G)	0	0.03	B
Nonlinearity (R)	0 K (λ_1)	0.4 K (λ_1)	B
	0 K (λ_2)	0.2 K (λ_2)	B
Repeatability (G)	0 K (λ_1)	0.2 K (λ_1)	A
	0 K (λ_2)	0.1 K (λ_2)	B
Reflectivity ρ (G)	0.9	0.03	B
Non-isothermal dT (R)	-10 K	10 K	B
Reflector focusing (G)	0 mm	0.5 mm	B
Temperature rise (R)	0 K	0.5 K	B
Diffusion coeff η (R)	0.98	0.02	B

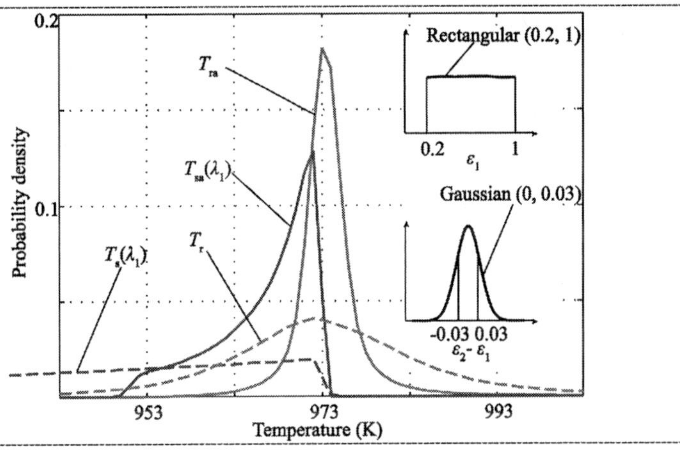

Figure 2. PDFs of the 4 methods calculated from 10^6 MCM at 973 K.

Figure 2 illustrates the Probability Distribution Functions (PDFs) of T_r, T_{ra}, $T_s(\lambda_1)$, and $T_{sa}(\lambda_1)$ calculated by MCM at 973 K. The PDFs were presented by histogram with 1 K scale. For concision, the distributions of $T_s(\lambda_2)$ and $T_{sa}(\lambda_2)$ were not drawn. Obviously, the PDF of $T_{sa}(\lambda_1)$ is greatly compressed compared with that of $T_s(\lambda_1)$ due to the multiple reflectance by the reflector, and the PDF of T_{ra} is also greatly compressed compared with that of T_r due to the difference $\Delta\varepsilon$ is shortened. On the whole, T_{ra} is much more superior to $T_{sa}(\lambda_1)$ in the aspects of standard uncertainty and expections. As the PDFs are asymmetric, the shortest 95% confidence intervals were calculated for expressions of the expanded uncertainties. The estimate, standard uncertainties, and shortest 95% coverage intervals are presented in Table 3.

Table 3. Uncertainty evaluation using MCM at 973K.

	Estimate (K)	Standard uncertainty (K)	Shortest 95% confidence interval (K)
$T_s(\lambda_1)$	937.5	25.2	[890.4, 973.4]

$T_{sa}(\lambda_1)$	966.7	5.5	[954.8, 973.5]
$T_s(\lambda_2)$	920.3	37.2	[852.4, 973.7]
$T_{sa}(\lambda_2)$	963.1	8.7	[945.1, 973.5]
T_r	974.0	14.2	[945.5, 1003.7]
T_{ra}	**973.9**	**4.0**	**[965.8, 982.0]**

6. Conclusions

The research results have shown that the proposed method solves the problem that the surface temperature measurement error caused by the in determination of emissivity which restricts the development of infrared thermometry. The instrument can be applied to the industry process where precise measurement of surface temperature is required.

7. Acknowledgments

This work was supported by the National Natural Science Foundation of China (61273178).

8. References

[1] Shirbhate N J and Dhongade G K 2014 *Int. J. Mech. Eng. Rob. Res.* **3** 289-292
[2] Rajora R and Dixit H K 2013 *Int. J. Mech. Eng. Rob. Res.* **2** 324-334
[3] Drury M D, Perry K P and Land T 1951 *J. Iron St. Inst.* **169** 245-250
[4] Ono A 1988 *Theory and Practice of Radiation Thermometry* ed. D P DeWitt and G D Nutter (NewYork: John Wiley &Sons), pp 607-614.
[5] Krapez J C, Belanger C and Cielo P 1990 *Meas. Sci. Technol.* **1** 857-864
[6] Hernandez D 2005 *Rev. Sci. Instrum.* **76** 024904
[7] Müller B and Renz U 2001 *Rev. Sci. Instrum.* **72** 3366-74
[8] Che X and Xie Z 2017 *IEEE Trans. Instrum. Meas.* **66** 1845-55
[9] Ciaurriz C, Hernandez D, Olalde G, Devilliers C and Salle M 1992 *Rev. Sci. Instrum.* **63** 3623-28
[10] Iuchi T and Hiraka K 2006 *Rev. Sci. Instrum.* **77** 114902
[11] Monte C and Hollandt J 2010 *Metrologia* **47** S172–S181
[12] BIPM 2008 *Document JCGM 101:2008*(http://www.bipm.org/en/publications/guides/gum.html)
[13] Muniz P R, Kalid R D A, Cani S P N and Magalhães R D S 2014 *Opt. Eng.* **53** 074101

Research on geometric properties of arbitrary beam sections

X L Chen[1], D Y Li[1] and L Zhang[1]

1 Chongqing University of Science and Technology, Chongqing 401331, China
E-mail: xlchen80@163.com

Abstract. The centroid, the moment of inertia, and the product of inertia for an arbitrary polygon and an arbitrary plane area, are formulated in algebraic forms and programmed with MATLAB software. Three numerical examples are shown and indicate the method can be easily computed by the computer for geometric properties of arbitrary beam sections in mechanics of materials with simply or multiply connected boundaries.

1. Introduction

Geometric properties of an area, including the centroid, the moment of inertia, the polar moment of inertia, and the product of inertia, are important quantities used in mechanics of materials, structural mechanics, fluid mechanics, and machine design, hence their computation by the computer is helpful for the engineer. Engineering analysis and design often use the properties of planar sections in calculations. For example, in the stress analysis of a beam under bending and torsional loads, cross-sectional properties are used to determine stress and displacement distributions in the beam cross section. It must be known of the area, centroid, and various moments and products of inertia of a cross section or a composed cross section in calculating the natural frequencies and shapes of a machine part. The centroid, moments and products of inertia of an area made up of several common shapes (rectangles usually) may thus be obtained by adding the moments of inertia of the component areas (U-shape, L-shape, C-shape, etc). This procedure is used to calculate the second moments of structural shapes in engineering practice because the determination of their moments of inertia is necessary for the design of structural components. Wen and Zhang [1] studied geometric properties of an arbitrary triangle and applied to convex polygons. Based on calculating the geometrical characteristic values of a triangle with a vertex on the coordinate origin, Cai [2] suggested a regular method for calculating geometrical and mechanical characteristic values of sections for structural bodies in a shape of prism. Carbonell et al. [3] presented examples of interactive simulations focusing on fundamental topics such as centroids, the properties of the moment of inertia, second moments of inertia with respect to two axes, principal moments of inertia and Mohr's Circle for plane stress. These learning tools feature animations, graphics and interactivity and were designed to encourage student participation and engagement in active learning activities, to effectively explain and illustrate course topics, and to build student problem-solving skills. Zhao et al. [4] presented a method for computing the sectional geometric properties of thin-walled elements with arbitrary polygon sections based on the theory of restraint torsion. Shi and Chen [5] presented the formulas of geometric parameters for the complicate thin section divided into finite segments. Xu et al. [6] gave the recount for the method and function of calculating distortion geometric nature of arbitrarily section in continuous box beam. Li and Zhou [7] discussed geometrical properties of section and arbitrary sections as well as calculation method, and presented a method parameter equation which can be established using basic geometrical elements line and curve. Zhou et al. [8] developed a finite element procedure of two-dimensional torsion and flexure problem for arbitrary cross-section by using the theorem of minimum potential energy, and the

IOP Conf. Series: Materials Science and Engineering **398** (2018) 012007 doi:10.1088/1757-899X/398/1/012007

cross-section was discretized with six-node and eight-node isotropic plane element to evaluate the torsional coefficient, shear center and the shear area correction coefficients o f any beam with irregular cross-section. The sectional area and inertial moment were also calculated with the numerical integration, and a finite element program was developed with this method. He [9] realized the computation for the moment of areas and moment of inertia opposite to the random axis based on the formula of parallel moves axis and revolution axis in plane graphics, the two-dimensional graphics software AutoCAD and the VBA development kit, thus simplified the computational process and reduced the duplicated work.

2. Computational method for geometric properties of an arbitrary polygon

Consider an arbitrary polygon, as shown in figure 1, which lies in the *x*-*y* plane. It should be noticed that the vertexes $A_i (i=1,2,...,n)$ must be continuously numbered with a counterclockwise direction, and A_{n+1} is assumed to coincide with A_1. x_i and y_i are the coordinates of the vertex A_i. S_i is the algebraic area of ΔOA_iA_{i+1} and can be expressed as in equation (1). Since the area of a polygon is the sum of S_i, substituting geometric properties of an arbitrary triangle shown in the reference [2] into the general integral definition of geometric properties, we have geometric properties of an arbitrary polygon shown as in equation (2)-equation (5), where x_C and y_C are the coordinates of the centroid C, I_x, I_y, and I_{xy} are the moment of inertia and the product of inertia respectively.

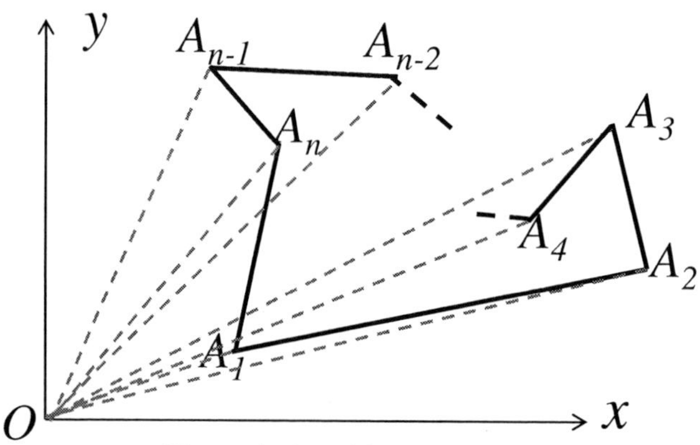

Figure 1. An arbitrary polygon

$$S_i = (x_i y_{i+1} - x_{i+1} y_i)/2 \quad (i=1,2,...,n) \tag{1}$$

$$x_C = \frac{\sum_{i=1}^{n} S_i \cdot \frac{x_i + x_{i+1}}{3}}{\sum_{i=1}^{n} S_i}, \; y_C = \frac{\sum_{i=1}^{n} S_i \cdot \frac{y_i + y_{i+1}}{3}}{\sum_{i=1}^{n} S_i} \tag{2}$$

$$I_x = \sum_{i=1}^{n} S_i \cdot \frac{y_i^2 + y_i y_{i+1} + y_{i+1}^2}{6} \tag{3}$$

$$I_y = \sum_{i=1}^{n} S_i \cdot \frac{x_i^2 + x_i x_{i+1} + x_{i+1}^2}{6} \tag{4}$$

IOP Conf. Series: Materials Science and Engineering **398** (2018) 012007 doi:10.1088/1757-899X/398/1/012007

$$I_{xy} = \sum_{i=1}^{n} S_i \cdot \frac{2(x_i y_i + x_{i+1} y_{i+1}) + x_i y_{i+1} + x_{i+1} y_i}{12} \tag{5}$$

2.1. MATLAB program

The MATLAB program for the computation of area, centroid, moment of inertia, polar moment of inertia, and product of inertia for an arbitrary polygon is as follows:

```
clear;clc;
xy=load('coordinates.txt') ;          % vertex coordinate matrix
n=length(xy);
xy=[xy;xy(1,:)];                       % An+1=A1
for i=1:n
s(i)=xy(i,1)*xy(i+1,2)-xy(i+1,1)*xy(i,2);
xci(i)=(xy(i,1)+xy(i+1,1));
yci(i)=(xy(i,2)+xy(i+1,2));
ixi(i)=xy(i,1)^2+xy(i,1)*xy(i+1,1)+xy(i+1,1)^2;
iyi(i)=xy(i,2)^2+xy(i,2)*xy(i+1,2)+xy(i+1,2)^2;
ixyi(i)=2*(xy(i,1)*xy(i,2)+xy(i+1,1)*xy(i+1,2))+xy(i,1)*xy(i+1,2)+xy(i+1,1)*xy(i,2);
end
S=sum(s)/2                             % area of an arbitrary polygon
xc=sum(s.*xci)/area/2/3                % the centroid
yc=sum(s.*yci)/area/2/3                % the centroid
Ix=sum(s.*iyi)/12                      % the moment of inertia about x-axis
Iy=sum(s.*ixi)/12                      % the moment of inertia about y-axis
Ip=Ix+Iy                               % the polar moment of inertia
Ixy=sum(s.*ixyi)/24                    % the product of inertia
```

2.2. Numerical examples

In figure 2(a), the coordinates of the vertexes are $A_1(-300,-50)$, $A_2(200,-50)$, $A_3(200,-350)$, $A_4(300,-350)$, $A_5(300,50)$, $A_6(-200,50)$, $A_7(-200,350)$, and $A_8(-300,350)$, where all the length units are mm. Substituting these coordinates into equation (1)-equation (5), geometric properties of the polygon can be solved by the above MATLAB program. Results are the area $S = 120000 mm^2$, the centroid $x_C = 0$ and $y_C = 0$, the moment of inertia $I_x = 2.9\text{e}9\,\text{mm}^4$ and $I_y = 5.6\text{e}9\,\text{mm}^4$, the polar moment of inertia $I_O = 8.5\text{e}9\,\text{mm}^4$, and the product of inertia $I_{xy} = \text{-}3\text{e}9\,\text{mm}^4$. In figure 2(b), the coordinates of the vertexes are $A_1(0,0)$, $A_2(9,-6)$, $A_3(9,0)$, $A_4(6,6)$, and $A_5(0,6)$, where all the length units are mm. Substituting these coordinates into equation (1)-equation (5), geometric properties of the polygon can be solved by the above MATLAB program. Results are the area $S = 72 mm^2$, the centroid $x_C = 4.625 mm$ and $y_C = 1 mm$, the moment of inertia $I_x = 648\,\text{mm}^4$ and $I_y = 1971\,\text{mm}^4$, and the product of inertia $I_{xy} = 81\,\text{mm}^4$ .The above results agree with the reference [10].

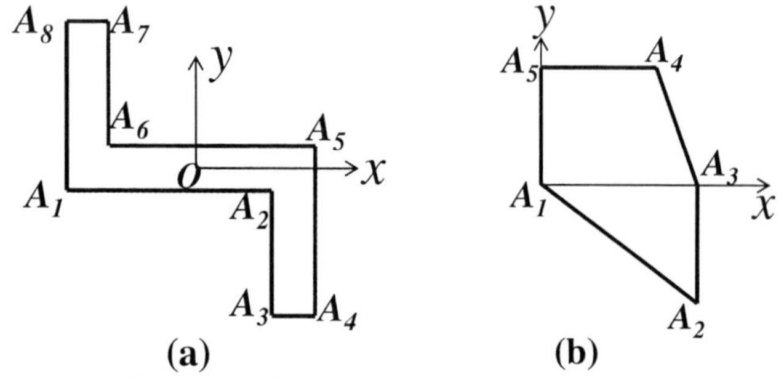

Figure 2. Polygonal cross-sectional areas

3. Computational method for geometric properties of arbitrary plane areas

As shown in figure 3, an arbitrary plane area can be easily meshed into a finite number of triangles by the software such as ANSYS, and the coordinates of nodes can be automatically obtained. Substituting geometric properties of an arbitrary triangle shown in the reference [1] into the general integral definition of geometric properties, we have geometric properties of an arbitrary plane area shown as in equation (6)-equation (10), where S_i is the area of meshed triangle $\Delta A_i A_j A_k$, x_C and y_C are the coordinates of the centroid C, I_x, I_y, and I_{xy} are the moment of inertia and the product of inertia respectively.

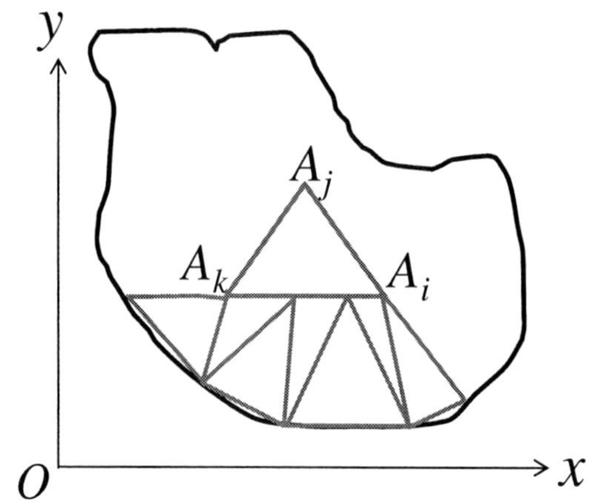

Figure 3. An arbitrary plane area

$$S_i = \frac{1}{2}\left| x_i(y_j - y_k) + x_j(y_k - y_i) + x_k(y_i - y_j) \right| \tag{6}$$

$$x_C = \frac{\sum_{i=1}^{n} S_i \cdot \frac{x_i + x_j + x_k}{3}}{\sum_{i=1}^{n} S_i}, \; y_C = \frac{\sum_{i=1}^{n} S_i \cdot \frac{y_i + y_j + y_k}{3}}{\sum_{i=1}^{n} S_i} \tag{7}$$

$$I_x = \sum_{i=1}^{n} S_i \cdot \frac{y_i^2 + y_j^2 + y_k^2 + y_i y_j + y_k y_j + y_i y_k}{6} \tag{8}$$

$$I_y = \sum_{i=1}^{n} S_i \cdot \frac{x_i^2 + x_j^2 + x_k^2 + x_i x_j + x_k x_j + x_i x_k}{6} \tag{9}$$

$$I_{xy} = \sum_{i=1}^{n} S_i \cdot \frac{2(x_i y_i + x_j y_j + x_k y_k) + x_i(y_j + y_k) + x_j(y_k + y_i) + x_k(y_i + y_j)}{12} \tag{10}$$

3.1. MATLAB program

With the triangular elements and coordinates of nodes obtained by the ANSYS software, the MATLAB Program for the computation of area, centroid, moment of inertia, polar moment of inertia, and product of inertia for an arbitrary plane area is as follows:

```
clear;clc;
node=load('node.txt');  % coordinates of nodes
b=load('ele.txt');      % triangular elements
n=length(b);
for m=1:n
  xi=node(b(m,7),2);yi=node(b(m,7),3);
  xj=node(b(m,8),2);yj=node(b(m,8),3);
  xk=node(b(m,9),2);yk=node(b(m,9),3);
s(m)=abs(xi*(yj-yk)+xj*(yk-yi)+xk*(yi-yj));
xci(m)=xi+xj+xk;
yci(m)=yi+yj+yk;
iyi(m)=yi^2+yj^2+yk^2+yi*yj+yj*yk+yk*yi;
ixi(m)=xi^2+xj^2+xk^2+xi*xj+xj*xk+xk*xi;
ixyi(m)=2*(xi*yi+xj*yj+xk*yk)+xi*(yj+yk)+xj*(yk+yi)+xk*(yi+yj);
end
S=sum(s)/2                  % area of an arbitrary plane area
xc=sum(s.*xci)/area/2/3     % the centroid
yc=sum(s.*yci)/area/2/3     % the centroid
Ix=sum(s.*iyi)/12           % the moment of inertia about x-axis
Iy=sum(s.*ixi)/12           % the moment of inertia about y-axis
Ip=Ix+Iy                    % the polar moment of inertia
Ixy=sum(s.*ixyi)/24         % the product of inertia
```

3.2. Numerical examples

In figure 4, a general cross-sectional area is meshed into 112 triangular elements by the ANSYS software, and the coordinates of nodes can also be obtained by the software. Substituting these coordinates into equation (6)-equation (10), geometric properties can be solved by the above MATLAB program. Results are the area $S = 7.5529\,mm^2$, the centroid $x_C = -4.29e-18 mm$ and $y_C = -1.1490 mm$, the moment of inertia $I_x = 20.3838 mm^4$ and $I_y = 2.0505 mm^4$, the polar moment of inertia $I_O = 22.4343\,\text{mm}^4$, and the product of inertia $I_{xy} = -2.2089e-16 mm^4$. While both the theoretical values of x_C and I_{xy} should be zero, it indicates numerical computation exits the rounding error. It must be noticed that the element type (plane stress or plane strain) and the material model are not used here.

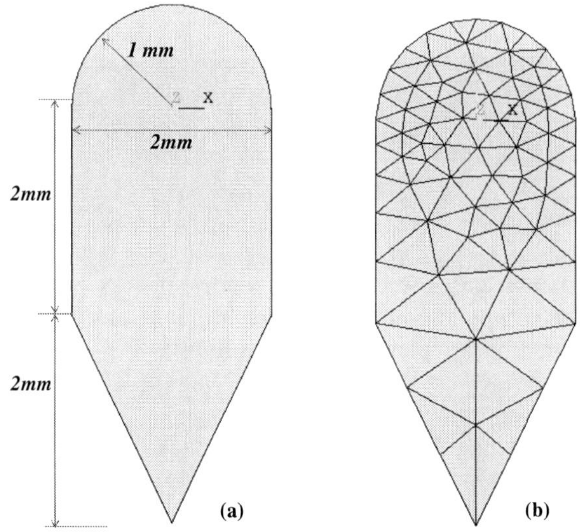

Figure 4. General cross-sectional areas

4. Conclusions

Equation (1)-equation (5) are exact for geometric properties of arbitrary simply connected polygons with straight edges. Equation (6)-equation (10) can be applied to compute beam sections with arbitrary boundaries. Both methods are easy to be executed by the computer program, however there may exits some little numerical errors caused by the triangular mesh or the computer operation. For the multiply connected polygon, it can be considered as a combination of some simply connected polygons and solved by equation (1)-equation (5), or solved directly by equation (6)-equation (10).

Acknowledgements

This work was supported by Scientific and Technological Research Program of Chongqing Municipal Education Commission (Grant No. KJ1713340) and the National Natural Science Foundation of China (Grant No. 11602043).

References

[1] Wen Z M and Zhang L S 2001 *J. Southern Institute Metall.* **22** 293–5
[2] Cai J B 1995 *Acta univ. agric. boreali-occidentalis* **23** 69–73
[3] Carbonell V, Romero C, Martinez E and Florez M 2013 *Eur. J. Phys.* **34** 991–1004
[4] Zhao Y, Zhou J O and Yan D H 2004 *Structural Engineers* **20** 38–41
[5] Shi Q and Chen C Y 2004 *Chin J Mech Eng* **40** 144–148
[6] Xu L H, Zhao Q and Jiang X 2005 *Forest Engineering* **21** 29–37
[7] Li N and Zhou X Y 2007 *Shanxi Architecture* **33** 58–60
[8] Zhou L Y, Li Q and Zhang S Z 2008 *J Comput Mech* **25** 634–639
[9] He H T 2011 *Tech Vocabulary* **215** 33–35
[10] Hibbeler R C 2004 *Engineering Mechanics: Statics* (Beijing: Higher Education Press)

IOP Conf. Series: Materials Science and Engineering **398** (2018) 012008 doi:10.1088/1757-899X/398/1/012008

Stiffness and Damping Formulations for Jointed Deployable Space Truss Structure

A Xi[1] W Zhang[1] and Gen Liu[1]

[1] College of Mechanical Engineering, Beijing University of Technology, Beijing, 100124, PR China

xiansu30@126.com

Abstract. In this paper, the dynamic behaviors of a elastic rod connected by a joint with clearance of deployable truss structures are investigated. A one-dimensional cantilever with joint support is examined in the deployable truss structure. These formulations consist of the nonlinear joint dynamic stiffness model and damping models are formulated and the dynamic characteristics of free-vibrating structures are estimated. The natural frequency transitions are investigated by considering the contact and noncontact conditions of the clearances, then the formulation of dynamic stiffness model is obtained. The friction models are mathematically formulated and simplified to make them easy to calculate. The impact damping models are formulated using the linear envelope of the time response and it can be negligible when the speed is small. The dynamic characteristics can be estimated by integrating these models and formulations using the energy loss factors. Experimental evaluations show the dynamic characteristics and demonstrate the validity of the formulations.

1. Introduction

The deployable space structures have characteristics of large size high precision [1-4]. The deployable structures not only satisfy the mission requirement but also to meet the good mechanical properties. With the characteristics of light weight, high storage ratio and enough stiffness, the truss become the preferred deployable structures[5,6]. The ring truss structure in the deployable antenna often need the hinge to connect each other, however, the hinge gap is inevitable. The gaps in the hinge make the truss structure to show a complex dynamic behaviour[7,8]. Therefore, it is important to study the characteristics of the joints in the ring truss support structure.

Figure 1. Truss antenna reflector.

In the research of hinge structure, scholars have done a lot of work. Hu et al. [9] put forward three scientific problems on the nonlinear dynamics and control of the large ring truss reflector structure. The effective of gaps in the nonlinear joint is studied. Yoshida [10] study the clearance hinge in the space structure, and the mechanical model of the hinge is established. Wang et al. [11] study on the nonlinear dynamic problems of two - bar structures with clearances. Lion et al. [12,13] uses the amplitude frequency response function to identify the dynamic parameters of the hinge. Crawley et al. [14-17] applies force state mapping method (FSM) to the parameter identification of nonlinear hinge, and uses the method to identify the model and parameters of spacecraft joint structure. The time history of the restoring force and the time history of the state variables are measured, and the least square method is used to fit the relationship between the restoring force and the state variable.

Stiffness and damping are important factor in determining the dynamic characteristics of hinge structure, this paper will analyze the radial dynamic characteristics of the planar revolute hinge on the connection of deployable ring truss antenna structure. The radial dynamic model of the hinge is established, and the dynamic stiffness and damping of the hinge are analyzed. Meanwhile, the relevant dynamic model is verified by experiments.

2. Radial dynamic stiffness of hinge

The ring truss deployable antenna is composed of a large number of bars and joints, and the whole structure is divided into mesh, the front net, ring truss tension ties and rear net, as shown in Figure2. The ring truss is the frame of the antenna, which plays a role in supporting surface and keeps the antenna structure stable.

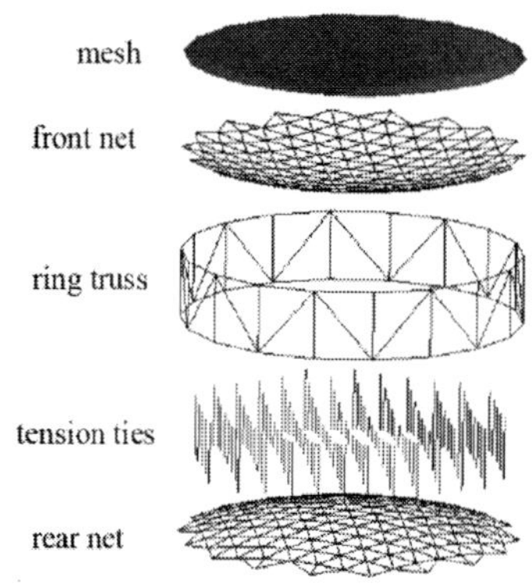

Figure 2. Sketch of circular truss

The stiffness of the joint is an important factor to determine the dynamic characteristics of the jointed structure. The assembly of the hinge and the pin provides the condition for the existence of the gap, and the gap between the hinge and the pin is inevitable in the process of manufacture, assembly and transportation.

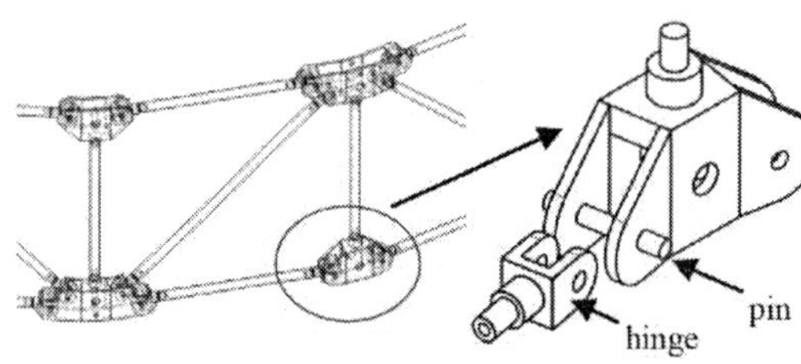

Figure 3. Sketch of rotary hinge

In this section, the radial dynamic characteristics of the planar revolute hinge in the annular truss antenna support structure are analyzed.

2.1. Static stiffness of hinge

Hertz contact theory is a theory of elastic non coordination of contact problems, the three-dimensional cylindrical contact is simplified to a two-dimensional contact, in Figure 4. The Hertz contact force

IOP Conf. Series: Materials Science and Engineering **398** (2018) 012008 doi:10.1088/1757-899X/398/1/012008

model is based on the complete elastic deformation, which mainly deals with the quasi-static contact problem of the elastic body, however, the energy loss during contact is not considered.

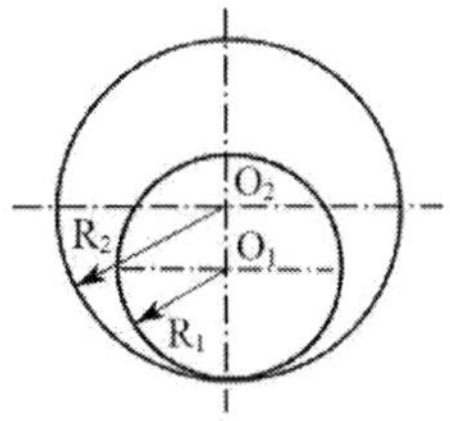

Figure 4. Contact state of rotary hinge

The Hertz contact force model is based on the complete elastic deformation, which mainly deals with the quasi-static contact problem of the elastic body, however, the energy loss during contact is not considered.

The Hertz contact force model can be described by a nonlinear spring

$$F_n = k_j \delta^n \tag{1}$$

where δ is the elastic deformation, k_j is the contact stiffness coefficient, the expression is as follows:

$$\mathrm{k_j} = \frac{4}{3\pi(\sigma_i + \sigma_j)} \left[\frac{R_1 R_2}{R_2 - R_1} \right]^{\frac{1}{2}} \tag{2}$$

where, σ_i , σ_j are contact modulus. $\frac{R_1 R_2}{R_2 - R_1}$ is the equivalent radius of contact.

$$\sigma_i = \frac{1-\upsilon_i^2}{\pi E_i}, \sigma_j = \frac{1-\upsilon_j^2}{\pi E_j} \tag{3}$$

The static stiffness is the relative stiffness at the moment of contact between the hinge and the pin.

2.2. Dynamic stiffness of hinge

The stiffness of the hinge under external excitation is studied, it is necessary to establish the dynamic stiffness model of clearance hinge. The natural frequency and dynamic stiffness of the hinge is derived. In figure 5, the motion of the clearance hinge can be divided into three states: gap state, transition state and contact state.

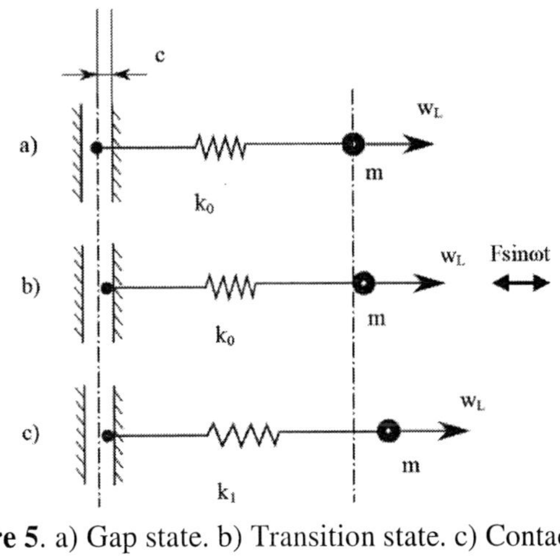

Figure 5. a) Gap state. b) Transition state. c) Contact state

In the contact and noncontact conditions, the equations of motion for the equivalent systems are given as

$$m\ddot{w}_L + k_0 w_L = 0 \ (w_L \le \varepsilon)$$

(4)

$$m\ddot{w}_L + k_1 w_L = (k_1 - k_0)\varepsilon \ (w_L > \varepsilon) \tag{5}$$

where $\ddot{w}_L$ is the acceleration of m, w_L is the deflection of m.

In the state, the displacement of the mass can be set

$$w_L = C_1 \sin \omega_0 t + C_2 \cos \omega_0 t \tag{6}$$

where C_1, C_2 are coefficient related to initial condition. ω_0 is the circular frequency of the hinge in the gap state. The initial condition of the hinge motion is assumed to be $t = 0$, $w_L = 0$, $\dot{w}_L = v_0$, $C_1 = v_0 / \omega_0$, $C_2 = 0$. When the hinge is in the transition state, according to the displacement continuity condition, $t = t_1$, $x = c$. Moreover, according to the law of conservation of energy, when the vibration displacement peak value is A, the potential energy is equal to the kinetic energy of the initial system

$$mv_0^2 / 2 = k_1 A^2 / 2 \tag{7}$$

The duration of the gap state is calculated

$$t_1 = \frac{1}{\omega_0} \arcsin \frac{\omega_0 c}{\omega_1 A} \tag{8}$$

where ω_1 is circular frequency of the contact state,

$$\omega_1 = \sqrt{\frac{k_1}{m}} \tag{9}$$

The duration of the contact state is

$$t_2 = \frac{1}{\omega_1} \arccos \frac{c}{c + A\lambda^2} \tag{10}$$

where λ is the natural frequency ratio of contact state and gap motion state.

$$\lambda = \frac{\omega_1}{\omega_0} \tag{11}$$

According to the literature [15], the natural frequency of the gap hinge structure is obtained by the superposition of the contact state and the gap state

$$f = \frac{1}{4}\left(\frac{1}{\omega_0} \arcsin \frac{\omega_0 c}{\omega_1 A} + \frac{1}{\omega_1} \arccos \frac{c}{c + A\lambda^2} \right)^{-1} \tag{12}$$

In the process of the radial movement of the hinge, when the hinge is in the gap state, $k_0 = 0$, and the natural frequency of the hinge can be expressed as

$$f = \frac{1}{4} \lim_{\omega_0 \to 0} \left(\frac{1}{\omega_0} \arcsin \frac{\omega_0 c}{\omega_1 A} + \frac{1}{\omega_1} \arccos \frac{c}{c + A\lambda^2} \right)^{-1} \tag{13}$$

$$f = \frac{\omega_1}{2\pi} \left(\frac{2c}{\pi A} + 1 \right)^{-1} \tag{14}$$

By Eq.14, the natural frequency of the hinge structure is directly proportional to the natural frequency of the contact section, meanwhile, the increase of the gap leads to the decrease of the natural frequency of the structure, furthermore, the natural frequency increases with the increase of amplitude.

The amplitude under external excitation is

$$w_{\max} = \frac{F / k}{\sqrt{[1 - (\omega / \omega_n)^2] + [2\zeta(\omega / \omega_n)]^2}} \tag{15}$$

where the circular frequency of the gap hinge structure is ω_n, k is static stiffness, ω is the circular frequency of external excitation, ζ is damping ratio. By Ref. [16], because the ζ is small, when the exciting force is small, the denominator of Eq.15 is simplified, and the dynamic stiffness k_d is obtained

$$k_d = kc_d = k\left|1-\left(\frac{\omega}{\omega_n}\right)^2\right| = k\left|1-[\frac{\omega}{\omega_1}\left(\frac{2c}{\pi A}+1\right)]^2\right| \tag{16}$$

Can be obtained by Eq.16, the dynamic stiffness coefficient changes with the natural frequency and the external excitation frequency of the gap hinge structure as well as the dynamic stiffness coefficient changes with the gap size and amplitude. Figure6 and Figure7 show the trend of dynamic stiffness coefficient. The dynamic stiffness can be seen from the chart, the influence of the gap size on the dynamic stiffness coefficient is significant.

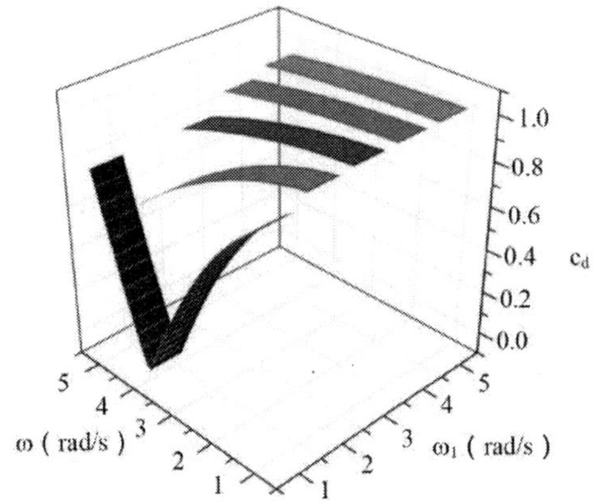

Figure 6. The trend of dynamic stiffness coefficient varying with frequency

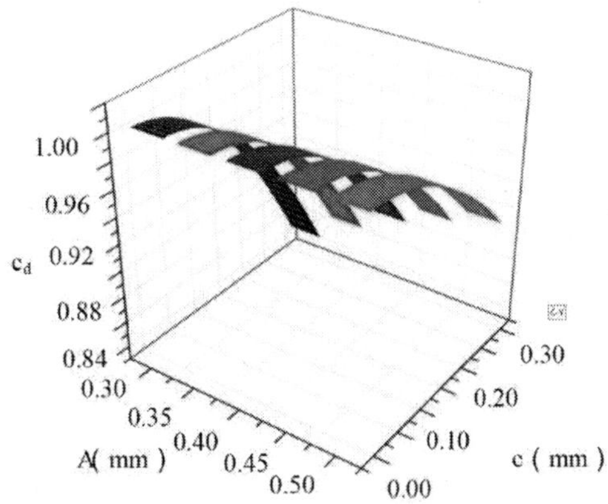

Figure 7. The trend of dynamic stiffness coefficient varying with amplitude and gap

3. Energy dissipation of the joint

The radial clearance of the pin-joint leads to the friction and impact when the vibration occurs. In order to model the joint damping in the hinge structure, it is necessary to analyze the radial energy dissipation.

The assembly error leads to the existence of sliding friction on the side of the hinge, and the energy dissipation caused by the sliding friction under the action of the F_p :

$$E_{slide} = \int_{t_1}^{t_2} c\dot{x}^2 dt = \int_{t_1}^{t_2} \frac{4\mu F_p}{\omega \pi A}\dot{x}^2 dt \tag{17}$$

where μ is the friction coefficient.

The friction force not only exists on the side of the hinge, but also occurs on the contact surface. The energy dissipation caused by contact friction can be obtained by analyzing the velocity of the contact surface. When the normal contact velocity is v, the contact half angle of the contact point is θ on the impact surface, the tangential velocity can be expressed in v_τ .

$$v_\tau = v\sin\theta \tag{18}$$

There are two contacts in the hinge in a period, so the energy dissipation caused by the contact friction is expressed as:

$$E_{conact} = 8\int_{t_1}^{t_2}\int_0^{\alpha} \mu N v \sin\theta d\theta dt \tag{19}$$

where N is the distributed contact force, α is the contact half angle, t_1 is the starting time of contact, t_2 is the termination time of contact.

Distributed contact force is:

$$N = \frac{E\sqrt{\delta - \theta^2 R}}{2\sqrt{R}} \tag{20}$$

where δ is contact depth, R is equivalent radius.

The velocity of the contact points in the contact area:

$$v = \frac{d\delta_w}{dt} = \frac{1}{2}\frac{\dot{\delta}\sqrt{\delta - \theta^2 R}}{\sqrt{\delta} + \sqrt{\delta}\sqrt{\delta - \theta^2 R}} \tag{21}$$

where $\delta_w = \sqrt{\delta}\sqrt{\delta - \theta^2 R}$.

The energy dissipation caused by contact friction can be expressed as:

$$E_{conact} = 8\int_{t_1}^{t_2}\int_0^{\alpha} \mu \frac{E\dot{\delta}}{4\sqrt{\delta R}}\left(2\delta - \theta^2 R\right)\sin\theta d\theta dt \tag{22}$$

The hinge clearance leads to the relative motion between the components, and when the two parts collide with each other the energy dissipated is

$$E_{impact} = \frac{1}{2}mv^2 \tag{23}$$

where v is velocity before impact.

Based on the above theory, the hinge radial energy dissipation produced by friction and impact can be expressed as:

$$E_{damp} = E_{slide} + E_{contact} + E_{impact} \tag{24}$$

The reduction of vibration caused by friction and collision can be measured by energy dissipation factor The energy dissipation factor is defined as the ratio of the energy dissipation to the total energy of a single cycle, so the energy dissipation factor D is:

$$D = \frac{E_{damp}}{E_{total}} \tag{25}$$

where E_{total} is total energy per cycle.

The relationship between the energy dissipation factor and the damping ratio is derived by combining the damping concept in the structural dynamics. The total energy of the structure at the beginning of deformation can be expressed as:

$$E_{total} = \frac{1}{2}ka_i^2 \tag{26}$$

where a_i is the displacement amplitude of periodic i, k is the stiffness of the structure. The energy dissipation expressed by displacement amplitude in a period is

$$E_{damp} = \frac{1}{2}k\left(a_i^2 - a_{i+1}^2\right) \tag{27}$$

Therefore, the energy dissipation factor can be expressed as

$$D = 1 - \left(\frac{a_{i+1}}{a_i}\right)^2 \tag{28}$$

The attenuation rate of structural vibration can be expressed as

$$\eta = \ln\frac{a_i}{a_{i+1}} = \ln(1 + \frac{a_i - a_{i+1}}{a_i}) \approx -1 + \frac{a_i}{a_{i+1}} \tag{29}$$

When the damping is small, the relationship between damping ratio η and energy dissipation factor D is obtained based on power series expansion

$$\xi = \frac{\eta}{2\pi} \approx \frac{0 + 1/2D}{2\pi} = \frac{E_{damp}}{4\pi E_{total}} \tag{30}$$

Therefore, the energy dissipation factor is proportional to the damping ratio, both of which can reflect the change of structural damping.

4. Radial dynamics experiment of the hinge

The dynamic experiment of the radial motion of the hinge structure is shown in Figure 8 and Figure 9. The external excitation of the hinge is sinusoidal, and laser sensor is used to acquire the displacement of hinge. The experimental part is the planar revolute hinge, the geometric dimensions as shown in Figure 10. The clearance is formed by matching the hinge size with the pin diameter, see table (1) for clearance data of hinge and column pin. Exert Sinusoidal excitation on the hinge, displacement of hinge radial vibration of will be acquired.

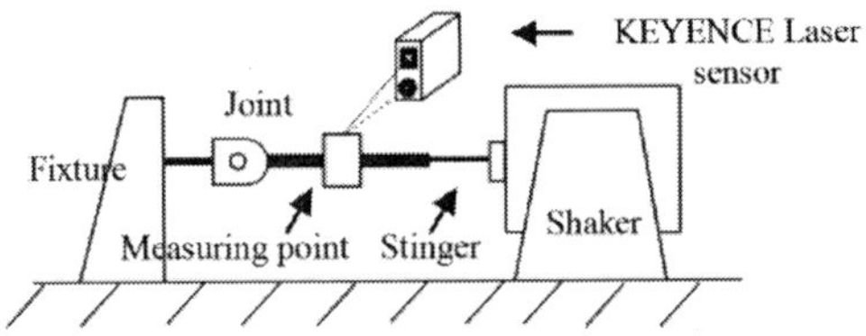

Figure 8. Experimental equipment layout

Figure 9. Experimental equipment

Table 1. Clearance value

R_2	R_1	c
3.0mm	2.1mm	0.45mm
	2.4mm	0.30mm
	2.5mm	0.25mm
3.2mm	2.1mm	0.55mm
	2.4mm	0.40mm
	2.5mm	0.35mm

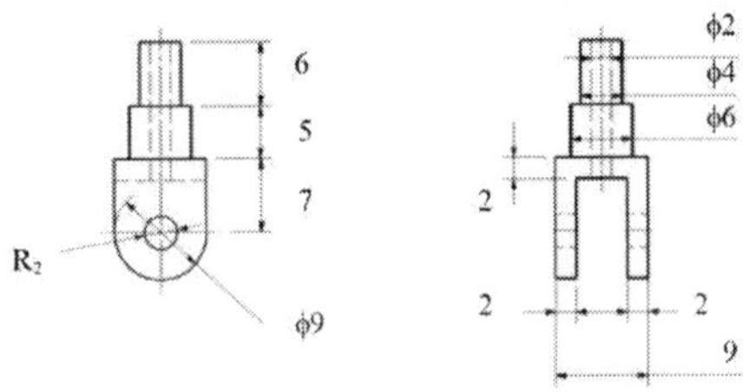

Figure 10. Geometry of hinge

When the excitation frequency is 5Hz, the clearance is 0.4mm, and the excitation amplitude is 0.604mm, the relationship between the external excitation amplitude and the hinge amplitude is shown in Figure11. From the chart, the external excitation is affected by the hinge motion and the amplitude is reduced; The time to reach the peak value of the hinge lags behind the external excitation. When the clearance decreases when the 0.3mm amplitude increased sharply as the Figure12, corresponding to the hinge and the dynamic stiffness coefficient will increase, and the theoretical calculation is consistent with the trend. The theoretical calculation is in good agreement with the experiment.

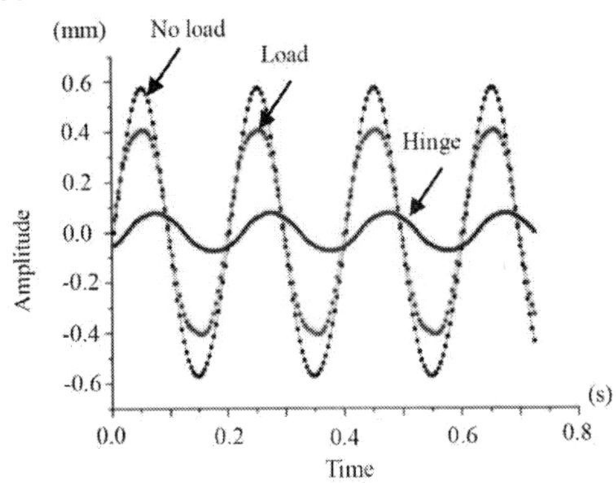

Figure 11. When the gap is 0.4mm, the hinge amplitude varies with the excitation

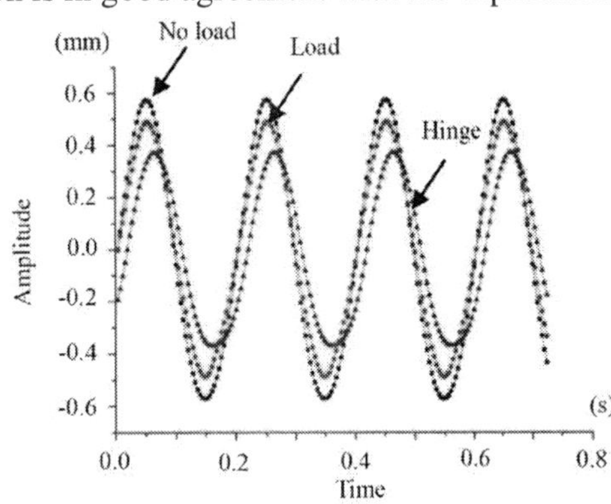

Figure 12. When the gap is 0.3mm, the hinge amplitude varies with the excitation

Apply a sweep excitation to the hinge, as shown in Figure13. The sweep frequency range is 1Hz-10Hz, the sweep time is 10s, and the clearance value is 0.4mm. The maximum amplitude of the external excitation is 0.604mm. In the initial stage, the input of 1Hz excitation, the amplitude of the hinge within a small range of shocks, when trigger sweep frequency, the amplitude of the hinge increases rapidly, thus it can be seen the excitation frequency promote the growth of the hinge amplitude, dynamic stiffness coefficient will decrease. Therefore, the external excitation increases the hinge dynamic stiffness coefficient and is consistent with the theoretical derivation.

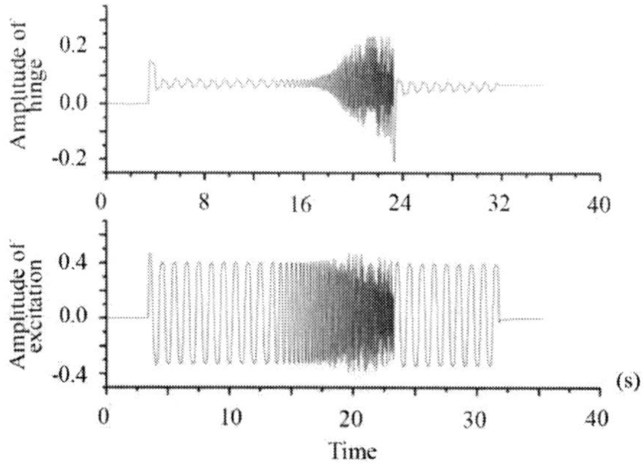

Figure 13. The amplitude variation of the hinge under swept frequency excitation

5. Conclusions

In this paper, the dynamic stiffness coefficient and damping of the clearance hinge structure in the large space deployable antenna ring truss structure under the radial motion are studied, the formulation of the radial clearance hinge dynamic expression of stiffness coefficient is carried out and verified by experiment, and meanwhile the damping characteristic is analysed. The size of the gap has a significant influence on the dynamic stiffness coefficient. When the gap is large, the dynamic stiffness coefficient increases obviously.

References

[1] Murphy D, McEachen M, Macy B, et al. *46th AIAA/ASME/ASCE/AHS/ASC Structures, Structural Dynamics and Materials Conference.* 2005: 2126.
[2] Chodimella S, Moore J, Otto J, et al. *47th AIAA/ASME/ASCE/AHS/ASC Structures, Structural Dynamics, and Materials Conference 14th AIAA/ASME/AHS Adaptive Structures Conference 7th.* 1916: 1603.
[3] Mitsugi J, Ando K, Senbokuya Y, et al. *ACTA ASTRONAUT*, 2000, 47(1): pp19-26.
[4] Tan G E B, Pellegrino S. Journal of Sound and Vibration, 2008, 314(3): pp783-802.
[5] Puig L, Barton A, Rando N. *ACTA ASTRONAUT*, 2010, 67(1): pp12-26.
[6] M.A. Brown. *Advances in Space Research*, 2011, 48(11): pp1747-1753.
[7] Moon F C, Li G X. *AIAA J*, 1990, 28(5): pp915-921.
[8] Bendiksen O O. *AIAA J*, 1987, 25(9): pp1241-1248.
[9] Hu H Y, Tian Q, Zhang W, et al. Advances in Mechanics, 2013, 43(4): pp390–414(in Chinese).
[10] Yoshida T. Journal of spacecraft and rockets, 2006, 43(4): pp771-779.
[11] Wang Y, Li F M. Applied Mathematical Modelling, 2015, 39(9): pp2518-2527.
[12] Lion C M. Journal of Vibration and Acoustics, 1991, pp113: 29.
[13] Wang J H, Liou C M. Journal of Sound and Vibration, 1990, 142(2): pp261-277.
[14] Crawley E F, Aubert A C. *AIAA J*, 1986, 24(1): pp155-162.
[15] Crawley E F, O'Donnel K J. *AIAA J*, 1987, 25(7): pp1003-1010.
[16] Ingham M D, Crawley E F. *AIAA J*, 2001, 39(2): pp331-338.
[17] Johnson K L. Contact mechanics. *Cambridge university press*, 1987.pp96-119.

Development of Under–anchorage Effective Prestress Detecting Instrument and Engineering Application

ZOU Zheng-qi[1] and GAO Hua-rui[2,*]

[1]CCCC First Highway Consultants Co., Ltd. , Xi'an, Shaanxi, China
[2]Geotechnical and Structural Engineering Research Center of Shandong University, Jinan, Shandong, China

*421334170@qq.com

Abstract. For detecting under–anchorage effective prestress of steel strand by lift–off test convenient. Revealing the testing principle of lift–off text and developing the testing in strument. The instrument consists of data acquisition card, oil pressure sensor, high precision displacement sensor and electric oil pump. Based on the graphical software platform LabVIEW, the test program is compiled by using G language. And further gives the program source code. Integrating the texting instrument (LTF 1.0) of under–anchorage effective prestress based on the software and hardware. So the engineering application is carried out. The research results show that the instrument can detect the under-anchorage effective prestress effectively by the load–displacement curve. Meanwhile it has the advantages of high precision, good safety and convenient operation. Through field application, the effective prestress of prestressed steel strand anchor was increased from 54.4% to 94.7%, and same strand dispersion was reduced from 10.2% to 1.7%.

1. Introduction

Prestressed anchoring technology is widely used in slope engineering, bridge engineering and geotechnical reinforcement projects. The tension quality of prestressed steel strands is closely related to the engineering quality [1] [2]. For the bridge project, if the prestressed anchor is too large, it will lead to deformation of the beam body. If the prestressing force of the anchor is insufficient, it will cause the cracking of the beam bottom and the deflection of the beam body, which will affect the safety of the project [3] [4]. Therefore, it is very necessary to carry out effective prestressed anchoring test.

Anchor prestressed test is divided into damage detection and non-destructive testing. Nondestructive testing is mainly through the ultrasonic [5], vibration and other techniques to measure the tension, but environmental factors will greatly affect the non-destructive testing, and thus the measurement accuracy is low. Although the damage detection has high accuracy, it will cause permanent damage to the structure and strand, and is not easy to operate. In contrast, the lift–off test, as a kind of micro-damage detection method [6], can not only accurately measure the effective prestressing force of steel strand under anchor [7], but also lessens the damage to steel strand and can be used in actual engineering [8] [9].

At this stage, there is fewer research results on the lift–off test, less specialized manufacturers can produce equipment used in lift–off test, and the corresponding criteria in the software are not rigorous enough. In order to obtain the tensile value of strand more accurately and apply this technique to engineering better, this paper presents an instrument based on LabVIEW software to detect the effective prestress of strand by using the lift–off test.

2. LIFT-OFF PRINCIPLE

Lift–off test, also known as the anti-pull method, tension the prestressed steel strand located in the un-grouting corrugated pipe, resulting in the prestressed strand work load. In the actual testing process, the relationship between the tensile test value and the displacement (tensile-elongation curve) is analyzed, and the effective prestress value under the anchor can be obtained.

In the testing process, the prestressing strand is affected by three forces: F_0, the effective prestress under the anchor; F_1, the reverse tension applied during the inspection; F_2, the reaction force of the anchor on the clamping piece. The relationships among the three are as follows:

$$F_1+F_2=F_0 \quad (1)$$

Under the ideal condition, the curve of the anchor effective prestressing force measured by the lift–off text is shown in Fig.1.

The first stage: When jacking the prestressing strand with tension, it can be regarded as an elastic body, and its equivalent stiffness k is expressed as follows.

$$k=F_1/\Delta l=EA/L \quad (2)$$

In the formula, E is the strand elastic modulus; A is the strand cross-sectional area; Δl is the strand elongation of the exposed section.

The tension is applied on the exposed strand, then the elongation changes linearly with the increase of F_1, as shown in Fig. 1 OA. At the same time, F_2 gradually decreases.

The second stage: when the jacking force F_1 increases to the same as the effective pre-stressing anchor F_0, the clip between the exposed section and the anchoring section is disengaged, and the two parts of the strand are simultaneously stressed. In equation (2), L increases, the equivalent stiffness of the strand decreases, and the slope of the tension-extension curve decreases, as shown in Figure 1, section AB. At this point, the tension at point A is the effective working stress under the anchor.

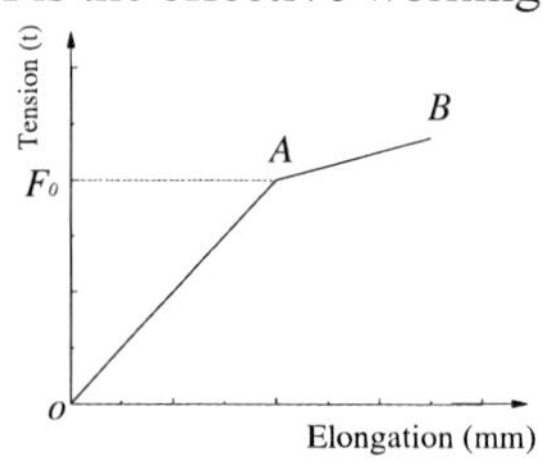

Figure 1. The tensile-elongation curve

3. ANCHOR EFFECTIVE PRESTRESS DETECTION SYSTEM RESEARCH AND DEVELOPMENT

3.1. System hardware and software environment

1) National Instruments (USA) develops LabVIEW software. LabVIEW is a program development environment that is significantly different from other computer languages in that LabVIEW uses a graphical editing language G to write programs that create charts and graphs instead of text lines. And LabVIEW uses data flow programming, and the data flow between nodes in the block diagram determines the order in which VIs and functions execute.

LabVIEW offers many controls similar in appearance to traditional instruments such as oscilloscopes and multimeters that can be used to easily create user interfaces. The user interface is called the front panel in LabVIEW. With icons and wires, you can programmatically control the objects on the front panel. The graphical source code for LabVIEW is somewhat similar to a flowchart, so it's called block diagram code.

LabVIEW has powerful data processing capabilities that give full play to your computer's capabilities, allowing users to define and assemble their own instruments to create more powerful instruments.

2) The chassis uses the NI cDAQ 9181 Ethernet chassis from NI, which accommodates four general-purpose 32-bit counters for use with the capture card.
System uses National Instruments NI 9215 acquisition card, the acquisition card has a sampling rate of 100kS / s, 16-bit resolution, can be 4 simultaneous sampling analog input.
Acquisition system also includes displacement sensor, electric oil pump and hydraulic pressure sensor. Acquisition hardware is shown in Figure 2. Among them, the hydraulic pressure sensor and the tubing are connected by using high-pressure thread tee.

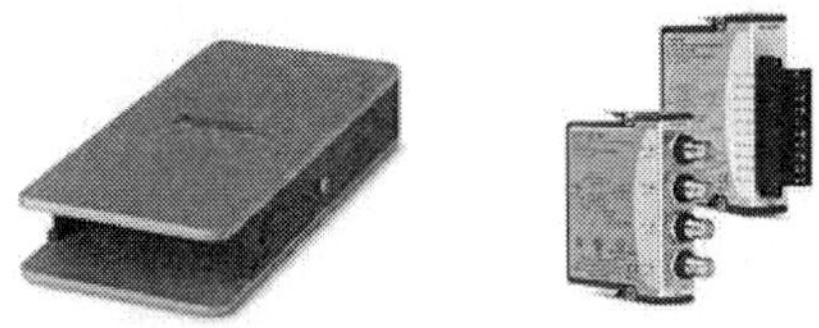

a) NI cDAQ 9181 chassis b) NI 9215 acquisition card

c) Electric oil pump d) Oil pressure sensor

Figure 2. Acquisition system hardware

3.2. Detection process

The system uses an electric oil pump to load the strand, the hydraulic pressure sensor fixed on the electric oil pump measures the oil pressure, and the displacement sensor fixed on the jack extends the measures elongation. The data is sent to the data acquisition card as an electrical signal, the acquisition card will be converted to a signal based on LabVIEW prepared by the data analysis program, the last tensile - extension curve through the LabVIEW front panel display.

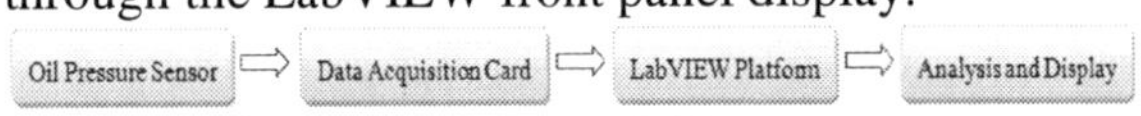

Figure 3. Data acquisition process

3.3. Data acquisition procedures

For hydraulic and displacement sensors, pressure and displacement changes are linear with the voltage. In order to ensure the accuracy of the actual test, the sensor can be calibrated through multiple measurements. The results are shown in Table 1 and Table 2.

Table 1. Hydraulic sensor calibration

Voltage（V）	0.68	1.36	2.04	2.66	3.35
Tension（t）	2	4	6	8	10

Table 2. Displacement sensor calibration

Voltage（V）	0.97	1.96	2.99	3.92	4.91
Elongation（mm）	10	20	30	40	50

After linear fitting the two sets of data, the voltage tension coefficient is 0.34 (V/t) and the voltage extension coefficient is 0.098 (V/mm). After the output voltage is input into the computer, the voltage needs to be analyzed and processed by LabVIEW.
We use hydraulic sensors and displacement sensors for signal acquisition. The LabVIEW program is shown in Figure 4.

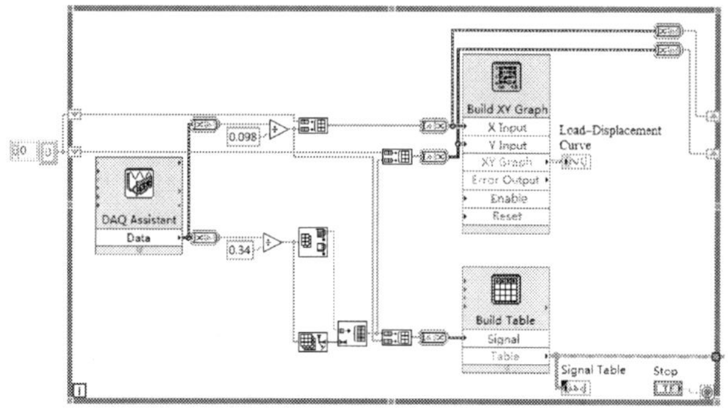

Figure. 4 Test program block diagram

The procedures for testing the following steps:

1) Configure the capture card. After the sensor is connected, then the data acquisition card should be configured. If the configuration is not reasonable, the work will be repeated and the acquisition process will be affected. To choose the correct measurement method, and the channel and related parameters are set to ensure the accuracy of the test.

2) Run LabVIEW virtual program. The sensor is connected to the acquisition card, and then run the program for data acquisition, the collected data is transmitted as an electrical signal to the computer through the DAQ Assistant. Enter the LabVIEW program for processing, the final output to be detected images. In the front panel, we can intuitively observe real-time detection of data.

3) The analysis and extraction of the data. When the detection image reaches the inflection point, click the "stop" button to complete the acquisition, and save the data for subsequent analysis and application.

4. PROJECT PROMOTION AND APPLICATION

4.1. Control passing rate

Based on the research software and test hardware, the anchor effective prestressing instrumentation (LTF 1.0) with lift-off test is integrated. Based on test equipment, we have carried out the application of engineering promotion. Choose to extract a 14th of the main bridge to explain. As shown in Fig. 5, three strands (19 root / bunch) of prestressed steel strands were randomly taken from this section and tested by using the instrument (LTF 1.0).

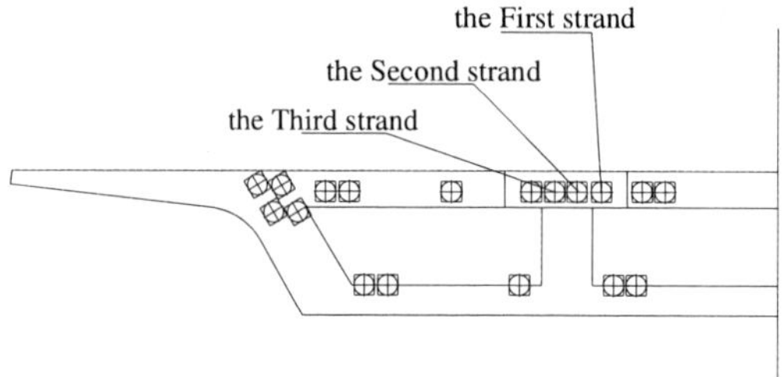

Figure 5. Test bridge cross-sectional view

The actual site test is shown in Figure 6. Three strands (randomly selected)'s single pass rate are only 54.4%, and the test results are shown in Figure 7. It can be seen that due to the quality of construction and other reasons, the tension of multiple strands can't meet the standard requirements, affecting the quality of the project.

This instrument can detect the problems in the actual on-site testing process and guide the on-site construction. In the subsequent construction, the bridge was tested again, the test results were shown in Figure 8.

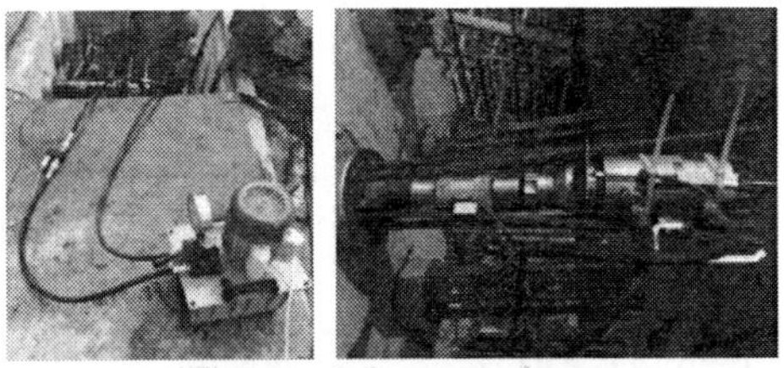

Figure. 6 the actual test

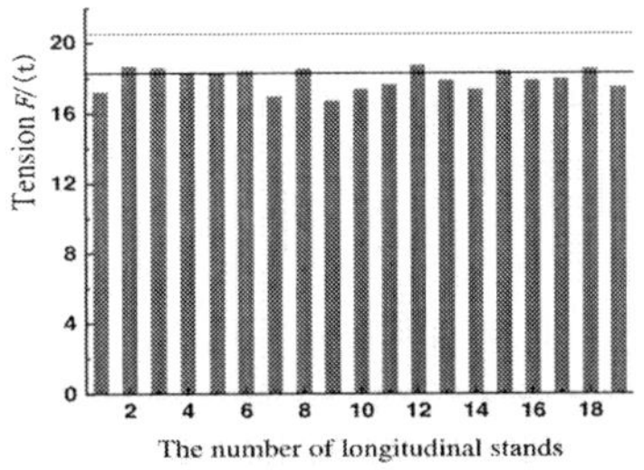

Figure 7. Strand tension (1)

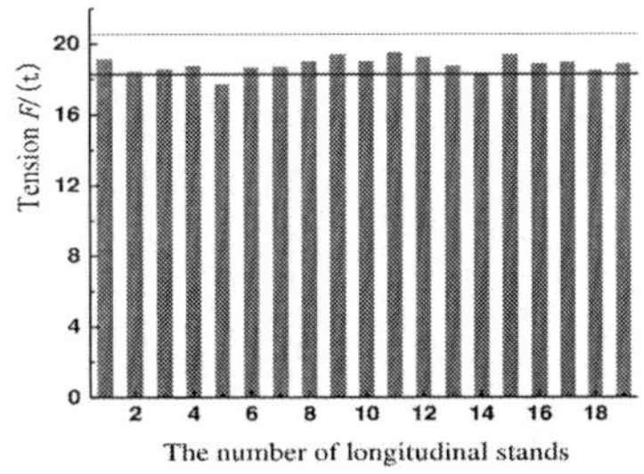

Figure 8. Strand tension (2)

By comparing the two tests can be seen, the first test strand average tension is 17.95t, did not meet the testing standards. The average tensile force of the second test strand is 18.68t, which reached the test standard, and the passing rate of single strand increases from 54.4% to 94.7%. The results show that the detection of tension control can make the strand average tension significantly improved.

4.2. Control the same strand dispersion

The same strand dispersion reflects the uniform distribution of tensile force. When the same strand dispersion is too large, the strand stress is not uniform, which will make a single strand break off, resulting in construction accidents. Therefore, it is necessary to control the same strand dispersion. Select the two sets of strand data before the tension control and after the tension control are compared. Before controlling the tension, the same strand dispersion of prestressing strand reaches 10.2%, in which the maximum tension of single strands in the strand reaches 21.94t, while the minimum is only 16.65t.The difference between the two reaches 5.29 t. This will seriously affect the mode of stress on the strand, or even cause wire breakage, causing serious engineering accidents. The distribution of tension in the test is shown in Figure 9.

In the subsequent construction, the steel strand was tested again. After guidance, the same strand dispersion significantly decreased, and the uniformity of the strand was guaranteed. Test tension distribution is shown in Figure 10.

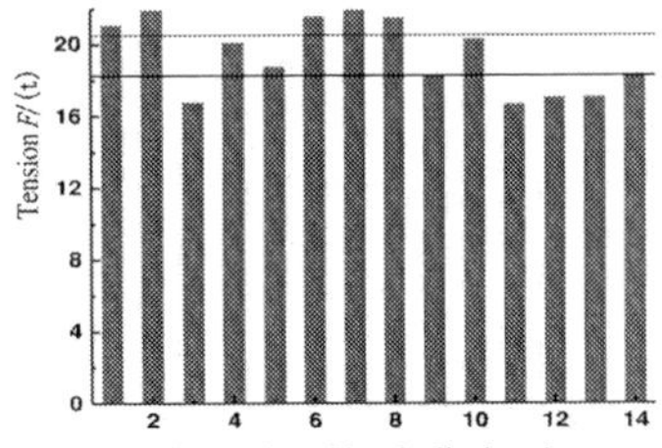

Figure 9. Strand tension (3)

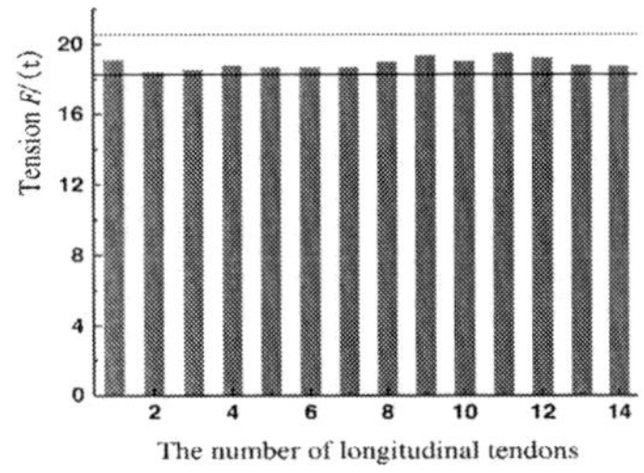

Figure 10. Strand tension (4)

It can be seen from the comparison that for the first time, the same strand dispersion is as high as 10.2% and does not reach the test control standard. Through the control of the tensioning process, the same strand dispersion has been reduced to 1.7% in the second test, which guarantees the construction quality. At the same time, it is shown that the same strand dispersion be effectively controlled by controlling the tension.

5. CONCLUSION

1) The principle of lift-off test is explained. We can tension the prestressed steel strand located in the un-grouting corrugated pipe, and analysis the tension - elongation curve to obtain the anchor effective prestress by using the lift-off test.

2) In this study, we developed the lift-off test equipment, which has the characteristics of high frequency, real-time, friendly software interface, easy to use and safe. After analyzing the tension - elongation curve, the anchor effective prestress can be detected accurately.

3) The testing instruments can be applied to the effective prestress test of prestressed anchorage structure in geotechnical engineering and bridge engineering. Comparing the results of before and after the tension control, the passing rate of effective tension increases from 54.4% to 94.7%, and the same strand dispersion decreases from 10.2% to 1.7%.Tension control can significantly improve the quality of the project, and has a wide range of promotional value.

References

[1] Fu Dan. Prestressed anchor working stress test method [D]. Beijing: Tsinghua University, 2010.

[2] Zhan Guichao, Liu Guanzhi.Analysis of the tensionof prestressed steel strand based on lift-off test[J]. Highway and Automotive Applications, 2016 (3): 191-193.

[3] Wang Jicheng. Bridge prestress and cable tension measurement and control technology[M]. China Communications Press, 2010.

[4] Yang Guoqiang, Xu Xiangfeng, Yao Chen, et al.Application of lift-off text in detecting effective tension under anchor [J] .Highway, 2016 (10): 81-84..

[5] Chen R H L, Wissawapaisal K. An ultrasonic method for measuring tensile forces in a seven-wire prestressing strand[J]. AIP Conference Proceedings, 2002, 615: 1295–1302.

[6] Fu Dan, Guo Hongxian, Cheng Xiaohui, et al. Working stress measurement of prestressed anchor cables: Detection mechanism and experimental study of lift-off test[J]. Rock and Soil Mechanics, 2012, 08: 2247–2252.

[7] Zhang Feng, Gao Lei, LiShucai et al. Test technology of under-anchorage effective prestress of steel stand in construction period[J].Journal of Northeastern University(Natural Science), 2017,38 (8): 1201-1205.

[8] Cheng Jianbo, Yao Chen, Zhang Feng.Model test research on lift-off test of prestressed steel strand working stress[J]. Highway and Automotive Applications, 2015 (3): 181-184..

[9] Xia Jing, Yuan Jianwei. Research on the effective prestressing detection method by the lift-off test[J]. Highway and Automotive Applications, 2016 (6): 210-212.

[10] Ministry of Communications of the People's Republic of China. JTG D62-2004 Code for design of highway reinforced concrete and prestressed concrete bridges and culverts[S]. Beijing: China Communications Press, 2004.

IOP Conf. Series: Materials Science and Engineering **398** (2018) 012010 doi:10.1088/1757-899X/398/1/012010

Simulation Analysis of Gas-solid Two-phase Flow for Heating Catalyst in Rotary Multi-cavity Kiln

Jingjing Xu[1] and Yuping Ma[1]

[1]School of Mechatronic Engineering and Automation,Shanghai University,Shanghai 200072, China

E-mail: xjj125@i.shu.edu.cn

Abstract. Based on CFD model framework, a simulation analysis of gas-solid two-phase flow in a multi-cavity kiln used for heating catalysts is presented. The heat transfer model includes radiation exchange among gas, refractory shell and catalyst, conduction inside the refractory shell, and convection heat transfer between the gas and the catalyst as well as the inner refractory shell wall. The contributions of heat transfer between the catalyst and the inner refractory shell wall by surface contact and heat transfer between the outer exposed refractory shell wall and surroundings have also been considered. The simulated results are in good agreement with experimental data.

1. Introduction

Rotary kilns are most commonly used in the metallurgical and chemical process industries for the thermal treatment of various granular solids. Catalyst heating production can also be performed in rotary kilns. Compared with a one-cavity kiln, the most significant advantage of a multi-cavity kiln is its remarkable ability in improving temperature uniformity of material to be heated and hence the productivity will be ultimately increased.

An analytical model was developed by Cook and Cundy [1] that described heat transfer processes between the rotating wall of a heated cylinder and an adjacent wet granular medium with an initial liquid water content equal to 2–7% of the mass of the dry solids. Manson and Unger [2] performed heat transfer processes in a rotary kiln incinerator and a detailed analysis could not be conducted. Sass [3] and Manitius et al. [4] developed heat transfer models for a rotary kiln dryer and an aluminum oxide rotary kiln, respectively. Ghoshdastidar has undertaken many studies about simulation of drying of materials in a rotary kiln. Ghoshdastidar et al [5] developed a detailed steady-state heat transfer model for the rotary kiln incinerator involving radiation, transient conduction and chemical reactions. P.S.Ghoshdastidar and Ankit Agarwal [6] presented a computer simulation and optimization study of heat transfer in a rotary kiln used for drying and preheating wood chips with superheated steam at 1 bar and conducted a parametric study in order to contribute to a good insight into the physics of the drying process in a rotary kiln. Koustubh Sinhal and P.S.Ghoshdastidar [7] reported a computer simulation of heat transfer inside a rotary kiln used for drying and preheating food products such as fruits and vegetables with superheated steam at 1 bar and conducted a detailed parametric study about the influence of controlling parameters on the axial solids and gas axial temperature profiles and the total predicted kiln length. Mastorakos et al. [8] performed a CFD-based model involving combustion, radiative heat transfer, conduction in the bed and walls and chemical reactions. Kaustubh S.Mujumdar and Vivek V. Ranade [9] reported a comprehensive computational fluid dynamics model to capture key

IOP Conf. Series: Materials Science and Engineering **398** (2018) 012010 doi:10.1088/1757-899X/398/1/012010

transport processes in rotary kilns. D.Sofialidis et al. [10] carried on a numerical simulation of the physico-chemical processes occurring inside a heated rotary kiln reactor.
The above literature review exhibits that no attempt has been made to simulate heating a catalyst in a rotary multi-cavity kiln, and there is little information about heat transfer inside a rotary kiln with the middle portion of the outer refractory shell wall acting as heat source. The present work aims to fill the gap in the existing literature.

2. Computational model

2.1. Model structure

The numerical simulation model is composed of three computational models, namely, the gas model, the catalyst model and the refractory shell model. The whole computational model is shown in Fig. 1 but the electrical heater and thermal insulation material are not included in the computational model. Dimensions of the rotary multi-cavity kiln are shown in Table 1.

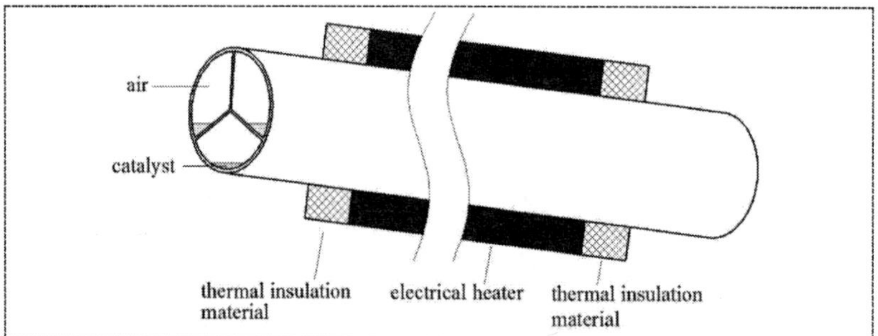

Figure 1. Schematic diagram of the rotary multi-cavity kiln.

Table 1. Dimensions of the rotary multi-cavity kiln.

Dimensions	Value
Length (m)	5.3
Inner diameter (m)	0.252
Outer diameter (m)	0.268
Baffle thickness (m)	0.006
Angle between baffles (degree)	120
Rotational speed (rpm)	0.7
Angle of inclination (degree)	5
Catalyst filling fraction in each cavity	7%

2.2 Hypothesis

- The moisture evaporation process is negligible in the article.
- Chemical reactions are not taken into account.
- Catalyst height and axial velocity in every cavity always remain unchanged.
- Interstitial gas among catalyst is neglected.
- Catalyst is treated as fluid.
- Kiln is assumed to be stationary.

2.3 Model equations

As mentioned above, the catalyst has been assumed as fluids, so the following equations of mass, momentum, and energy conservations are suitable for the catalyst and gas.
The continuity equation:

$$\frac{\partial \rho}{\partial t} + \nabla \cdot (\rho U) = 0 \tag{1}$$

The momentum equation:

$$\frac{\partial(\rho U)}{\partial t} + \nabla \cdot (\rho U \otimes U) = -\nabla p + \nabla \cdot \tau \tag{2}$$

The thermal energy equation:

$$\frac{\partial(\rho h)}{\partial t} - \frac{\partial p}{\partial t} + \nabla \cdot (\rho U h) = \nabla \cdot (\lambda \nabla T) + U \cdot \nabla p + \tau : \nabla U \tag{3}$$

Within the refractory shell, the conservation of energy equation can account for heat transport due to conduction.

$$\frac{\partial(\rho h)}{\partial t} = \nabla \cdot (\lambda \nabla T) \tag{4}$$

In the above equations, ρ is fluid density, p is fluid pressure, U is velocity vector, τ is stress tensor, h is static enthalpy and λ is thermal conductivity.
There is no need to specify thermal boundary conditions on each interface among the gas, the refractory shell wall and the catalyst, for these boundaries have temperatures and thermal fluxes calculated automatically.
The radiative transfer equation of the discrete transfer model can be written as:

$$\frac{dI_v(r,s)}{ds} = -(K_{av} + K_{sv}) I_v(\boldsymbol{r}, \boldsymbol{s}) + K_{av} I_b(v, T) + \frac{K_{sv}}{4\pi} \int_{4\pi} I_v(\boldsymbol{r}, \boldsymbol{s}') d\Omega' \tag{5}$$

Where v stands for frequency, $\boldsymbol{r}$ stands for position vector, $\boldsymbol{s}$ stands for direction vector, $\boldsymbol{s}'$ stands for scattering direction vector, s stands for path length, K_{av} stands for absorption coefficient, K_{sv} stands for scattering coefficient, I_b stands for blackbody emission intensity, I_v stands for spectral radiation intensity which depends on position and direction, T stands for local absolute temperature, and Ω stands for solid angle.
As no suspended materials exist inside the multi-cavity kiln, the scattering radiation effect of heat rays is neglected. Hence, equation (5) can be simplified as:

$$\frac{dI_v(r,s)}{ds} = -K_{av} I_v(\boldsymbol{r}, \boldsymbol{s}) + K_{av} I_b(v, T) \tag{6}$$

2.4 Model mesh

Compared with tetrahedral meshes, even if there are less hexahedral meshes, higher calculation accuracy, less solving time and more reliable results could also be achieved. As a result, hexahedral meshes are utilized as much as possible. In order to favorably capture gas flow characters of the velocity and temperature boundary layers, the inflation technology are employed for high quality meshes. The cross-sectional mesh viewed from the axis of kiln is shown in Fig. 2.

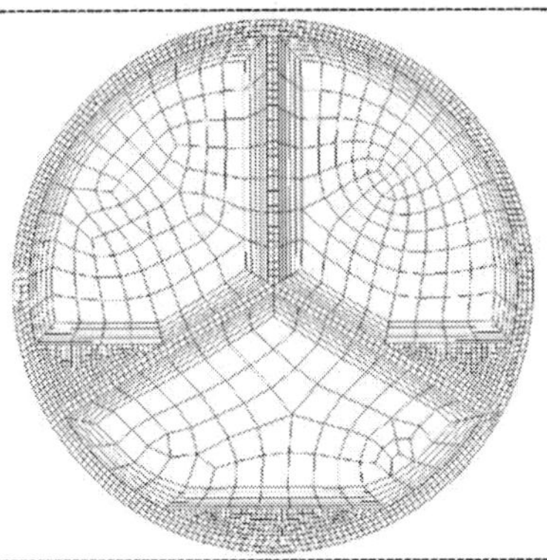

Figure 2. Cross-sectional mesh viewed from the axial direction.

IOP Conf. Series: Materials Science and Engineering **398** (2018) 012010 doi:10.1088/1757-899X/398/1/012010

3. Solution methodology

The software CFX has been applied to solve discretization equation to attain numerical results. Although governing equations were discretized by using first-order schemes (see Ref. [6]), high resolution schemes are taken to achieve higher accuracy and capture finer details for complex flow fields. Table 2 is the input data to the program.

Table 2. Input data to the program.

Refractory shell	
(a) Density (kg/m^3)	7500
(b) Specific heat (J/kg·K)	630
(c) Thermal conductivity (W/m·K)	25.5
(d) Emissivity	0.9
(e) Outer middle refractory shell wall temperature (K)	1118
(f) Convective heat transfer coefficient between the outer exposed shell wall and surroundings[a] (W/m2·K)	10
(g) Ambient temperature outside the multi-cavity kiln (K)	286
Catalyst	
(a) Stacking Density (kg/m^3)	1450
(b) Specific heat (J/kg·K)	3427
(c) Thermal conductivity (W/m·K)	0.4104
(d) Absorption coefficient (mm^{-1})	1
(e) Scatting coefficient (mm^{-1})	0
(f) Inlet velocity (m/s)	0.000184
(g) Inlet temperature (K)	286
(h) Outlet pressure relative to 1 bar (Pa)	0
Gas	
(a) Absorption coefficient (mm^{-1})	0.01
(b) Scatting coefficient (mm^{-1})	0
(c) Inlet velocity (m/s)	0.194
(d) Inlet temperature (K)	343
(e) Outlet pressure relative to 1 bar (Pa)	-50

[a]Obtained through engineering experience and provided by SINOPEC CATALYST COMPANY SHANGHAI DIVISION.

The termination criterion for iteration is:

- residual values for a convergence run nearly reach below 10-5;
- monitored interesting flow quantities such as the temperature and mass flow at all outlets approximately stay the same level;

The computations are performed on the DELL PRECISION T3500 workstation. The typical CPU time used for the steady state solution is about 19 hours.

4. Results and confrontation with experiments

4.1 Results and Discussion

The results (figures not shown) indicate that the circumferential and radial temperature distributions for the refractory shell are uniform because of the fact that the thermal conductivity of the refractory shell is large and that the baffle thickness is small, which will contribute to enhancing heat transfer by conduction.

The results (figures not shown) also reveal that, in respect of the catalyst in each cavity, circumferential and radial temperature variation is very small, and hence the catalyst temperature in every cavity could be considered uniformly distributed. This may be explained by the fact that the axial catalyst velocity is very low and the refractory shell temperature distributions are uniform axially and circumferentially. Three catalyst locations at three cavities are shows in Fig. 3. Fig. 4 shows axial average temperatures of catalyst in three cavities and implies that the temperatures are almost identical, especially in the high-temperature region.

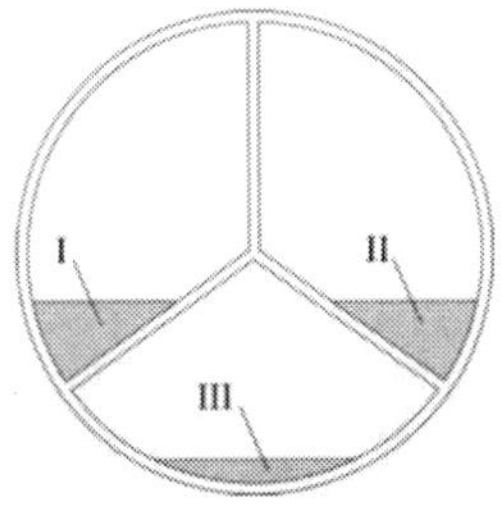

Figure 3. Catalyst locations in three cavities.

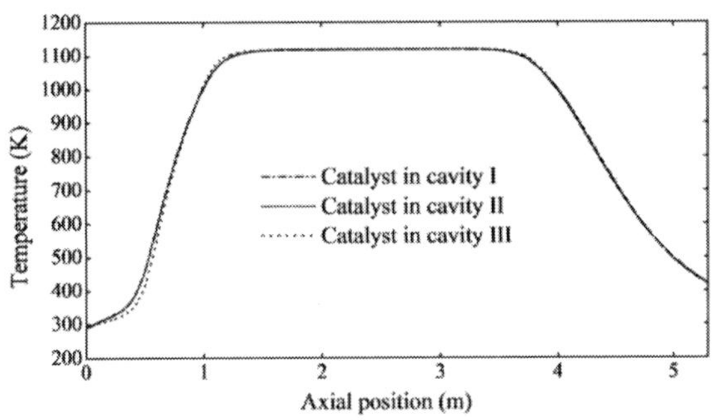

Figure 4. Axial temperatures of catalyst in three cavities.

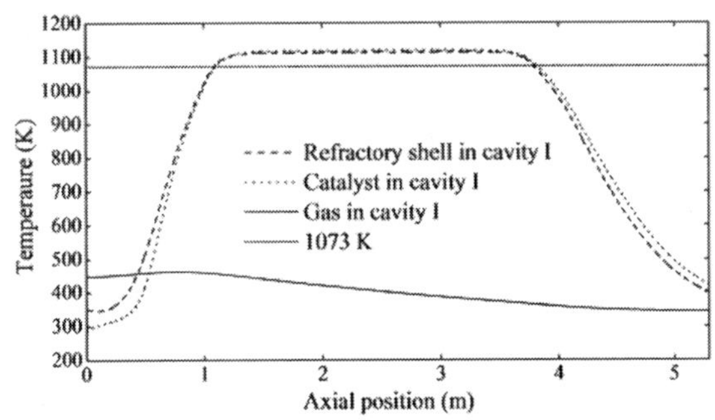

Figure 5. Axial temperature distributions of the refractory shell, catalyst and gas in cavity I.

Since the temperature distributions in three cavities for the gas, catalyst and refractory shell are nearly same, the cavity I has just been taken into account. Axial average temperature distributions of the refractory shell, catalyst and gas in cavity I are shown in Fig. 5. At the axial position 1.5~3.4 m, the axial average temperature of the refractory shell keeps constant 1118 K. When the axial position ranges from 1.5 m to 0 m and from 3.4 m to 5.3 m, the temperature becomes to decrease which can be explained by the fact that in the absence of the electrical heater, the outer exposed refractory shell wall has to transfer heat to surroundings by convection and radiation. At the axial position 0~1.0 m, the catalyst temperature is lower than the refractory shell, and the maximum difference is 35 K. This can be explained as follows. The catalyst inlet temperature is low and gradually absorbs thermal energy from the refractory shell. The catalyst temperature is almost the same as the refractory shell during the axial position 1.0~3.7 m. The catalyst temperature is higher than the refractory shell at the axial position 3.7~5.3 m, and the maximum difference is 20 K, which could be attributed to the direct heat transfer from the outer exposed refractory shell wall to surroundings through convection and radiation and indirect heat loss for the catalyst. When the gas flows into the multi-cavity kiln, the temperature increases slowly at the axial position 5.3~4 m and then goes up fast from the axial position 4 m to 0.7 m. This is because that the low-temperature gas needs to absorb thermal energy from the catalyst and the refractory shell and therefore the higher the temperatures for the catalyst and the refractory shell are, the faster the gas temperature rises. The gas temperature gradually decreases when the gas flow close to the outlet, because of the fact that heat transfer occurs from the gas to low-temperature catalyst and the refractory shell.

In order to satisfy the catalytic requirement of good performance, catalyst temperature must stay above 1073 K for at least 3.8 hours. From the simulation results, at the axial length ranging from about 1.10 m to 3.81 m, the catalyst temperature maintains higher than 1073 K, and the corresponding residence time in the high-temperature region is 4.09 hours (see Table 3) that satisfies the residence time requirement.

It will be essential to check the thermal energy distributions in the rotary multi-cavity kiln that could contribute to energy-saving analysis. Table 4 shows thermal energy distributions indicating that the overall thermal energy (heat source) provided by the middle portion of the outer refractory shell wall is 6578 W, heat loss from the outer exposed shell wall to surroundings is 3390 W, and the consumed thermal energy for heating the gas and the catalyst is 2805 W and 372 W respectively. The energy imbalance by iterative error is 0.17%.

Table 3. Axial position and residence time in the high-temperature region for the catalyst.

Parameters	Value
Axial position (m)	1.10~3.81
Residence time (h)	4.09

Table 4. Thermal Energy Distributions inside the Rotary Multi-cavity Kiln.[a]

	Heat source	Outer exposed shell wall to surroundings	Gas	Catalyst	Energy imbalance by iterative error (%)
Thermal energy (W)	+6578	-3390	-2805	-372	0.17

[a]The sign "+" means energy provided by the electrical heater and "-" means consumed energy.

4.2 Confrontation with Experiments

In order to verify the correctness of the simulation model, the relevant experiment was carried out. For the reason of intractability, the gas and catalyst temperatures were not acquired and just the temperature on one baffle surface could be measured. Fig. 6 shows two rows experimental points on the baffle surface viewed from the axial direction, named Row 1 and Row 2. Each row has 22 experimental points where thermocouple probes of contact sensor were located. The temperature at each experimental point was measured twice and then the average temperature was acquired.

From simulation results, the temperatures of 500 points at each row were obtained. It is apparent from Fig. 7 that the maximum deviation between experimental points and simulated points in row 1 and in row 2 are 11.8% and 17.8% at the axial position 0.65 m, respectively. The other deviations are all below

10%. Hence, the experimental temperatures coincide with the simulated results and then the simulated method can be confirmed feasible.

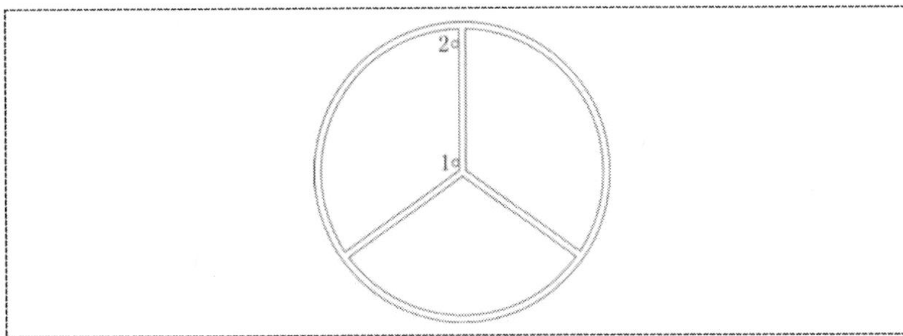

Figure 6. Experimental locations viewed from the axial direction.

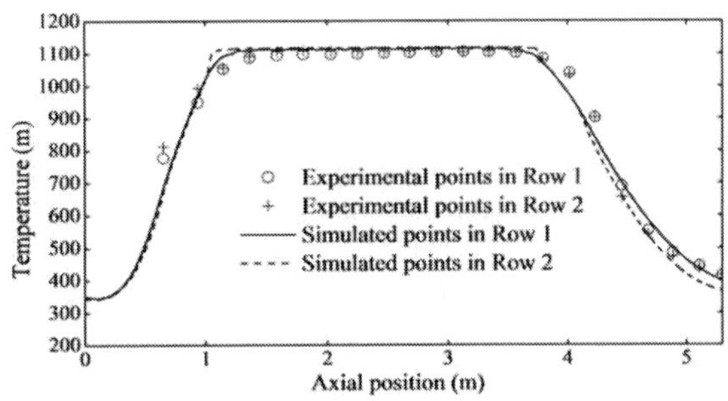

Figure 7. Temperature distributions of test points on the baffle surface: Comparison between experiment and simulation.

5. Conclusions

A comprehensive framework of a CFD-based model, composed of the gas, the catalyst and the refractory shell which are coupled together by mass and energy communication through common boundaries, has been addressed to simulate heat transfer in the rotary multi-cavity kiln where the catalyst is heated. The radiation, convection and conduction among the gas, the catalyst and the refractory shell are considered. The temperature on one baffle surface is in fine accordance with measured data. Although the rotary multi-cavity kiln in the paper is just a pilot kiln, the reults could serve as a guide of an industrial rotary multi-cavity kiln design that will be the further efforts to make. By using the above numerical stimulation method, the temperature and flow field during heating catalyst in an industrial model of a multi-chamber rotary kiln can be simulated and improved to meet catalyst performance, save energy, and provide valuable guidance for designing reasonably a gas-burning multi-chamber rotary kiln and technological parameters.

6.Acknowledgements

The authors acknowledge financial support from SINOPEC CATALYST COMPANY SHANGHAI DIVISION.

7.References

[1] Cook, Charles A.; Cundy, Vic A. Heat Transfer between a Rotating and a Moist Granular Bed. International Journal of Heat and Mass Transfer 1995, **38 (3)**, 419–432.

[2] Manson, L.; Unger, S. Hazardous Material Incineration Design Criteria. EPA Report 1979, No. 600/2-79-198.

[3] Sass, A. Simulation of the Heat-Transfer Phenomena in a Rotary Kiln. Industrial & Engineering Chemistry Product Research and Development 1967, **6 (4)**, 532–535.

IOP Conf. Series: Materials Science and Engineering **398** (2018) 012010 doi:10.1088/1757-899X/398/1/012010

[4] Manitius, A.; Kureyuz, E.; Kawecki, W. Mathematical Model of the Aluminum Oxide Rotary Kiln. Industrial & Engineering Chemistry Product Research and Development 1974, **13 (2)**, 132–142.

[5] Ghoshdastidar, P.S.; Rhodes, C.A.; Orloff, D. I. Heat Transfer in a Rotary Kiln During Incineration of Solid Waste, 23rd ASME National Heat Transfer Conference, Denver, Colorado, 1985, Aug. **4–7**, ASME Paper No.85-HT-86.

[6] Ghoshdastidar P.S.; Agarwal, Ankit. Simulation and Optimization of Drying of Wood Chips with Superheated Steam in a Rotary Kiln. Journal of Thermal Science and Engineering Applications 2009, **1**, 1-13.

[7] Koustubh Sinhal; Ghoshdastidar P.S. Computer Simulation of Drying of Food Products with Superheated Steam in a Rotary Kiln. Journal of Thermal Science and Engineering Applications 2012, **4**, 1-13.

[8] E Mastorakos; A Massias; C.D Tsakiroglou; D.A Goussis; V.N Burganos; A.C Payatakes. CFD Predictions for Cement Kilns Including Flame Modelling, Heat Transfer and Clinker Chemistry. Applied Mathematical Modelling 1999, **23(1)**, 55–76.

[9] Kaustubh S. Mujumdar; Vivek V. Ranade. CFD modeling of rotary cement kilns. Asia-Pacific Journal of Chemical Engineering 2008, **3**, 106-118.

[10] Sofialidis, D.; Faltsi, O.; Sardi, K.; Skevis, G.; Skodras, G.; Kaldis, S.P.; Sakellaropoulos, G. Modelling low-temperature carbonisation of solid fuels in a heated rotary kiln for clean fuel production. Fuel 2005, **84**, 2211–2221.

Study on Optimization Target of Camber of Independent Suspension

Yishan Sun and Zhicheng Wu

School of Mechanical Engineering, Beijing Institute of Technology, Beijing, China

E-mail: sys1996127@163.com

Abstract. For vehicles with independent suspension, the kinematic characterization and optimization of the wheels has a great influence on the handling stability of the vehicle. The target of the optimization is an important factor of assuring the good performance of the suspension after optimization. Through analyzing the difference between the target of wheel camber optimization while vehicles running in straight-line and curve, and comparing the difference between the target of wheel camber optimization while the suspension has linear and nonlinear stiffness, a design range of camber optimization was put forward according to actual running conditions of a vehicle, a method of confirming the target of camber optimization based on weighting method was carried out.

1. Introduction

Conventional studies about the target of suspension kinematics optimization are mainly focused on the wheel alignment parameters.

A researcher optimized the wheel alignment parameters in order to increase the handling stability of a vehicle, the goal of the optimization includes the variety range of camber is 1°, the variety range of caster angle is from 4°to 6°, the variation of inclination angle is small, the scrub radius is small and the toe angle is minimum while wheels jumping. [1]

A researcher determined the weight of every factor of the wheel alignment parameters and carried out a general function equation according to the weight. [2]

A researcher defined two general function equations according to the weight of every factor of the wheel alignment parameters. The camber and toe angle were assigned to a group and the inclination angle and caster angle were assigned to a group. Then the goal of the optimization was defined as two groups of goal. [3]

A researcher hoped to make the variation of wheel alignment parameters minimum while the wheel jumping and to assure the toe angle is negative number while the wheel jumping and toe angle is positive number while the wheel bounding, the toe angle characteristic of rear suspension is opposite to the front suspension and the variation of camber is negative number. [4]

A researcher optimized the characteristic of suspension through changing the position of the hard points and optimizing the wheel alignment parameters. [5]

A researcher built a model, analyzed the different factor of the camber and improved the driver's driving comfort by optimizing the camber.[6]

In our study, we analyzed the difference between the target of wheel camber optimization while vehicles running in straight-line and curve, and compared the difference between the target of wheel camber optimization while the suspension has linear and nonlinear stiffness. We put forward a design

IOP Conf. Series: Materials Science and Engineering **398** (2018) 012011 doi:10.1088/1757-899X/398/1/012011

range of camber optimization according to actual running conditions of the vehicle and carried out a method of confirming the target of camber optimization basing on the weighting method.

2. The difference between the target of camber optimization while vehicle travelling in straight line and curve

An ideal kinematic characterization of wheel has a positive impact on the running stability of the vehicle, among which the camber decides the contact between wheel and ground. Ideal camber change regulation should ensure that the wheel has good contact with the ground, which is to say, the wheel should be perpendicular to the ground.

2.1. The target of camber optimization while vehicle travelling in straight

In this condition, the car body will have vertical vibration by large margin when the vibration frequency is close to the resonant frequency and the roll of the car body can be ignored. In order to get good contact between the wheel and the ground, the variety range of camber should be assured as small as possible.

2.2. The target of camber optimization while vehicle travelling in curve

In this condition, the car body will roll by large margin. If the target of optimization is the same with the target when vehicle drives in straight, the wheel will roll with the car body. It will cause bad contact between the wheel and the ground and reduce the attachment coefficient and the tire life, as well as having a bad impact on the stability of vehicle as shown in Figure 1.

Figure 1. The wheel will roll with the car body when the suspension is designed conventionally

When vehicle drives in curve, the car body will roll. In order to get good contact between the wheel and ground, the wheel should be perpendicular to the ground, which is to say, the camber should be equal to the body roll angle φ_r ,the amber angle of inner wheel should be positive number and the amber angle of outer wheel should be negative number. The condition when the body rolls is shown as Figure 2.

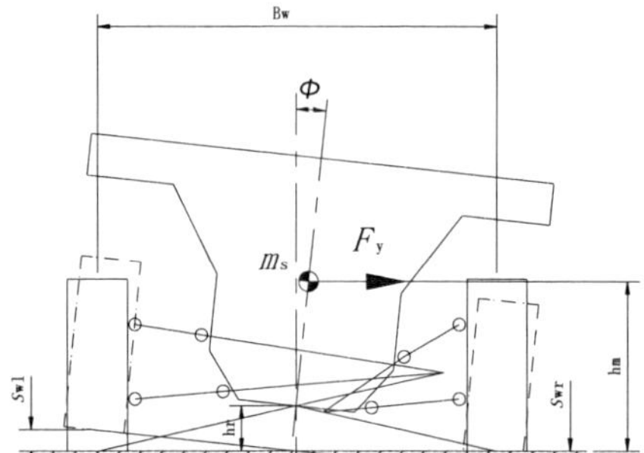

Figure 2. The body rolls and the wheel is perpendicular to the ground

The study above clearly shows that there is difference between the targets of camber optimization while vehicle travelling in straight and curve. So the target of camber optimization should be confirmed taking the actual running condition into consideration.

3. The difference between the target of wheel camber optimization while the suspension has linear and non-linear stiffness

3.1. Suspension of linear stiffness

As shown in Figure 2, the roll angle φ_r can be calculated as follows

$$\varphi_r = arctan\left(\left(s_{wl} - s_{wr}\right) / B_w\right) \tag{1}$$

s_{wl},s_{wr} represents the wheel travel of left suspension and right suspension, respectively. They are decided by the stiffness characteristic of suspension. B_w represents the wheel-base.

For suspension of linear stiffness, the lateral movement of barycenter caused by the roll of body can be ignored, so the wheel travel of left suspension and right suspension can be considered equal in the figure and opposite in the direction. So the relationship between the roll angle and wheel travel can be simplified as follows

$$\varphi_r = arctan\left(2s_{wl} / B_w\right) = -arctan\left(2s_{wr} / B_w\right) \tag{2}$$

The relationship diagram can be inferred from that equation which is shown in Figure 3.

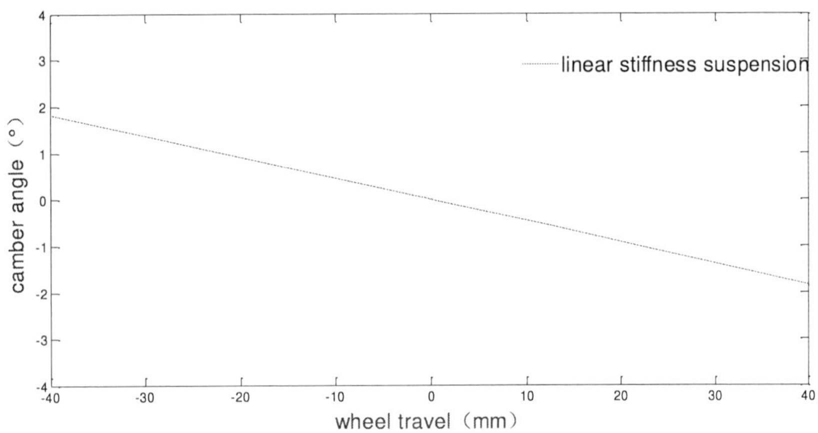

Figure 3. The relationship between wheel travel and camber in linear stiffness suspension

3.2. Suspension of non- linear stiffness

For suspension of nonlinear stiffness, the wheel travel of left wheel and right wheel is not the same. So the relationship between roll angle and wheel travel is complicated and it can't be confirmed by simple expression. So we took the speed as intermediate variable and inferred the relationship expression between roll angle and speed, as well as the expression between wheel travel and speed. Then the relationship expression between roll angle and wheel travel can be confirmed.

3.2.1.The expression between roll angle and speed. The roll angle can be calculated as follows

$$\varphi_r = M_{\varphi r} / K_{\varphi r} \tag{3}$$

$M_{\varphi r}$ represents roll moment, and $K_{\varphi r}$ represents roll angle stiffness.

$M_{\varphi r}$ has three parts, $M_{\varphi r1}$ is caused by the centrifugal force of sprung mass, $M_{\varphi r2}$ is caused by the gravity force of sprung mass, $M_{\varphi r3}$ is caused by the centrifugal force of un-sprung mass. So roll moment can be calculated as follows

$$M_{\varphi r} = M_{\varphi r}1 + M_{\varphi r}2 + M_{\varphi r}3 \tag{4}$$

*3.2.1.1.*Roll moment($M_{\varphi r1}$) caused by the centrifugal force of sprung mass. $M_{\varphi r1}$ which is caused by the centrifugal force of sprung mass can be calculated as follows

$$M_{\varphi r}1 = F_y L_r \tag{5}$$

F_y represents centripetal force, L_r represents the swing arm.

$$F_y = m_s v_a^2 / R_a = a_y G_s \tag{6}$$

a_y represents lateral acceleration, G_s represents the gravity force of sprung mass, L_r represents the difference between the barycenter of sprung mass(h_m) and the roll center(h_r),which means that

$$L_r = h_m - h_r \tag{7}$$

3.2.1.2. Roll moment($M_{\varphi r2}$) caused by the gravity force of sprung mass. When the body rolls, it will rotate around the roll center, and the barycenter will move. The distance of movement is L_e . so $M_{\varphi r2}$ can be calculated as follows

$$M_{\varphi r}2 = G_s L_e \approx G_s L_r \varphi_r \tag{8}$$

3.2.1.3. Roll moment($M_{\varphi r3}$) caused by the centrifugal force of un-sprung mass. $M_{\varphi r3}$ which is caused by the centrifugal force of un-sprung mass can be decided by the structure and the size of the suspension. As the un-sprung mass is much less than sprung mass, and the swing arm is almost equal to the radius of wheel, this part of moment has a little influence on the moment so it can be ignored in the process. Roll angle stiffness $K_{\varphi r}$ has two parts, which is angle stiffness $K_{\varphi r1}$ caused by the spring and angle stiffness $K_{\varphi r2}$ caused by the lateral stabilizer bar. So roll angle stiffness $K_{\varphi r}$ can be calculated as follows

$$K_{\varphi r} = K_{\varphi r}1 + K_{\varphi r}2 \tag{9}$$

3.2.1.4. Angle stiffness $K_{\varphi r1}$ caused by the spring. It is decided by the suspension stiffness (K_{su}) and stiffness of spring (K_{sp}).

$$K_{\varphi r}1 = 0.5 K_{sp} B_w^2 / i_s^2 \tag{10}$$

i_s represents the lever ratio of spring which is decided by the structure of suspension guide mechanism.

3.2.1.5. Angle stiffness $K_{\varphi r2}$ caused by the lateral stabilizer bar. It is decided by torsional rigidity of lateral stabilizer bar ($K_{\varphi b}$), the length of horizontal stabilizer bar's u-shaped arm (L_u) and the distance between horizontal stabilizer bar's post fulcrums(B_b). Angle stiffness $K_{\varphi r2}$ can be calculated as follows

$$K_{\varphi r}2 = K_{\varphi b}\left(\frac{B_b}{L_u}\right)^2 \tag{11}$$

So when the vehicle which has nonlinear stiffness suspension travels in curve, the relationship expression between roll angle and the speed can be derived as follows

$$\varphi_r = \frac{m_s v_a^2 (h_m - h_r) / R_a}{\dfrac{K_{sp} \cdot B_w^2}{i_s^2} + \dfrac{K_{\varphi b} \cdot B_b^2}{L_u^2} - m_s g (h_m - h_r)} \tag{12}$$

3.2.2. The relationship expression between wheel travel and speed. For independent suspension which has oil gas spring, the stiffness is nonlinear, and the stiffness increases as the wheel travel increases, the load characteristic of oil gas spring can be described as follows.

$$F_s = F_{sp} \cdot \frac{V_{sp}^k}{\left(V_{sp} - \pi\left(\dfrac{d_s}{2}\right)^2 \cdot L_s\right)^k} \tag{13}$$

F_s represents dynamic load, F_{sp} represents the spring load when the suspension is in static equilibrium position, V_{sp} represents the volume of the gas when the suspension is in static equilibrium position, d_s represents the diameter of the hydro-cylinder, L_s represents dynamic travel, k represents gas polytropic exponent.

Assume $F_s = F_{sp} + \Delta F_s$, ΔF_s is the increased load of oil gas spring, L_s can be carried out as follows.

$$L_s = \frac{V_{sp} - \dfrac{V_{sp}}{\sqrt[k]{\dfrac{F_{sp} + \Delta F_s}{F_{sp}}}}}{\pi\left(\dfrac{d_s}{2}\right)^2} \tag{14}$$

The wheel travel can be described as follows.

$$S_{w1} = i_s \cdot L_{sl}$$
$$S_{w2} = i_s \cdot L_{sr} \tag{15}$$

L_{sl}、L_{sr} represents the amount of spring compression of left and right suspension, respectively.
If there is no horizontal stabilizer bar, oil gas spring will bear all the increased load of the suspension (ΔF_z).

$$\Delta F_{sl} = i_s \cdot \Delta F_{zl}$$
$$\Delta F_{sr} = i_s \cdot \Delta F_{zr} \tag{16}$$

The increased load of the suspension(ΔF_z) can be calculated as follows.

$$\Delta F_{zl} = -\Delta F_{zr} \approx F_y h_m / B_w = m_s v_a^2 h_m / B_w R_a \tag{17}$$

m_s represents the sprung mass, v_a represents the speed, h_m represents the height of the barycenter, R_a represents the steering radius.
If there is horizontal stabilizer bar, it will bear part of the increased load of the suspension. It can be described as follows.

$$\Delta F_{sl} = \Delta F_{zl} - \Delta F_b$$
$$\Delta F_{sr} = \Delta F_{zr} + \Delta F_b \tag{18}$$

$\triangle F_b$ represents the load on horizontal stabilizer bar.

$$\Delta F_b = K_{\varphi b} B_b tan\varphi_r / L_u^2 \tag{19}$$

So the relationship expression between the wheel travel and the speed can be carried out as follows.

$$S_w = \frac{V_{sp} - \dfrac{V_{sp}}{\sqrt[k]{\dfrac{F_{sp} \pm m_s v_a^2 h_m / R_a B_w \mp K_{\varphi b} \cdot B_b / L_u^2 \cdot \tan[\dfrac{m_s v_a^2 (h_m - h_r) / R_a}{\dfrac{K_{sp} \cdot B_w^2}{i_s^2} + \dfrac{K_{\varphi b} \cdot B_b^2}{L_u^2} - m_s g (h_m - h_r)}]}{F_{sp}}}}}{\pi \cdot \left(\dfrac{d_s}{2}\right)^2} \cdot i_s \tag{20}$$

3.2.3. The relationship expression between wheel travel and roll angle. According to the equations above, when the vehicle with suspension of nonlinear stiffness travels in curve, the relationship expression between the wheel travel and roll angle can be calculated as follows.

$$S_w = \frac{V_{sp} - \dfrac{V_{sp}}{\sqrt[k]{\dfrac{F_{sp} \pm \varphi_r [\dfrac{K_{sp} \cdot B_w^2}{i_s^2} + \dfrac{K_{\varphi b} \cdot B_b^2}{L_u^2} - m_s g (h_m - h_r)] h_m / [B_w (h_m - h_r)] \mp K_{\Phi b} \cdot B_b \tan\varphi_r / L_u^2}{F_{sp}}}}}{\pi \cdot \left(\dfrac{d_s}{2}\right)^2} \cdot i_s \tag{21}$$

The relationship expression between the camber and wheel travel while running in curve can be shown in the Figure 4.

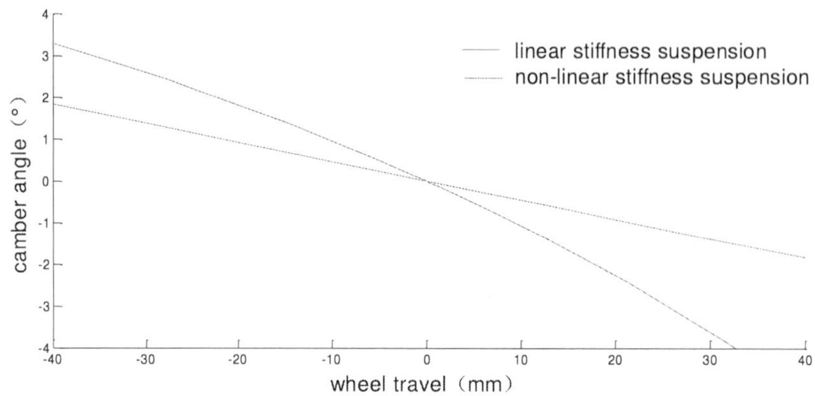

Figure 4. The relationship between wheel travel and camber in suspension of linear and nonlinear stiffness

4. The target of camber optimization settled by weight method

Taking the actual running condition of the vehicle into consideration, the design range should be the area between the target when vehicle travels in straight and the target when vehicle travels in curve. The shade shows the range as Figure 5.

If the vehicle travels in straight line more, the camber kinematic characterization should be more close to the straight line optimization objectives, if the vehicle travels in curve line more or the vehicle rolls a lot, the camber kinematic characterization should be more close to the curve optimization objectives. It should be decided according to the actual running condition.

So confirming the target of camber optimization should be based on a lot of experience and experiment data, it also needs a lot of statistics and analysis. In one of our tests about vehicles, we tracked a vehicle while running. After our analysis about the travelling time and road, we found that this vehicle travelled in curve 70 percent of the time, so the target of camber optimization can be confirmed according to the weight as shown in Figure 5.

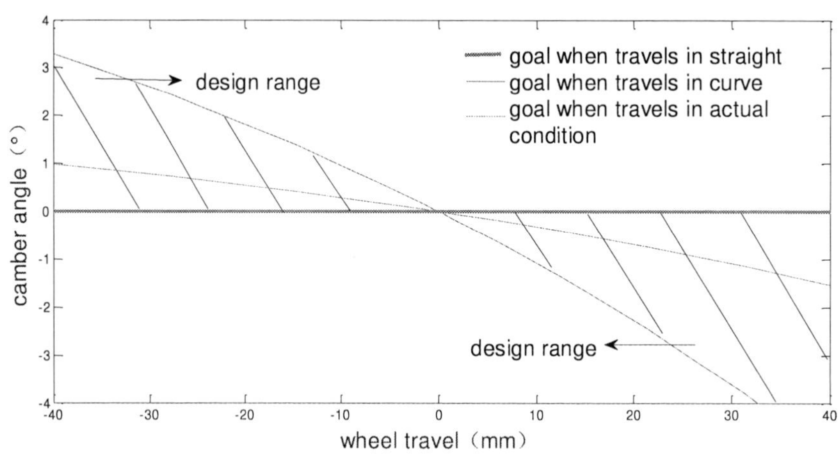

Figure 5. The design range when vehicle travels in actual condition

5. Conclusions

During the comparison and analysis, we come to the following conclusions:

(1) Through the analysis of different working conditions during straight travel and curve travel, the difference of the camber optimization target under different working conditions was obtained.

(2) Through analyzing the relationship expression between the camber and wheel travel under the curve driving condition, the difference between the target of wheel camber optimization while the suspension has linear and nonlinear stiffness was pointed out.

(3)According to the difference of the optimization target, the optimal design range of the wheel camber was given considering the curve and straight driving condition. The method of determining the target of camber optimization was proposed according to the weighted ratio of straight driving and curve driving under the comprehensive driving conditions

6. References

[1] Yu, HF The simulation of suspension's impact on handling stability based on ADAMS. [D]Dalian University of Technology 2007.DOI:10.7666/d.y1092407.

[2] Li YT, Zhao ZG, Zhang T Multi objective optimization of wishbone suspension based on Adams [J] Machine Design and Manufacturing Engineering. 2009, 38(17): 30-34,48.DOI:10.3969 /j.issn.1672-1616.2009.17.008

[3] Tao JJ. The design and optimization of mining dump truck's suspension[D]Hunan University,2015

[4] Liao YS, Liu YJ, Chen J The kinematic optimization of Macpherson suspension based on Adams. [J]Science Technology and Engineering 2012,12(20):4935-4940,4950.DOI:10.3969 /j.issn.1671-1815.2012.20.017.

[5] Lee H G, Won C J, Kim J W. Design sensitivity analysis and optimization of McPherson suspension systems[C] Proceedings of the World Congress on Engineering. 2009, 2: 1-3.

[6] Mahmoodi-Kaleibar M, Javanshir I, Asadi K, et al. Optimization of suspension system of off-road vehicle for vehicle performance improvement [J]. Journal of Central South University, 2013, 20(4): 902-910.

IOP Conf. Series: Materials Science and Engineering **398** (2018) 012012 doi:10.1088/1757-899X/398/1/012012

Modal Analysis of a Compressor Disk

G Liu[1], W Zhang[2] and A Xi[3]

[1] College of Mechanical Engineering, Beijing University of Technology, Beijing, 100124, PR China
[2] College of Mechanical Engineering, Beijing University of Technology, Beijing, 100124, PR China
[3] College of Mechanical Engineering, Beijing University of Technology, Beijing, 100124, PR China

liugen1991@hotmail.com

Abstract. The compressor disk is one of the key parts of the aeroengine. Its structure and strength play an important role in the reliability of the compressor. Most of the damage are caused by resonance stress. At the same time, the dynamic design and analysis of the key components are required because of the rotating state. When the compressor disk is in normal operation, the disk rotates around the rotating shaft in a large scale. Centrifugal inertia force will affect the deformation of the structure, and induce a certain change of natural frequency when compared to the static condition, which has increased complexity to modelling and numerical calculation. In this paper, the effect of dynamic stiffening on the vibration characteristics of disc is studied. In order to derive the equations of motion, Hamilton's principle is used, and the boundary conditions are obtained at the same time. We studied the natural frequency of the isotropic plate and laminate plate by Ritz method. Using the popular finite element analysis software ANSYS, we carry on simulation computation of vibration modes on a certain type of compressor in the working process. Besides, by investigation of the influence of structure parameter and rotating speed on the forward travelling wave and back travelling wave, we studied the critical speed of the compressor disk, which is essential in the rotator.

1. Introduction

Rotating structure, especially high speed rotating structure, is vulnerable under external disturbance and strong vibration. Sometimes a very small external perturbation would induce structural fracture, resulting in a serious accident. In recent years, many accidents have occurred in domestic and foreign aviation engines due to excessive vibration caused by the vibration of the disk fatigue damage. Therefore, research on transverse vibration characteristics of rotating circular plate under high speed has very important engineering value.

A perturbation technique had been used by Park et al. [1] to investigate the free vibration behaviour of rotating disks, and by Baddour and Zu [2] to investigate the nonlinear dynamic behaviour of rotating disks. The study of transverse vibration behaviour of spinning circular disks having one of the various models, such as rectangular, cylindrical or polar, has also been carried out [3-4].

Powmya et al. [5] presented solutions, which were based on principle of collocating the equations of motion at Chebyshev zeroes for the free vibration analysis of laminated plate. Hashemi et al. [6] dealt with the free vibration behavior of laminated transversely isotropic circular plates with axisymmetric rigid core attached to the center. Sujata et al. [7] showed that the fatigue crack of dimension as measured on the fracture surface is unstable, and it would lead to instantaneous fracture of the disk under the prevailing engine operating conditions. Debabrata et al. [8] studied the effects of the individual loadings and their combination on the free vibration dynamic behavior. Zhou [9] used the structural module of finite element of ANSYS to calculate the stress, deformation, natural frequency and mode shapes of gas compressor disc. The results show that the stress concentration occurred on balance hole.

It is evident from the literature review that research on large amplitude dynamic analysis of high rotating speed disc is scarce. At the same time, research on rotating disk of variable structure parameters is not available in the literature in spite of the tremendous potential in industrial applications. In this paper, the effect of dynamic stiffening on the vibration characteristics of disc is studied.

2. The modeling of compressor disk

2.1. Potential energy

The dynamic behaviour of a rotating annular disk having inner radius r_1 and outer radius r_2, rigidly mounted on a shaft, is investigated. The disk is considered to be flexible while the shaft is taken as a rigid body, as shown in Figure 1.

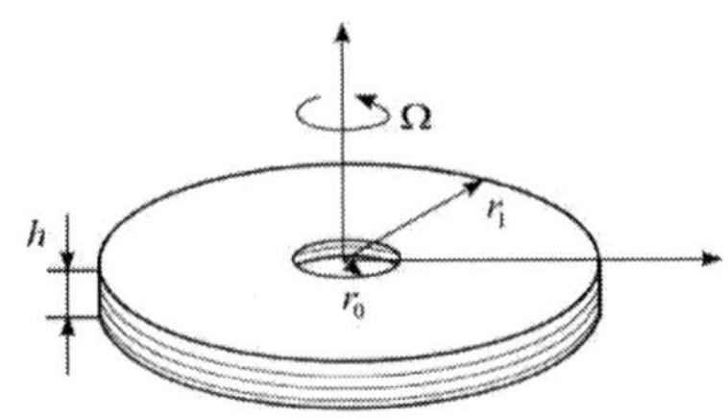

Figure 1. Schematic diagram of rotating disk

Also, the disk material is assumed to be isotropic and homogeneous and to follow linear elastic material behaviour.

In thin plate theory, it is usually assumed that the linear filaments of the plate initially perpendicular to the middle surface remain straight and perpendicular and do not contract or extend. This assumption leads to a relationship between the displacement of an arbitrary point and the displacements of the middle surface. They are given by,

$$u_r(\mathrm{r},\theta,\mathrm{z},\mathrm{t}) = u(r,\theta,t) + z\varphi_1(r.\theta,t) \tag{1a}$$

$$u_\theta(\mathrm{r},\theta,\mathrm{z},\mathrm{t}) = v(r,\theta,t) + z\varphi_2(r,\theta,t) \tag{1b}$$

$$u_z(\mathrm{r},\theta,\mathrm{z},\mathrm{t}) = w(r,\theta,t) \tag{1c}$$

Where u_r, u_θ, u_z denote the displacements of any point on the middle surface and φ_1,φ_2 are the rotations of the normal to the midplane about h, r axes, respectively.

The disk thickness is considered to be small compared to its outer radius in order to have the plane stress assumption for the present formulation valid and in order to neglect the effect of shear deformation and rotary inertia. When the transverse shear effects are neglected,

$$\varphi_1(\mathrm{r},\theta,\mathrm{t}) = -\frac{\partial w}{\partial r}, \varphi_2(r,\theta,t) = -\frac{\partial w}{r\partial\theta} \tag{2}$$

The von-Karman plate theory can be shown to lead to the following nonlinear stain-displacement expressions

$$\varepsilon=\begin{Bmatrix}\varepsilon_r\\ \varepsilon_\theta\\ \varepsilon_{r\theta}\end{Bmatrix}=\begin{Bmatrix}\dfrac{\partial u_r}{\partial r}\\ \dfrac{u_r}{r}+\dfrac{1}{r}\dfrac{\partial u_\theta}{\partial \theta}\\ \dfrac{1}{2r}\left(\dfrac{\partial u_r}{\partial \theta}-u_\theta+r\dfrac{\partial u_\theta}{\partial r}\right)\end{Bmatrix}+\begin{Bmatrix}\dfrac{1}{2}\left(\dfrac{\partial u_z}{\partial r}\right)^2\\ \dfrac{1}{2r^2}\left(\dfrac{\partial u_z}{\partial \theta}\right)^2\\ \dfrac{1}{2r}\dfrac{\partial u_z}{\partial r}\dfrac{\partial u_z}{\partial \theta}\end{Bmatrix} \tag{3}$$

Substitute equations (1a-1c) to (3), we can obtain

$$\varepsilon=\varepsilon_0+\varepsilon_1=\begin{Bmatrix}\dfrac{\partial u}{\partial r}+\dfrac{1}{2}\left(\dfrac{\partial w}{\partial r}\right)^2\\ \dfrac{u}{r}+\dfrac{1}{r}\dfrac{\partial v}{\partial \theta}+\dfrac{1}{2r^2}\left(\dfrac{\partial w}{\partial \theta}\right)^2\\ \dfrac{1}{r}\left[\dfrac{\partial u}{\partial \theta}-u+r\dfrac{\partial v}{\partial r}+\dfrac{\partial w}{\partial r}\dfrac{\partial w}{\partial \theta}\right]\end{Bmatrix}+z\begin{Bmatrix}-\dfrac{\partial^2 w}{\partial r^2}\\ -\dfrac{\partial^2 w}{r^2\partial \theta^2}-\dfrac{\partial w}{r\partial r}\\ -2\dfrac{\partial w}{\partial r}\dfrac{\partial w}{r\partial \theta}+\dfrac{\partial w}{r^2\partial \theta}\end{Bmatrix} \tag{4}$$

Where $\varepsilon_0,\varepsilon_1$ are linear and nonlinear strain vectors

According to the shear deformation theory, the constitutive equations for the *kth* layer of a polar orthotropic laminated plate can be written in the following form in polar co-ordinates

$$\begin{Bmatrix}\sigma_r\\ \sigma_\theta\\ \sigma_{r\theta}\end{Bmatrix}=\begin{bmatrix}\bar{Q}_{11} & \bar{Q}_{12} & 0\\ \bar{Q}_{21} & \bar{Q}_{22} & 0\\ 0 & 0 & \bar{Q}_{66}\end{bmatrix}\begin{Bmatrix}\varepsilon_r\\ \varepsilon_\theta\\ \varepsilon_{r\theta}\end{Bmatrix} \tag{5}$$

$$\begin{Bmatrix}\bar{Q}_{11}\\ \bar{Q}_{12}\\ \bar{Q}_{22}\\ \bar{Q}_{66}\end{Bmatrix}=\begin{bmatrix}C^4 & 2C^2S^2 & S^4 & 4C^2S^2\\ C^2S^2 & C^4+S^4 & C^2S^2 & -4C^2S^2\\ S^4 & 2C^2S^2 & C^4 & 4C^2S^2\\ C^2S^2 & -2C^2S^2 & C^2S^2 & (C^2-S^2)^2\end{bmatrix}\begin{Bmatrix}Q_{11}\\ Q_{12}\\ Q_{22}\\ Q_{66}\end{Bmatrix} \tag{6}$$

where $C=\cos\theta, S=\sin\theta$, $Q_{11,}Q_{12,}Q_{22,}Q_{66}$ are the elastic constants are expressed in terms of material constants in the plate co-ordinates as

$$Q_{11}=\frac{E_1}{1-\mu_{12}\mu_{21}}, Q_{12}=\mu_{12}\frac{E_1}{1-\mu_{12}\mu_{21}},\ Q_{22}=\frac{E_2}{1-\mu_{12}\mu_{21}}, Q_{66}=G_{12}=\frac{E}{2(1+\mu)}$$

where E is Young's modulus of elasticity , μ is Poisson's ratio.

The stress resultants acting on a laminate are obtained as,

$$U=\frac{1}{2}\iiint\left(\sigma_r\varepsilon_r+\sigma_\theta\varepsilon_\theta+\sigma_{r\theta}\varepsilon_{r\theta}\right)rdrd\theta dz \tag{7}$$

2.2. *Kinetic energy*

The original position of a particle is given by

$$\boldsymbol{r}_0=r\boldsymbol{e}_r+z\boldsymbol{e}_z \tag{8}$$

The deformed position of the same particle is given by

$$\boldsymbol{r}=\boldsymbol{r}_0+\boldsymbol{u}=(r+u_r)\boldsymbol{e}_r+u_\theta\boldsymbol{e}_\theta+(z+u_z)\boldsymbol{e}_z \tag{9}$$

The velocity of this particle is obtained

$$v=\frac{d\boldsymbol{r}}{dt}=\left(\dot{u}_r-\Omega u_\theta\right)\boldsymbol{e}_r+\left[\dot{u}_\theta+\Omega(r+u_r)\right]\boldsymbol{e}_\theta+\dot{u}_z\boldsymbol{e}_z \tag{10}$$

IOP Conf. Series: Materials Science and Engineering **398** (2018) 012012 doi:10.1088/1757-899X/398/1/012012

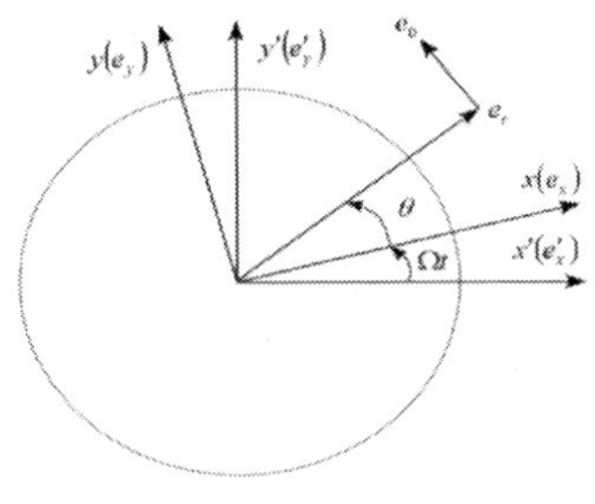

Figure2. Diagram of fixed coordinate system and dynamic coordinate system

According to the Figure 2, we can obtain

$$\frac{d\mathbf{e_r}}{dt}=\Omega\mathbf{e_\theta},\frac{d\mathbf{e_\theta}}{dt}=-\Omega\mathbf{e_r},\frac{d\mathbf{e_z}}{dt}=0 \tag{11}$$

Since the unit vectors are orthonormal, the square of the speed can be written as

$$\mathbf{v}\cdot\mathbf{v}=\left(\dot{u}_r-\Omega \mathrm{u}_\theta\right)^2+\left[\dot{u}_\theta+\Omega\left(\mathrm{r}+\mathrm{u}_\mathrm{r}\right)\right]^2+\dot{u}_z{}^2 \tag{12}$$

$$T=\frac{1}{2}\rho\iiint\left(v\cdot v\right)rdrd\theta dz \tag{13}$$

In order to derive the equations of motion, Hamilton's principle is used

$$\begin{aligned}\rho h&\left[\frac{\partial^2 u}{\partial t^2}-v\frac{\partial\Omega}{\partial t}-\Omega^2(r+u)-2\Omega\frac{\partial v}{\partial t}\right]=D_0\left[\frac{1+\mu}{2r}\frac{\partial^2 v}{\partial r\partial\theta}+\frac{\partial^2 u}{\partial r^2}-\frac{3-\gamma}{2r^2}\frac{\partial v}{\partial\theta}-\frac{1+\mu}{2r^3}\left(\frac{\partial w}{\partial\theta}\right)^2-\frac{u}{r^2}+\frac{\partial u}{r\partial r}\right]\\&+D_0\left[\frac{\partial w}{\partial r}\frac{\partial^2 w}{\partial r^2}+\frac{1+\mu}{2r^2}\frac{\partial w}{\partial r}\frac{\partial^2 w}{\partial r\partial\theta}\right]+G\frac{h}{r}\left[\frac{1}{r}\frac{\partial^2 u}{\partial\theta^2}+\left(\frac{\partial w}{\partial r}\right)^2+\frac{1}{r}\frac{\partial w}{\partial r}\frac{\partial^2 w}{\partial r\partial\theta}\right]\end{aligned} \tag{14}$$

$$\begin{aligned}\rho h&\left[\frac{\partial^2 v}{\partial t^2}+(r+u)\frac{\partial\Omega}{\partial t}-\Omega^2 v+2\Omega\frac{\partial u}{\partial t}\right]=D_0\left[\frac{1+\mu}{2r}\frac{\partial w}{\partial r}\frac{\partial^2 w}{\partial r\partial\theta}+\frac{3-\mu}{2r^2}\frac{\partial u}{\partial\theta}+\frac{1+\mu}{2r}\frac{\partial^2 u}{\partial r\partial\theta}+\frac{1}{r^2}\frac{\partial^2 v}{\partial\theta^2}+\frac{1}{r^3}\frac{\partial w}{\partial\theta}\frac{\partial^2 w}{\partial\theta^2}\right]\\&+\frac{Gh}{r}\left[r\frac{\partial^2 v}{\partial r^2}+\frac{1}{r}\frac{\partial w}{\partial\theta}\frac{\partial w}{\partial r}+\frac{\partial v}{\partial r}-\frac{v}{r}+\frac{\partial w}{\partial\theta}\frac{\partial^2 w}{\partial r^2}\right]\end{aligned} \tag{15}$$

$$\begin{aligned}\rho h&\left[\frac{\partial^2 w}{\partial t^2}++\frac{h^2\Omega^2}{3}\nabla^2 w-\frac{h^2\partial^2}{3\partial t^2}\nabla^2 w\right]=D_0\left[-\frac{h^2}{3}\nabla^4 w+\frac{3}{2r^4}\left(\frac{\partial w}{\partial\theta}\right)^2\frac{\partial^2 w}{\partial\theta^2}+\frac{1}{r^3}\frac{\partial^2 v}{\partial\theta^2}\frac{\partial w}{\partial\theta}-\frac{1}{2r^3}\left(\frac{\partial w}{\partial\theta}\right)^2\frac{\partial w}{\partial r}\right]\\&+D_0\frac{1+\mu}{2}r\frac{\partial w}{\partial\theta}\frac{\partial u}{\partial\theta}+D_0\frac{\partial w}{\partial\theta}\left[+\frac{1+\mu}{2}r^2\frac{\partial^2 u}{\partial r\partial\theta}+\frac{1-\mu}{2}rv\right]+D_0\frac{\partial w}{\partial\theta}\left[\frac{r^2}{2}\frac{\partial w}{\partial\theta}\frac{\partial^2 w}{\partial r^2}+r^2\frac{\partial w}{\partial r}\frac{\partial^2 w}{\partial r\partial\theta}+\frac{1-\mu}{2}r^3\frac{\partial^2 v}{\partial r^2}-\frac{1-\mu}{2}r^2\frac{\partial v}{\partial r}\right]\\&+\frac{1}{r^4}\frac{\partial w}{\partial r}\left[r^2\frac{\partial w}{\partial\theta}\frac{\partial^2 w}{\partial r\partial\theta}+(1+\mu)\mathrm{r}^3\frac{\partial u}{\partial r}+\frac{1+\mu}{2}r^3\frac{\partial^2 v}{\partial r\partial\theta}\right]+\frac{1}{r^4}\frac{\partial w}{\partial r}\left[\frac{1-\mu}{2}r^2\frac{\partial^2 u}{\partial\theta^2}+\frac{r^3}{2}\left(\frac{\partial w}{\partial r}\right)^2+\frac{r^2}{2}\frac{\partial w}{\partial r}\frac{\partial^2 w}{\partial\theta^2}+r^4\frac{\partial^2 u}{\partial r^2}+\frac{3r^4}{2}\frac{\partial w}{\partial r}\frac{\partial^2 w}{\partial r^2}\right]\\&-\frac{1-\mu}{2}r^2\frac{\partial v}{\partial\theta}\frac{1}{r^4}\frac{\partial w}{\partial r}+\frac{1}{r^4}\frac{\partial^2 w}{\partial\theta}\left[vr\frac{\partial u}{\partial r}+r(\mathrm{u}+\frac{\partial v}{\partial\theta})\right]+\frac{\partial^2 w}{\partial r^2}\left[\frac{\partial u}{\partial r}+\frac{v}{r}(\mathrm{u}+\frac{\partial v}{\partial\theta})\right]+\frac{1-\mu}{r^2}\frac{\partial^2 w}{\partial r\partial\theta}\left[\frac{\partial u}{\partial\theta}-v+r\frac{\partial v}{\partial r}\right]\end{aligned} \tag{16}$$

Where

$$D_0=\frac{Eh}{\left(1-\mu^2\right)},\nabla^2=\frac{\partial}{\partial r^2}+\frac{1}{r}\frac{\partial}{\partial r}+\frac{1}{r^2}\frac{\partial^2}{\partial\theta^2}$$

The boundary term obtained is given below

$$\frac{\partial}{\partial r}\left(\nabla^2 w\right)-\frac{1-\mu}{r^2}\frac{\partial^2}{\partial\theta^2}\left(\frac{\partial w}{\partial r}-\frac{w}{r}\right)=0 \qquad \frac{\partial^2 w}{\partial r^2}+\frac{\gamma}{r}\frac{\partial w}{\partial r}+\frac{\gamma}{r^2}\frac{\partial^2 w}{\partial\theta^2}=0$$

$$\frac{\partial u}{\partial r}+\frac{1}{2}\left(\frac{\partial w}{\partial r}\right)^2+\frac{\mu}{r}\left(\frac{\partial v}{\partial \theta}+u\right)+\frac{\mu}{2}\left(\frac{\partial w}{r\partial \theta}\right)^2=0 \qquad \frac{\partial u}{r\partial \theta}+\frac{\partial v}{\partial r}-\frac{v}{r}+\frac{1}{v}\frac{\partial^2 w}{\partial r\partial \theta}=0$$

Dynamic characteristics of transverse vibration of rotating circular plate can be described by the "wave theory" and "critical speed" theory. The vibration of a section is composed of two traveling waves in the opposite direction: forward traveling wave and backward traveling wave,

$$f_f=f_d+n\Omega \tag{17}$$

$$f_b=f_d-n\Omega \tag{18}$$

The n is the number of nodal radius. Apparently, when n=0, the frequency of forward travelling wave and backward travelling wave are the same. While when n$\geqslant$1, the frequency of forward traveling wave is bigger than backward traveling wave. When fb = 0, which means Ω= fb/n is the critical speed.

When a thin circular plate rotates at constant angular velocity, the equilibrium equation of the plate in the radial direction is expressed as

$$d\left(hr\sigma_r\right)/dr-h\sigma_\theta+h\rho r^2\Omega^2=0 \tag{19}$$

The radial and circumferential stress components in the equation are written as

$$\sigma_r=\frac{E}{1-\mu^2}\left(\mu\frac{u}{r}+\frac{du}{dr}\right),\sigma_\theta=\frac{E}{1-\mu^2}\left(\frac{u}{r}+\mu\frac{du}{dr}\right) \tag{20}$$

Substituting u and v into yields the equation

$$r^2\frac{d^2u}{dr^2}+r\frac{du}{dr}-u=\frac{1-\mu^2}{E}\rho r^3\Omega^2 \tag{21}$$

The general solution of equation (21) is obtained as

$$u=Ar+Br^{-1}-\frac{1-\mu^2}{8E}\rho r^3\Omega^2 \tag{22}$$

Combining the boundary conditions, the unknown coefficients can be obtained as follows

$$A=\frac{\left(1-\mu^2\right)\rho\Omega^2}{8E\left[r_0^2(1+\mu)+r_1^2(1-\mu)\right]}\left[r_0^4(3+\mu)+r_1^4(1-\mu)\right] \tag{23}$$

$$B=-\frac{\left(1-\mu^2\right)\rho\Omega^2 r_0^2 r_1^2}{8E\left[r_0^2(1+\mu)+r_1^2(1-\mu)\right]}\left[r_0^2(3+\mu)-r_1^2(1+\mu)\right] \tag{24}$$

3. Solution

To set up the eigen value problem for determination of free
vibration frequencies and the corresponding mode shapes in the present work, we can write the displacements in the form as follows

$$u(r,\theta,t)=\sum_{m=0}^{M}C_{rmn}V_{rmn}e^{i(n\theta+\lambda t)} \tag{25a}$$

$$v(r,\theta,t)=\sum_{m=0}^{M}C_{\theta mn}V_{\theta mn}e^{i\left(\frac{\pi}{2}n\theta+\lambda t\right)} \tag{25b}$$

$$w(r,\theta,t)=\sum_{m=0}^{M}C_{wmn}V_{wmn}e^{i\left(n\theta+\lambda t\right)} \tag{25c}$$

Where m, n are the number of the node cycle and the number of node diameter of the disk mode, M is the truncation order.

By substituting equations (7) and (13) into the Lagrange function, neglecting nonlinear terms

$$L=T_{\max}-U_{\max} \tag{26}$$

According to the Ritz method,

$$\frac{\partial L}{\partial C_{rmn}}=0,\frac{\partial L}{\partial C_{\theta mn}}=0,\frac{\partial L}{\partial C_{wmn}}=0 \tag{27}$$

The following frequency equations are derived

$$\left(K-\lambda^2 M\right)S=0 \tag{28}$$

Where K is the stiffness matrix, M is the mass matrix and S is the coefficients matrix, which can be written

$$S=\left\{C_{r01}\cdots C_{r1n}\cdots C_{rmn},C_{\theta 01}\cdots C_{\theta 1n}\cdots C_{\theta mn},C_{w01}\cdots C_{w1n}\cdots C_{wmn}\right\}$$

4. Numerical results and discussions

4.1. Modal analysis of the disk

The example model in this study is from the first-stage disk of the low-compressor in some type of aeroengine. The model parameters of isotropic plate are listed in Table 1.

Table 1. Parameters of isotropic plate

Structural parameters	Symbols	Values
Inner radius	r_0	30mm
Outer radius	r_1	90mm
Thickness	h	8mm
Density	ρ	7840kg/m^3
Young's modulus	E	210Gpa
Poisson's ratio	μ	0.3

Laminate plate parameters are: the number of layer k=8, the orientation is [0/45/-45/90], Ex=25x106N/m2, Ey=1x106N/m2, Gxy= Gxz =0.5x106N/m2, Gyz =0.2x106N/m2. The truncation order M=5, which can lead to good convergence.

The natural frequency and corresponding nodal diameter and nodal cycle are listed in Table 2.

Table 2. The natural frequency of a static and rotating disk

Mode		**Ω**=0		**Ω**=12000r/min	
(*m*, *n*)		isotropic	laminate	isotropic	laminate
(0.0)	FTW	919.80.	919.10	943.74	942.10
	BTW	919.80	919.10	943.74	942.10
(0.1)	FTW	954.27	953.96	967.21	967.10
	BTW	954.27	953.96	961.10	959.32
(0.2)	FTW	1107.81	1009.27	1121.43	1110.21
	BTW	1107.81	1009.27	1119.27	1109.10
(0.3)	FTW	1817.32	1815.31	1991.30	1971.24
	BTW	1817.32	1815.31	1984.32	1965.31
(0.4)	FTW	2183.81	2181.23	2278.43	2241.32
	BTW	2183.81	2181.23	2213.74	2201.78
(1.0)	FTW	2651.73	2649.54	2741.24	2710.37
	BTW	2651.73	2649.54	2720.34	3694.38
(1.1)	FTW	2783.49	2781.78	2841.32	2819.42
	BTW	2783.49	2781.78	2810.17	2801.38

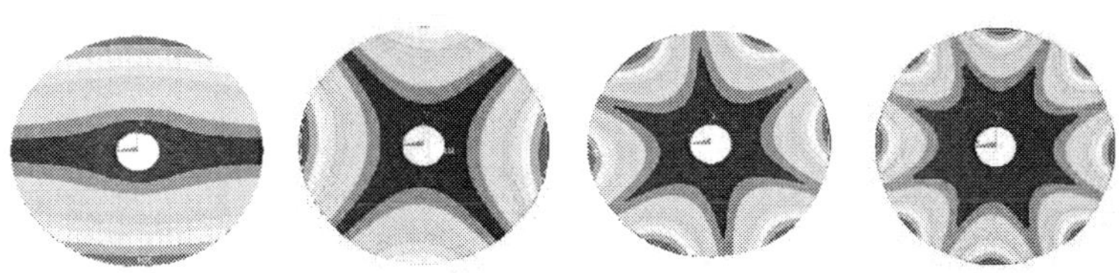

Mode (0, 1) Mode (0, 2) Mode (0, 3) Mode (0, 4)

Figure 3. Modal analysis of a compressor disk with pre-stress

From the Table 2, we can find that the dynamic frequency of the disc is larger than that of the static frequency due to the stress stiffening effect generated by the rotation. Besides, it is not difficult to find that with the increase of the node diameter, the amplitude of the dynamic frequency increases, and the amplitude is enlarged for the same node diameter.

4.2. The influence of structure parameter on natural frequency of the disk

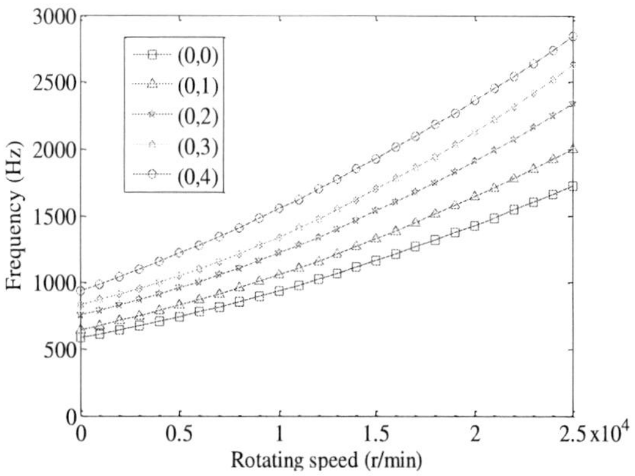

Figure 4

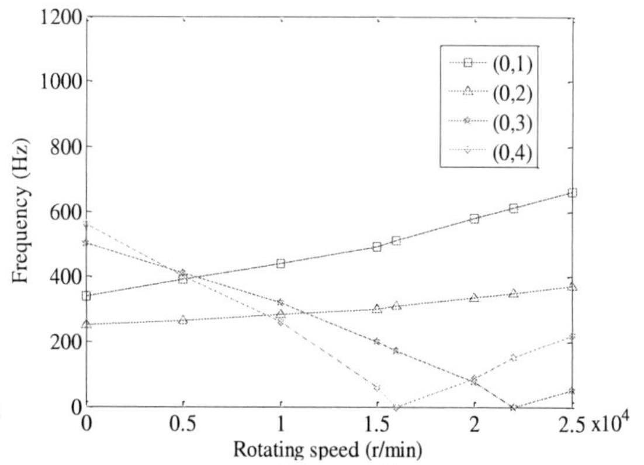

Figure 5

Figure 4. The natural frequency of isotropic plate with rotating speed (radius ratio=3, Backward travelling wave)

Figure 5. The natural frequency of isotropic plate various rotating speed (radius ratio=3, Forward travelling wave)

Since the critical speed is the angular velocity when the back traveling wave frequency is zero, the critical speed is the intersection of the back traveling wave frequency and the horizontal axis.

From the above two figures, the forward traveling wave frequency increases with the rotating speed increases, the back traveling wave frequency reduces with the rotating speed increases. Note that the critical speed of n=3 is greater than that of n=4.

There is non-intersection point of the back traveling wave and the horizontal axis when the number of node is 1 and 2, that is, there exists no critical speed. And for other higher-order modes, their vibration frequency increases at first with the increase of rotation speed and then decreases to zero, then the frequency is further increasing. That is to say, for laminated circular plate and single plate, there is also exit dynamic buckling phenomenon and has dynamic buckling critical speed.

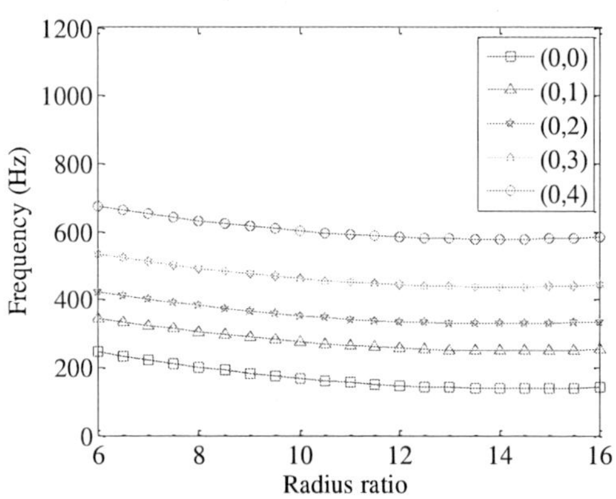

Figure 6

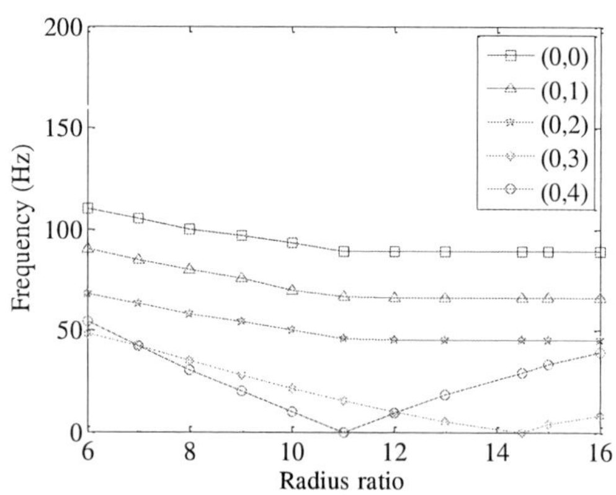

Figure 7

Figure 6. The natural frequency of laminate plate with various radius ratio (Ω=1200r/min, Forward travelling wave)

Figure 7. The natural frequency of laminate plate with various radius ratio (Ω=1200r/min, Backward travelling wave)

From the above two figures, we can obtain that when the radius ratio is over 10, the curve is relatively flat, which demonstrate the influence of radius ratio on the modal frequency of the front traveling wave is more and more small with the increase of the radius ratio. While the change trend of vibration frequency of back traveling wave is that with increase of the radius ratio, the frequency decreases to zero first and then increases. The reason could be due to that the increase of the radius ratio decreases the total bending stiffness of laminated circular plate, then makes the vibration frequency decrease.

4.3. Critical speed of the disk

The critical speed of dynamic instability is the most concerned parameter, which determines whether the whole vibration system is safe or not. Using the research of the influence of geometric parameter to circular plate, we could optimize the design and improve the system critical speed, in turn made the rotating circular plate work more stable. Here, we discussed the critical speed of laminated circular plate with different radius ratio.

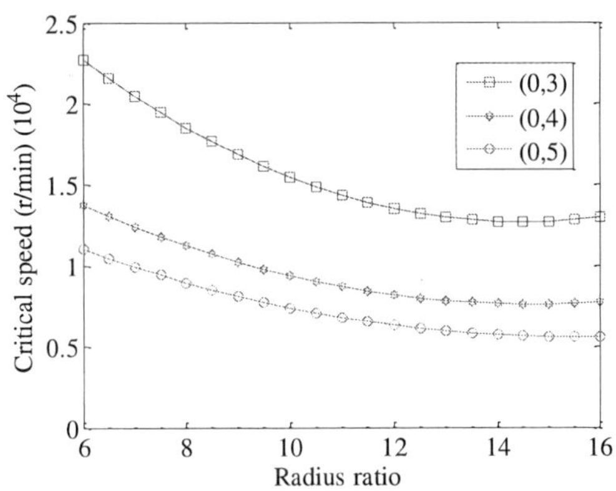

Figure 7. The critical speed of laminate plate with various radius ratio

Our results here show that for lower order modes n=0, n=1 and n=2, there does not appear the case that the back traveling wave vibration frequency is zero, that is, the three modes are stable. This figure also shows the change of the critical speed of 3 kinds of high order modes n=3, n=4 and n=5. From the graph, we can note that with the radius ratio increases, critical speed of each mode decrease. This is due to that the increase of radius ratio reduces the bending stiffness of layer laminated circular plates, which make the natural frequency decrease, then decreases the critical speed.

5. Conclusions

This paper established dynamic equation of the rotating circle plate and got the corresponding boundary conditions. As the Lagrange method is a good solution to solve the forward and back travelling wave mode identification problem, the effect of geometric parameters on the critical speed of rotating laminated circular plate are discussed. Conclusions are drawn as follows,

(1) Composite materials and isotropic materials have different material properties, such as elastic modulus, density and so on, so their natural frequency characteristics are different. According to the results obtained in Figure 10, to improve the dynamic performance of the structure and reduce the plate deformation with high rotating speed, composite structures are more appropriate than isotropic structures in experiencing the same motion.

(2) By numerical calculation, we confirm that the rotating circular plate has traveling wave vibration characteristics, namely with the increasing of the speed, forward travelling wave vibration frequency increases, while back travelling wave vibration frequency decreases to zero and then increases, which indicate that the there is a dynamic instability critical speed of the compressor disk.

(3) We discuss in detail the effect of structural parameters of laminated circular plate on the critical speed. With increase of radius ratio of the laminated circular plate, the forward traveling wave vibration frequency and critical speed decreased.

References

[1] Park J S, Shen I Y, Ku C P R, 2003 *J. appl. mech-t. asme*, 70, pp. 299–310.
[2] Natalie Baddour, Jean W. Zu, 2007 *Appl math model*, 31, pp.54-77.
[3] S.H.Hashemi, S.Farhadi, S.Carra, 2009 Journal of Sound and Vibration, 323, pp.366–384.
[4] Khoshnood A and Jalali M A, 2008 *J sound vib,* 314, pp. 147–160.
[5] Powmya A and Narasimhan M C, 2015 *Internatioal Jounal Advanced Structure Engenieering*, 7.pp.129-141.
[6] Hosseini-Hashemi Sh and Rezaee V, 2012 *J sound vib*, 331, pp.5581–5596.
[7] Sujata M and Jagannathan N, 2013, *Journal of Failure Analysis and Prevention*, 13, pp.437–444.
[8] Debabrata Das, Prasanta Sahoo, and Kashinath Saha, 2012 *International Journal for Computational Methods in Engineering Science and Mechanics*, 13,pp.37–59..
[9] Zhou peng, Hu Yu-li, 2005 *Machinery Design and Manufacture*,06(2),pp.06.

Research on Regenerative Braking Control Strategy and Simulink Simulation for 4WD Electric Vehicle

Yi-bing Xie[1], **Shou-cheng Wang**[1]

[1] College of Electromechanical Engineering, Qingdao University of Science and Technology, Qingdao 266061, China

xieyibing@126.com

Abstract. With the reduction of energy and environmental degradation, pure electric vehicles are a breakthrough point to solve the global energy and environmental crisis. However, due to the current battery technology limitations, the mileage cannot meet the use requirements. Regenerative braking is based on the existing structure and components of the electric vehicle. The braking torque is generated simultaneously by controlling the motor to generate electricity, and the braking energy is recovered at the same time. The realization of regenerative braking can compensate for the shortage of continuously driving mileage of pure electric vehicles, and maximize the energy saving advantages and social benefits. The electro hydraulic braking system is the intermediate product of traditional brake system and brake by wire system, electro-hydraulic brake system of pure electric vehicle braking energy recovery can not only improve the effect of energy recovery based on, but also can realize the braking performance. Based on Simulink, a pure electric vehicle model of electro-hydraulic braking system is established. Based on this model, the regenerative braking control strategy of electro-hydraulic braking system of 4WD Electric vehicle is developed.

1. Introduction

With the gradual reduction of global available energy and the worsening situation of the environment, mankind needs a safer and low-carbon energy system and environment. Major automobile manufacturers in the world have invested heavily in the development of new energy vehicles, and the government departments have gradually introduced relevant policies to encourage the development of new energy vehicles. Compared with traditional vehicles, electric vehicles have an absolute advantage, you can recover part of the braking energy[1]. Regenerative braking energy recovery, the vehicle is in the process of braking, motor control and power generation, power generation in the process of mechanical energy is used to vehicle performance, at the same time the motor power will produce braking torque through the transmission, transmitted to the wheels by the transmission, braking and charging the battery, to achieve the braking energy recovery however, the inertia energy braking process in traditional vehicle are converted into heat. Braking energy recovery can prolong the vehicle mileage, improve the energy efficiency of the brake friction brake by the bear is reduced, and the regenerative braking can also reduce brake heat load, reduce friction wear, improve the safety of vehicles and use economy. Urban traffic requires frequent parking and deceleration, and braking energy recovery has more important significance and considerable economic and social benefits. The establishment of Simulink electro hydraulic braking system of electric vehicle based on the model, on the basis of the model development of 4WD Electric Vehicle Regenerative Braking Control Strategy of electro hydraulic braking system, and provide a theoretical basis for the electric vehicle drive and brake type selection, provide control strategy for electric vehicle regenerative braking system development industry[2].

2. Related work

A regenerative braking system is an uncertainty system, it has nonlinear parameter and serious disturbances[3]. The braking mode of EV includes hydraulic braking mode, electric braking mode and hybrid braking mode. The traditional friction braking mode is also called hydraulic braking mode, it lets vehicle decelerate or stop with friction plates. The kinetic energy is converted into thermal energy during the braking process, which is very wasteful because the energy cannot be recovered at all[4]. In contrast, energy can be recovered in fully electric braking mode, so it is also called regenerative braking mode[5]. In the braking process, electric braking mode can recover energy, but it needs a relatively long braking time due to the limitation of motor. However, the braking force of the hydraulic braking mode is large and the stability is better. The hybrid braking mode is to combine these two modes and allocate them rationally, so as to recover as much energy as possible. The key question is how to allocate torques between hydraulic braking mode and electric motor braking mode under the premise of baking safety. There are some papers about regenerative braking control system and several control strategies have been presented. Paper[6] presented a nonlinear model predictive controller for regenerative braking system, which improved the recovery efficiency by distributing the braking torque between front and rear axis respectively. However, it did not allocate torques between the hydraulic braking mode and the electric braking mode, which limited the energy recovery efficiency. Paper[7] presented a control strategy based on fuzzy control, which used the displacement of the brake pedal, the variation rate of the brake pedal, the vehicle speed and the wheel slip rate error as input. And the method in paper[8] used the vehicle speed and the achievable force of motor as input. However, the above research did not take the State of Charge (SOC) of energy storage device into consideration, which is a significant point that affects energy recovery efficiency, and it may cause damage to the battery.

3. Regenerative Braking Controller Design

3.1. Vehicle Dynamics Model

As lateral performance is neglected in this study, considering the vehicle dynamics model adopted here has five degrees of freedom for longitudinal motion and the rotational movement of the four wheels. A single wheel braking model is shown in Figure 1.

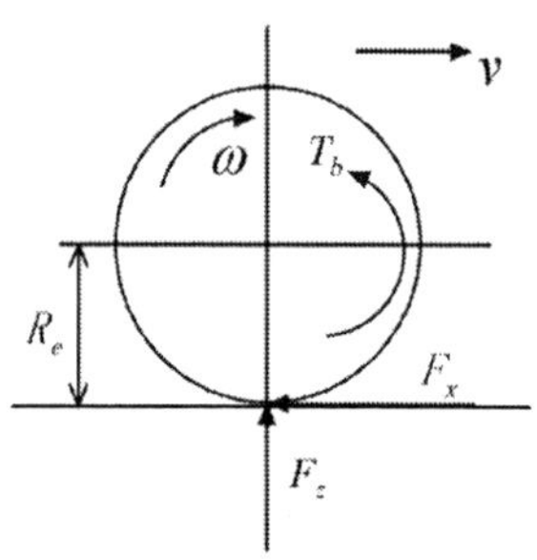

Figure 1. Single wheel braking model

3.2. Braking Torque Allocation of Front and Rear

Wheels Safety is the premise of energy recovery of the whole braking process. Experience has shown that, vehicle would be unstable if rear wheel locks before front wheel. A shock rotation or tail flick might appear while vehicle is driven in high speed on low friction coefficient road. In order to ensure vehicle in a safe state, a large adhesion between tire and road is expected. As the mutative tire load ratio of front and rear axis have the staple impact on tire-road adhesion, we consider allocating the brake torque based on the mutative tire load ratio of front and rear axis. According to the previously mentioned tire model, we can easily get the mutative tire load ratio of front and rear axis based on normal force equations.

$$\frac{F_{zf}}{F_{zr}} = \frac{l_r g - h a_x}{l_f g + h a_x} \tag{1}$$

Therefore, the brake torque coefficient of front and rear axis K can be described as

$$K = \frac{F_{zf}}{F_{zr}} = \frac{T_{mf}}{T_{mr}} = \frac{T_{hf}}{T_{hr}} \tag{2}$$

3.3. Torque Allocation of Hydraulic Brake and Electric rake Based on MPC

In this section, a model predictive controller is designed for torque allocation of hydraulic brake and electric brake. First, the controller design model is given based on vehicle dynamics equations. Then the cost function is established for the purpose of braking safety and energy Optimization, next the control constraints are introduced. Finally, the approach for solving the cost function is presented. According to the current measurement information obtained, model predictive control method solve the finite horizon open loop optimization problems online at each sampling time, and the first element of the obtained control vectors from optimal solution would be applied to the electric vehicle model. On the basis of the new measurements, this procedure is repeated at next sampling time in optimization process.

3.1.1 Controller design model. According to the vehicle dynamics model, the state equation for hydraulic and electric braking distribution can be stablished. The state and control variables are defined as follows,

$$X = \begin{bmatrix} x_1 \\ x_2 \end{bmatrix} = \begin{bmatrix} \omega_{mf} \\ \omega_{mr} \end{bmatrix} \tag{3}$$

$$U = \begin{bmatrix} u_1 \\ u_2 \end{bmatrix} = \begin{bmatrix} T_m \\ T_h \end{bmatrix} \tag{4}$$

Then based on above equation, the state equations can be described as follows,

$$\dot{x}_1 = \frac{g_0}{2J}[\mu_f F_{zf} R_e - \frac{K}{1+K}(g_0 u_1 + u_2)] \tag{5}$$

$$\dot{x}_2 = \frac{g_0}{2J}[\mu_r F_{zr} R_e - \frac{1}{1+K}(g_0 u_1 + u_2)] \tag{6}$$

Note that, brake torque Tm and Th are both defined positive. K is the ratio of front wheel brake torque to rear wheel brake torque, which is defined in the above section.

3.1.2 Cost function. Due to the goal that recovering as much energy as possible by distributing the brake torque between hydraulic and electric brake, several cost function terms are need to be established. Actually, the braking mode switching is determined by the characteristics and constraints of the electric motor. First, in order to meeting the demand of total required brake torque and maximizing energy recovery, the electric motor braking mode is implemented as far as possible. If motors have achieved the maximum torque corresponding the current motor speed, then hydraulic braking mode compensate the remaining part of total required braking demand. Secondly, as the vehicle velocity decreases, energy recovery power of electric motors would be lower than its internal copper loss power, then the motors cannot produce regenerative current. Therefore, the electric motor braking mode is completely converted into hydraulic braking mode while the motor speed achieve the certain value.

This section propose a brake torque allocation controller for regenerative braking system of electric vehicle with four in-wheel motors. The regenerative braking system include several modules, brake torque demand calculation module, brake torque distribution module, brake torque tracking module, and battery management unit. The structure of regenerative braking control system is shown in Figure 2. In this article, we focus on the research of brake torque distribution module. This module mainly consists of two parts work, one aspect is to allocate the brake torque on front and rear axis, and the other is distributing the brake torque between hydraulic braking mode and electric braking mode. Note that the brake torque demand calculation and tracking module, and battery management unit are assumed useful and beneficial, we will conduct research on these issues in the future.

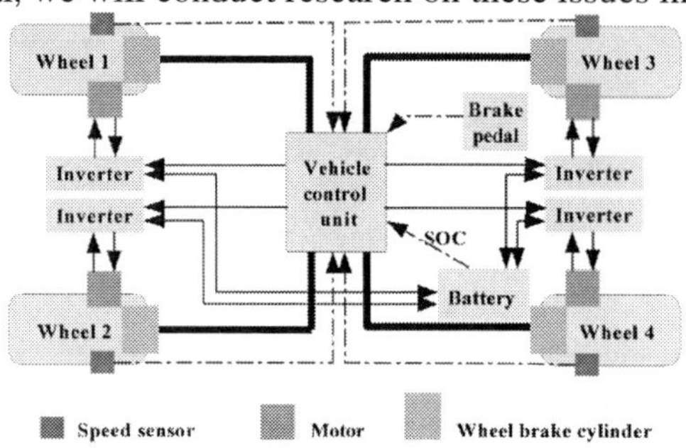

Figure 2. Structure of regenerative braking control system

4. Simulations and Analysis

In this paper, the regenerative braking control system is implemented through the combined simulation. The vehicle model with four individually driven in-wheel motors is established in AMESim software, and the fuzzy controller is designed in Simulink software. This paper only study the situation of straight driving, and the motor used is a permanent magnet brushless DC motor. The model is divided into six modules, including vehicle controller, vehicle module, motor module, tire module, driver module and battery module. In the AMESim, the parameters are set as follows: the car quality is 1360 kg, the maximum torque of the motor is 118 N m, the wheel radius is 0.275 m, the peak voltage of the battery is 450V, the rest are all default values. The driver model uses automatic catch and simplifies the transmission, thinking that there is only one gear and that it has a fixed transmission efficiency and that the reversing situation is ignored. The simulation is carried out in NEDC (New European Driving Cycle) and the simulation time is 1200 s. The vehicle braking and acceleration are relatively stable, similar to the actual situation in the city. And the actual vehicle speed can be well tracked the controlled speed, this indicates that the controller can ensure that the vehicle is in a stable state corresponding to the driver expectation. The regenerative braking torques of the motors on front and rear axles are shown in Figure3 and Figure4. The hydraulic braking torques on front and rear axles are shown in Figure5 and Figure6.

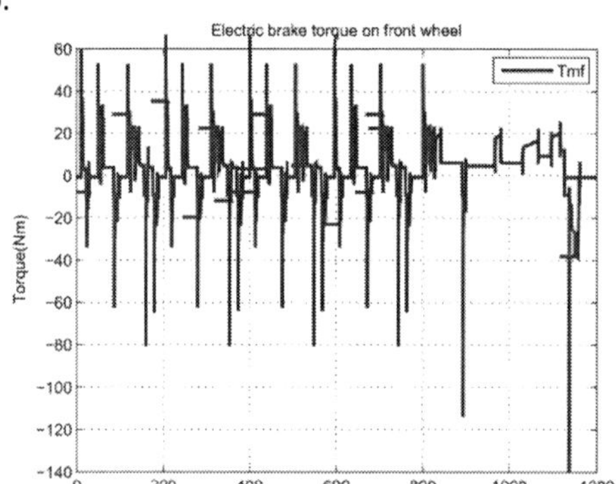

Figure 3. Electric brake torque on front wheel

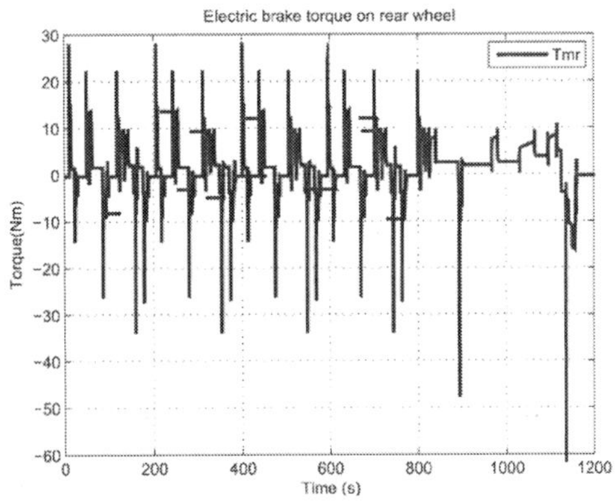

Figure 4. Electric brake torque on rear wheel

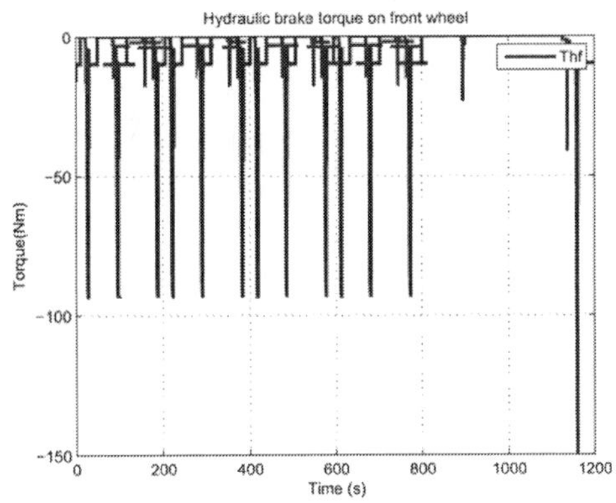

Figure 5. Hydraulic brake torque on front wheel

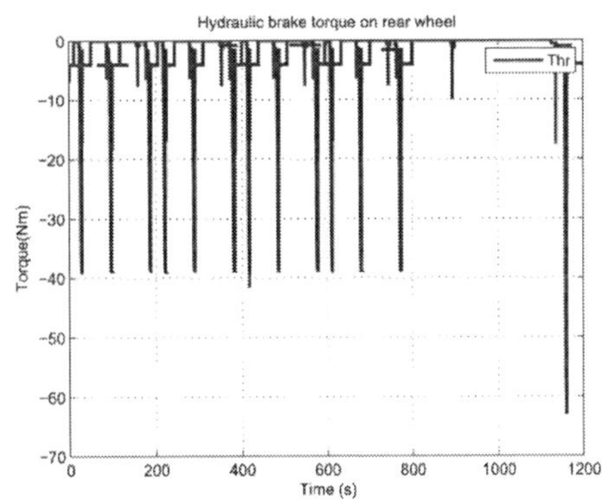

Figure 6. Hydraulic brake torque on rear wheel

5. Conclusions

The establishment of Simulink electro hydraulic braking system of electric vehicle based on the model, on the basis of the model development of 4WD Electric Vehicle Regenerative Braking Control Strategy of electro hydraulic braking system, including front and rear axle braking force distribution method and electro-hydraulic hybrid brake coordination control method, control strategy for the development of regenerative braking system. Finally the establishment of the front axle drive and four wheel drive electric vehicle model, selecting basic braking and city driving cycle simulation, braking energy recovery comparison of different drive in the form of pure electric vehicle driving rate and economy. The simulation results show that the four wheel drive electric vehicle with front axle drive electric vehicles are energy, the best braking energy recovery of each control algorithm, the braking

IOP Conf. Series: Materials Science and Engineering **398** (2018) 012013 doi:10.1088/1757-899X/398/1/012013

energy recovery rate, and provide a theoretical basis for the electric vehicle drive and brake type selection, control strategy for the development of regenerative braking system.

6. References

[1] Gao H, Gao Y and Ehsani M 2001 A neural network based SRM drive control strategy for regenerative braking in EV and HEV. Electric Machines and Drives Conference *J. IEEE International* **2** 571-575

[2] Junzhi Z, Yutong L, Chen L and Ye Y 2014 New regenerative braking control strategy for rear-driven electrified minivans *J. Energy Conversion and Management* **82** 135-145

[3] Xu G, Li W, Xu K and Song 2011 An intelligent regenerative braking strategy for electric vehicles *J. Energie* **4** 1461-1477

[4] Li X, Xu L, Hua J, Li J and OUYANG M 2008 Regenerative braking control strategy for fuel cell hybrid vehicles using fuzzy logic. Electrical Machines and Systems, 2008. ICEMS 2008 *J. International Conference* **2** 2712-2716

[5] Xu G, Xu K, Zheng C, Zhang X and Zahid T 2011 Fully electrified regenerative braking control for deep energy recovery and maintaining safety of electric vehicles *J. IEEE Transactions on Vehicular Technology* **65**1186-1198

[6] Huang X and Wang J 2012 Model predictive regenerative braking control for lightweight electric vehicles with in-wheel motors *J. Proceedings of the Institution of Mechanical Engineers, Part D: Journal of Automobile Engineering* **26** 1220-1232

[7] Chu L, Yao L, Yin J.-K, Chao L.-B and WeI W 2011 Study on the braking force allocation dynamic control strategy based on the fuzzy control logic *J. Industrial Engineering and Engineering Management (IE&EM)* **12** 635-639

[8] Zhang Z, Xu G, Li W and Zheng L 2010 Regenerative braking for electric vehicle based on fuzzy logic control strategy *J. Mechanical and Electronics Engineering (ICMEE)* **1** 319-323

IOP Conf. Series: Materials Science and Engineering **398** (2018) 012017 doi:10.1088/1757-899X/398/1/012017

Method for Measuring Clearance of Automobile Body Panel with Double Structured Light System

Ruochen Liu[1], Lei Guo[1], Shibin Yin[1], Haqing Liu[1], Qingbao Wang[1]

[1] School of Electronic Engineering and Optoelectronic Technology, Nanjing University of Science and Technology, Nanjing 210094,China

Abstract. According to the changes of the autobody measuring clearance often face the shooting angle problem caused by internal decline caused by the lack of data acquisition accuracy, a high precision and robust measurement of binocular gap line structured light and the system parameter calibration method of system integration. First the mathematical measurement model of the system is established according to the Zhang Zhengyou camera calibration model to calculate the camera parameters, the extrinsic parameters of the camera calibration of the light plane, while the introduction of auxiliary line structured light laser by solving the intersection of two camera coordinate system, finally gives the calculation model of the door gap and combined with the calibration data for the gap size. Compared with a three-dimensional scanning measuring device, the system can measure only two images at the same time. The experimental results show that when the shooting angle is less than 45 degrees at the working distance of 400mm and 50*50 field of view, the accuracy reaches 0.08mm, which meets the needs of fast and real-time measurement gap.

1. Introduction

With the rapid development of China's automobile industry in the intelligent manufacturing, automobile production in all aspects of the detection of measurement is particularly important. Which in the car body matching quality link, the body outside the board cover gap measurement is an important means of evaluation. Its measurement accuracy will directly affect the appearance of the vehicle, the car enclosed, the size of the external noise and switch the effect of the door. These indicators will directly affect the user's driving experience. At present, in the vast majority of automobile plants and parts factory or the use of traditional manual measurements, such as the feeler, Cubing model and coordinate machine and other means to measure. These are measured by contact method, only after the parts are tested after the match, not only can not achieve the process of matching parts of the gap in real-time measurement, and time-consuming and laborious, the measurement of substandard products can only be abandoned The So how to use advanced vision Detection and laser measurement technology, objectively reflect the actual state of the gap has become an important issue. In order to solve this problem, Chen Yong and others proposed the use of monocular structure of the optical system to measure, but taking into account the loss of field of view of the reasons for the amount of data, the use of multiple viewing angle and always get the sensor pose, Up to ± 0.1mm [1]. Therefore, in this measurement system, whether the complete capture of the entire contour of the gap to become the key to the measurement.

For the blind problem of measurement, Li Xinghua uses the binocular line structure to measure the size of the round bar, and the use of high-precision gauge and guide rails on the two light plane angle calibration, the average absolute error of 0.059mm[2]. Xiao Hong use the system to measure the use of limit matching to find the feature points by calculating the parallax measurement of the brake shoe thickness, accuracy of ± 2mm, which is due to binocular feature points corresponding to the error caused by excessive[3]. Therefore, the use of binocular structure of light system, how to binocular measurement of the point cloud data is also the key to determine the accuracy of the measurement accuracy. Liu Hui use the projector to obtain the corresponding point to solve the two-point cloud between the RT matrix, and the use of ICP algorithm for further point cloud registration. Wang Hua out of the two points in the frequency domain to register, the measurement accuracy of ± 0.05mm[5]. For the gap of the body cover, this paper presents the use of binocular structure of light system for measurement, the measurement system is a difficulty is the system calibration, the traditional calibration methods require more sophisticated calibration equipment or in the acquisition[7]. Three-dimensional data after the use of point cloud registration algorithm for registration, expensive, large amount of computing, can not meet the needs of real-time computing. A point cloud registration method is proposed for this problem. It is only necessary to perform a one-step RT conversion to complete the registration, without the need of complex point cloud registration algorithm.

2. Binocular structure light measurement system

As the gap measurement and shooting angle, and monocular vision may exist blind spot[8]. The measurement system consists of two cameras, two lines of light structure, the relative position of the relationship shown in Figure 1.

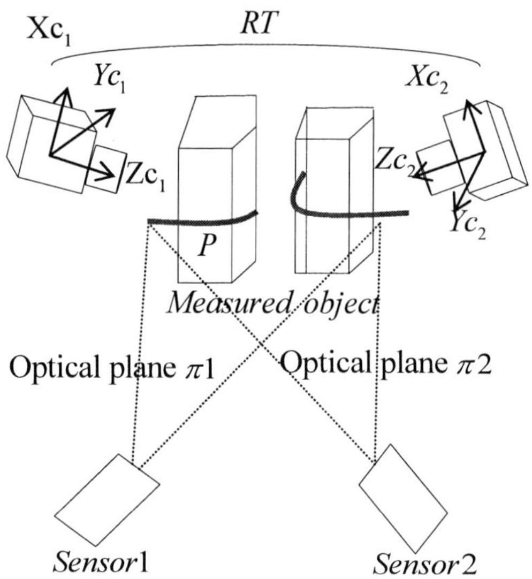

Figure 1. System structure.

After the two light beams of the laser are adjusted, the measured objects are irradiated from the left and right ends, and the camera is also connected to both ends of the measured object to facilitate the collection of the internal information of the gap of the measured object. Where the camera 1 coordinate system is, the camera 2 coordinate system is the center of the laser light point in the camera 1 coordinate system coordinates, according to the ideal aperture imaging model available

$$P_{z1}\begin{bmatrix} u_1 \\ v_1 \\ 1 \end{bmatrix} = \mathbf{A}\begin{bmatrix} P_{x1} \\ P_{y1} \\ P_{z1} \end{bmatrix} \tag{1}$$

Which is the parameter matrix in the camera $\begin{bmatrix} f_x & 0 & c_x \\ 0 & f_y & c_y \\ 0 & 0 & 1 \end{bmatrix}$

(u_1, v_1) For the camera coordinates corresponding to the next point.
According to formula (1) available

$$f_x P_{x1} + (C_x - u_1) P_{z1} = 0 \tag{2}$$

$$f_y P_{y1} + (C_y - v_1) P_{z1} = 0 \tag{3}$$

$$AP_{x1} + BP_{y1} + CP_{z1} + D = 0 \tag{4}$$

Therefore, the three-dimensional coordinates of the point under the camera 1 can be obtained by the formulas (2) to (4)

$$\begin{cases} P_{x1} = \dfrac{u_1 - C_x}{f_x} P_{z1} \\ P_{y1} = \dfrac{v_1 - C_y}{f_y} P_{z1} \\ P_{z1} = \dfrac{Cf_x f_y}{f_x f_y - Af_y(u_1 - C_x) - Bf_x(v_1 - C_y)} \end{cases} \tag{5}$$

Because both cameras meet the theory of aperture imaging, we can obtain the three-dimensional coordinates of all the data points on the light bar in the two camera coordinate systems by acquiring the parameters of the two cameras in the light plane.
In addition, if the relative position of the two cameras and the two lasers is the same, we can transform the two camera coordinate systems by rigid body transformation

$$\begin{bmatrix} P_{x1}' \\ P_{y1}' \\ P_{z1}' \end{bmatrix} = \mathbf{R}_b \begin{bmatrix} P_{x2} \\ P_{y2} \\ P_{z2} \end{bmatrix} + \mathbf{T}_b \tag{6}$$

Among them, for the rotation matrix and the translation matrix, you can point the camera 2 coordinates into the camera 1, the calculation.

3. Parameter calibration

First, an object plane is defined so that the rotation matrix is reduced by a column vector. According to Eqs. (1) to (5), the imaginary relationship of the image point to the spatial point

$$\begin{bmatrix} u \\ v \\ 1 \end{bmatrix} = sM \begin{bmatrix} r_1 & r_2 & \mathbf{t} \end{bmatrix} \begin{bmatrix} P_x \\ P_y \\ 1 \end{bmatrix} \tag{7}$$

Which M is the parameter matrix in the camera, is any non-zero scale factor. Set the coherence matrix H, ie

$$H = \begin{bmatrix} h_1 & h_2 & h_3 \end{bmatrix} = sM \begin{bmatrix} r_1 & r_2 & \mathbf{t} \end{bmatrix} \tag{8}$$

By the projective matrix with the satisfaction of the orthogonal characteristics (with the dot product and the vector length equal), available

$$h_1^T B h_2 = 0 \tag{9}$$

$$h_1^T B h_1 = h_2^T B h_2 \tag{10}$$

Where $B = M^{-T} M^{-1}$ 。

(9) and (10), we can obtain the closed solution of the parameters in the camera from the matrix at that time $K \geq 3$.

$$\begin{aligned}
f_x &= \sqrt{\lambda / B_{11}} \\
f_y &= \sqrt{\lambda B_{11} / \left(B_{11}B_{22} - B_{12}^2\right)} \\
c_x &= -B_{13} f_x^2 / \lambda \\
c_y &= (B_{12}B_{13} - B_{11}B_{23}) / (B_{11}B_{22} - B_{12}^2) \\
\lambda &= B_{33} - \left[B_{13}^2 + c_y \left(B_{12}B_{13} - B_{11}B_{23}\right)\right] / B_{11}
\end{aligned} \tag{11}$$

The laser strip is irradiated on a plane, assuming that the coordinates of the three points on the plane in the world coordinate system are $(0,0,0)$, $(1,0,0)$, $(0,1,0)$, then the coordinates in the camera coordinate system can be determined $P_{c1\sim3} = \mathbf{R}P_{b1\sim3} + \mathbf{t}$ from three points to determine the plane equation The Therefore, only need to know the plane relative to the camera coordinate system under the external parameters and then according to formula (2), (3) can be calculated on the laser bar point in the camera coordinate system under the three-dimensional coordinates. And then reverse the need to obtain the plane of the unilateral should be known through the internal reference to the external parameters, that is,

$$\begin{aligned}
r_1 &= \lambda M^{-1} h_1 \\
r_2 &= \lambda M^{-1} h_2 \\
r_3 &= r_1 \times r_2 \\
t &= \lambda M^{-1} h_3 \\
\lambda &= 1 / \left\| M^{-1} h_3 \right\|
\end{aligned} \tag{12}$$

In theory, more than two positions of the external solution to solve the plane.

Since all the points are in the light plane, it is possible to establish the coordinate system on the light plane and convert the point of the camera coordinate system to the coordinate system by using the rigid body transformation, which can effectively convert the 3D data into two-dimensional plane data, The calculation.

From the light plane available $Ax + By + Cz + D = 0$, set $lx = -\frac{D}{A}, ly = -\frac{D}{B}, lz = -\frac{D}{C}$

The unit vector $\mathbf{R}_x$ of the coordinate system TCP axis can be considered to be parallel to the axis of the camera coordinate system

$$\mathbf{R}_x = [0 \quad \frac{ly}{\sqrt{ly^2 + lz^2}} \quad -\frac{lz}{\sqrt{ly^2 + \mathrm{lz}}}] \tag{13}$$

$$\mathbf{R}_x = [0 \quad \frac{ly}{\sqrt{ly^2 + lz^2}} \quad -\frac{lz}{\sqrt{ly^2 + \mathrm{lz}}}] \tag{14}$$

The unit vector TCP of the coordinate system axis is

$$\mathbf{R}_x = \mathbf{R}_y \times \mathbf{R}_z \tag{15}$$

$$\mathbf{R}_s = \begin{bmatrix} \mathbf{R}_x^T & \mathbf{R}_y^T & \mathbf{R}_z^T \end{bmatrix} \tag{16}$$

$$\mathbf{T}_s^T = \begin{bmatrix} 0 & 0 & -\frac{d}{c} \end{bmatrix} \tag{17}$$

The auxiliary laser is used to intersect the optical plane by the auxiliary laser beam, and the camera 1 acquires the image simultaneously with the camera 2. Fig. Extract the sub-pixel center of the light bar through the matrix. Considering that the matrix will only solve the maximum eigenvalue at the intersection of the light bars, the normal direction at the intersection of the light bars can not be determined. We through the template matching method to remove the light bar part of the intersection, the image is divided into four single-line area.

$$d = \sum_{i=1}^{N} \left[y_i - (a + bx_i) \right]^2 \to \min \tag{18}$$

Respectively a , b , for partial alignment and solving equations can be fitted by two straight lines to solve the intersection coordinates. Finally, through (6) to solve the two camera coordinate system in the laser bar between the location of the transition relationship.

4. Size calculation

The clearance detection position on the body cover is shown in Fig. First of all, through the matrix to extract the light bar center, a reasonable choice of extraction area will greatly reduce the calculation of time[9]. So first through the image segmentation to find the location of the gap, only the light bar bending part of the convolution, the rest of the linear part of the gray-scale centroid method can be calculated. Considering that the gap internal information may be elliptical. The objective function of the polygon approximation algorithm [1] is changed to an elliptic function

$$Ax^2 + Bxy + Cy^2 + Dx + Ey + F = 0 \tag{19}$$

If the fitting error is less than the straight line fitting error, it is considered ellipse, that is,

$$f(A,B,C,D,E,F) = \sum_{i=1}^{m} (y(x_i) - y_i)^2 < f(k,b) \tag{20}$$

So you can get four inflection points $P_1 \sim P_4$, fitting two elliptical centers, the direction between the measurement direction, minus the corresponding ellipse radius is the gap measurement

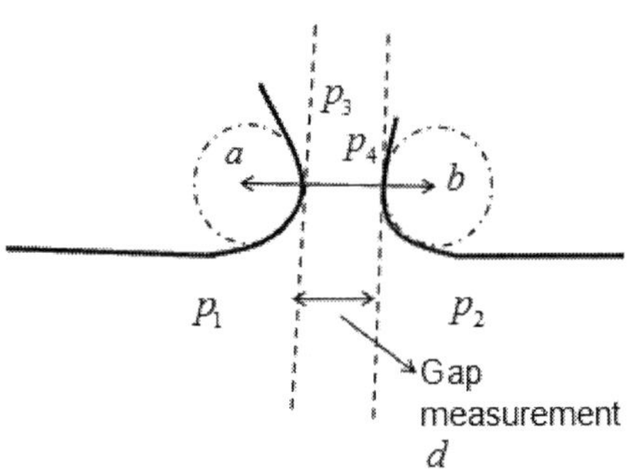

Figure 2. Gap measurement.

IOP Conf. Series: Materials Science and Engineering **398** (2018) 012017 doi:10.1088/1757-899X/398/1/012017

5. Experiment and discussion

We hope you find the information in this template useful in the preparation of your submission.
The experiment uses Basler's 1300-60gm camera with 35mm lens, and a 10mm pitch circular target. The working distance is 400mm, and the camera 1 is the global coordinate system. EXPERIMENTAL PROCEDURE 1) 6 images were taken by different orientations before placing the plane target in front of the camera. 2) The two lasers are turned on, so that the laser strip is irradiated on the target. In order to facilitate the extraction of the feature, the laser strip can not hit the circular feature point. 3) Turn off the external light source, open the auxiliary line structure light, intersect the laser, and collect the picture. Which need to collect the laser irradiation in four different locations of the target picture 4. Each position of the laser strip intersects the picture 4 sheets. Through the 10 target images to establish the camera internal parameters to optimize the calibration function

$$f(f_x, f_y, c_x, c_y, k_{1\sim2}, p_{1\sim2}, \mathbf{R}, \mathbf{T}) = \sum_{i=1}^{N} d^2\left(X_d^{'}, X_d\right) \tag{21}$$

Where t$d\left(X_d^{'}, X_d\right)$he distance from the projection point to the actual projection point is calculated. The calculated back projection error is 0.03mm and the optical plane fitting error is 0.02mm, which can meet the measurement requirement. The measured gap measured by the experiment is 1.4 mm for the feeler measurement with an accuracy of 0.02 mm.
As the two lasers for the mechanical adjustment coincidence, and can not guarantee that the two laser light plane completely coincide. So the object will be measured before and after the move, not in the working distance, the laser pattern thicker, but does not meet the Gaussian distribution, but there are different levels of up and down as shown in Figure 3, and the laser out of the pupil of the angle there is a certain relationship , It is necessary to use the straight line segment in the image of the camera 1 to solve the point of the straight line segment in the image of the camera 2 to its average distance as the correction matrix to complete the point cloud registration as shown in Fig. 4 (c).

(a) (b)

Figure 3. (a)Image coordinates before registration. (b)Three dimensional coordinate before registration.

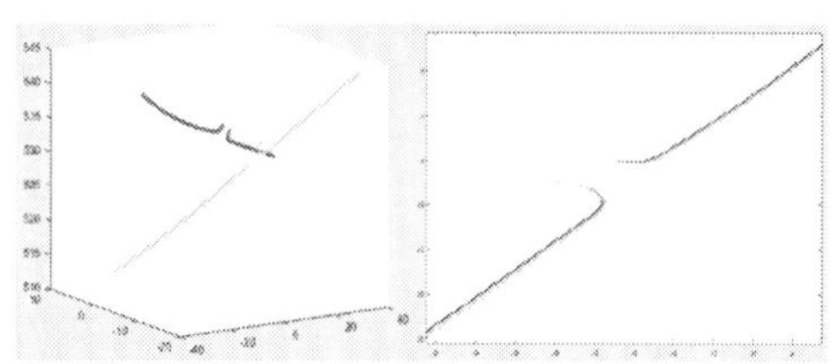

(a) (b)

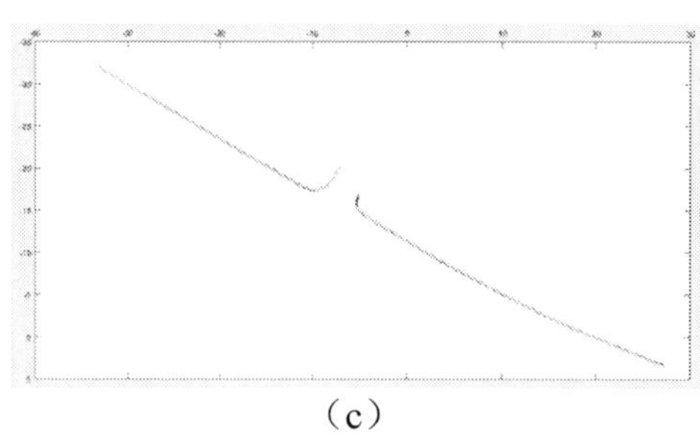

(c)

Figure 4. (a) before registration, (b) after

The square sum of the Eu-distance distance between the correction matrix in the different positions in the field of view and the corrected two-point cloud is as follows

Table 1. Use the correction matrix to correct the effect.

location	Fix the T matrix	Average distance
1	0.01， -0.16	0.36
2	0.03， -0.18	0.42
3	0.03， -0.18	0.38
4	-0.03， 0.16	0.57
5	0.04， 0.18	0.33
6	0.05， -0.10	0.48
7	0.09， -0.14	0.71
8	0.14， 0.01	0.45
9	0.02， 0.20	0.61
10	0.02， 0.20	0.63

Because the gap can not guarantee a certain picture of the gap and the gap between the surface, so at different angles to capture the gap picture, as shown in Table 2. The experimental results show that the average error can be less than 0.1mm in the shooting range of 45 ° angle Table 2 measurement results from different angles.

Table 2. Measurement results from different angles.

Angel	Monocular measurement	error	Binocular measurement	error
0	1.4820	0.082	1.3686	-0.0314
5	1.5200	0.12	1.3663	-0.0337
10	1.6488	0.2488	1.3733	-0.0267
15	1.7150	0.315	1.3903	-0.0097
20	1.76191	0.36191	1.4186	0.0186
25	1.8128	0.4128	1.3714	0.0286
30	1.8790	0.479	1.3778	0.0222
35	1.7631	0.3631	1.4519	0.0519
40	1.8944	0.4944	1.4258	0.0258

45	1.9141	0.5141	1.4358	0.0358
50	2.2220	0.822	1.5323	0.1323
55	2.2750	0.875	1.5685	0.1685
60	2.3823	0.9823	1.7013	0.3013

6. Conclusion
In this paper, a binocular structured light measurement system is proposed based on the high precision and high robustness measurement of the gap of the body cover. And the principle of the system is introduced. The three-dimensional data is transformed into two dimensions by not only reducing the complexity of the calculation. The conversion relationship. And the correction method of the mechanical adjustment error due to the laser is given. Eventually completed the gap measurement. The system is measured at different angles at different angles, with a measurement accuracy of 0.1 mm within the 45 ° shooting angle and good robustness. In the future work, the construction of two lasers can spread the model, you can more accurately coordinate system 1, the other transplanted to the embedded in the production of small hand-held detection system is one of the future direction of development.

7. References
[1] Yong C 2015 Research on key technology of laser clearance measurement for autobody panel. *J. Jiangsu University* **24** 11-12
[2] Xinghua L, Jingmei Z and Peifen C 2015 Research on on-line measurement system of double structured light vision sensor. *J. Computer engineering and application* **14** 175-178
[3] Hong X, and Xiao, X 2016 Measurement of the thickness of brake shoes based on binocular structured light. *J. Optics and optoelectronic technology* **3** 54-57
[4] Hui L, Boxiong W and Huaiyi R 2012 ICP algorithm in point cloud matching of binocular structured light system applications. *J. Journal of Tsinghua University (NATURAL SCIENCE EDITION)* **7** 946-950
[5] Hua W, Chunqi Xing and Jin G 2017 The alignment of twin line structured light wheel pairs using point cloud frequency domain registration. *J. Ptics and precision engineering* **3** 616-624
[6] Hui C, Feng Z, and Zhan S 2014 A practical means for the calibration of a binocular structured light system, IEEE International Conference on Information Science and Technology. *J. IEEE* **12** 615-618
[7] An Y, Bell T and Li B 2016 Method for large-range structured light system calibration. *J. Appl Opt* **33** 9563-9572
[8] Guowei F, Xiao Z and Fei K 2014 Calculation method for gap clearance of automotive parts based on laser measurement. *J. laser and infrared* **4** 370-373
[9] Wei L, Peng G and Yang Z 2017 High precision measurement method for 3D surface of large parts. *J. Infrared and laser engineering* **9** 3

IOP Conf. Series: Materials Science and Engineering **398** (2018) 012016 doi:10.1088/1757-899X/398/1/012016

The Decoupling Control of ARS and DYC Based on the Neutral Network Inverse System Method

Xiwen Tang[1], Rui Dong[1], Shilong Wu[1], Guang Xia[2] and Hai Xie[2]

[1]School of Radar Confrontation, National University of Defense, Hefei, China
[2]Institute of Automotive Engineering, Hefei University of Technology, Hefei, China

shelley7983@163.com, eei_shlkf@sina.com , rui_jadeite@sina.com, xiaguang008@163.com and 78791428@qq.com

Abstract. According to the interference and coupling characteristics among sub-systems of the chassis of the forklift truck, the the neutral network inverse system method of the non-linear system is applied for the decoupling control of the active rear steering (ARS) and the direct yaw moment control (DYC) of the forklift truck. Based on the analysis of the reversibility of the chassis system, a back propagation neutral network reverse system model is set to decouple the chassis system into two independent pseudo-linear system; A PD closed loop controller is designed to comprise a composite controller with the neutral network reverse system and the simulation is performed for verification. The simulation result shows that the decoupling control strategy with the neutral network reverse system can eliminate the interference and coupling among sub-systems of the chassis and thus enhances the status tracking and operating stability of forklift trucks.

1. Introduction

Recent years, the application of integrated control over the sub-systems of the chassis of vehicles to exploit the functional potentials of sub-systems has become a hotspot in the study of the vehicle dynamics[1-3]. The sub-systems of the chassis of vehicle are designed to realize certain functions. The active rear steering (ARS) and the direct yaw moment control (DYC) can both enhance the lateral stability[4-5]; therefore, the cross-interference of the control function occurs, and, inevitably, the interfering coupling occurs, debasing the performance of the vehicle system and the reliability. If decoupling is performed for the integrated chassis system whose channels are inter-coupled to eliminate the interactions among dynamic sub-systems during the design of the control system, additional functions that independent sub-systems cannot be provided may be realized and therefore the comprehensive performance of the vehicle is enhanced.

References [6-7] study the decoupling control of the sub-systems of the chassis. This paper mainly discusses the decoupling control on the ARS and the DYC of forklift trucks with the neutral network reverse system method of the non-linear system based on the non-linear dynamic characteristics of vehicles caused by non-linearity of tires, and uses Matlab/Simulink to perform simulation analysis on the control system to verify the effects of the decoupling control on the control system.

2. Reference model of the control system

The reference model of the control system is a 2-DOF dynamic model. The ARS and the DYC respectively control the yaw and steering stability laterally and vertically and applies a 2-DOF vehicle model of lateral and yaw as the common control reference model, as shown in Figure 1.

In Figure 1, v, β and ω_r are respectively the vertical velocity, slidslip angle and yaw velocity of the vehicle; m is the mass of the forklift truck; a, b are respectively the front and rear wheel bases; Iz is the moment of inertia; Caf and Carare respectively cornering stiffness of the front and rear wheels; δ_r, δ_c and T are the turning angle of the rear wheels applied by the driver, the turning compensation angle of the rear wheels and the yaw control torque.

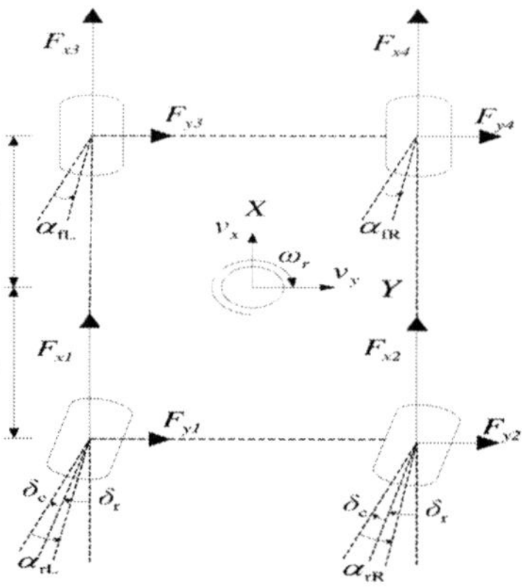

Figure 1. Reference model of the control system

The status variable of the system is defined as $x=(\beta \ \ \omega_r)^T$, the control input variable as $u=(\delta_c \ \ T)^T$ and the control output variable as $y=(\beta \ \ \omega_r)^T$, then the state equation of the vehicle can be described as follows:

$$\begin{cases} M\dot{x} = Kx + Nu + Q\delta_r \\ y = Cx \end{cases} \tag{1}$$

wherein:

$$M = \begin{pmatrix} m\upsilon & 0 \\ 0 & I_z \end{pmatrix},$$

$$K = \begin{pmatrix} -2(C_{af}+C_{ar}) & -mv + \dfrac{2(aC_{af}-bC_{ar})}{v} \\ 2(aC_{af}-bC_{ar}) & -2\dfrac{a^2C_{af}+b^2C_{ar}}{v} \end{pmatrix}, N = \begin{pmatrix} 2C_{ar} & 0 \\ 2C_{ar} & 1 \end{pmatrix}, Q = \begin{pmatrix} 2C_{ar} \\ 2bC_{ar} \end{pmatrix}, C = \begin{pmatrix} 1 & 0 \\ 0 & 1 \end{pmatrix}.$$

Obviously as shown above, the reference model (1) of the vehicle system is a typical 2-input and 2-output multi-variable system.

3. Design of the decoupling control system

3.1. Reversibility analysis

For the multi-input and multi-output integrated chassis control system (1) described by the above state equation, the Interactor algorithm is usually used to verify the system reversibility: calculate the derivative of each order of output variable **y** relative to time until the derivative equation of the output explicitly contains the input variable ***u***, according to (1), the following can be obtained:

$$\begin{aligned} \dot{y}_1 = \dot{\beta} &= -2(C_{af}+C_{ar})\beta / mv \\ &+2\omega_r(aC_{af}-bC_{ar})/mv^2 - \omega_r + 2C_{ar}(\delta_r+\delta_c)/mv \end{aligned} \tag{2}$$

Recorded as $Y_1 = \dot{y}_1$, its rank of the Jacobian matrix relative to the input variable u is:

$$t_1 = rank(\frac{\partial \boldsymbol{Y}_1}{\partial u^T}) = rank\begin{pmatrix} \frac{\partial \dot{y}_1}{\partial \delta_c} & \frac{\partial \dot{y}_1}{\partial T} \end{pmatrix} = 1 \tag{3}$$

$$\dot{y}_2 = \dot{\omega}_r = 2(aC_{af} - bC_{ar})\beta / I_z - 2(a^2C_{af} + b^2C_{ar})\omega_r / I_z v + 2bC_{ar}(\delta_r + \delta_c)/I_z + T/I_z \tag{4}$$

Recorded as $Y_2 = (\dot{y}_1 \quad \dot{y}_2)^T$, its rank of the Jacobian matrix relative to the input variable u is:

$$t_2 = rank(\frac{\partial \boldsymbol{Y}_2}{\partial u^T}) = rank\begin{pmatrix} \frac{\partial \dot{y}_1}{\partial \delta_c} & \frac{\partial \dot{y}_1}{\partial T} \\ \frac{\partial \dot{y}_2}{\partial \delta_c} & \frac{\partial \dot{y}_2}{\partial T} \end{pmatrix} = 2 \tag{5}$$

Due to nonnegative integer $\alpha_1 = 1$, $\alpha_2 = 1$ makes t_2 the number of the outputs of system (1), then the vector relative degree o the system is:

$\boldsymbol{a} = (\alpha_1 \quad \alpha_2)^T = (1 \quad 1)^T$; meanwhile, $\sum_{i=1}^{2} \alpha_i = n = 2$ (n is the degree of the system). According to the implicit function theorem, the reverse system of the integrated system (1) does exist and the output u (the input of the original system) of the reverse system can be described as follows:

$$\boldsymbol{u} = \varphi(\boldsymbol{x}, \dot{y}_1, \dot{y}_2) = \varphi(\boldsymbol{x}, \boldsymbol{v}) \tag{6}$$

wherein $\boldsymbol{v} = (v_1 \quad v_2)^T = (\dot{y}_1 \quad \dot{y}_2)^T$, $\varphi(\square)$ is the nonlinear mapping relation between the output and the input of the reverse system. Set $z_1 = y_1$ and $z_2 = y_2$, and $v_1 = \dot{y}_1$ and $v_2 = \dot{y}_2$ as the inputs of the reverse system, then the standard form of the reverse system of the integrated system (1) can be described as:

$$\begin{cases} \dot{z}_1 = v_1 \\ \dot{z}_2 = v_2 \\ \boldsymbol{u} = \bar{\varphi}(z_1 \quad z_2 \quad v_1 \quad v_2) \end{cases} \tag{7}$$

According to (7), the pseudo-linear system comprised by connecting the reverse system (7) in series before the original system (1) is equivalent to two first order integral linear sub-systems. Therefore, the control of the close coupling multi-variable system (1) with the ARS and the DYC can be transformed into the control on the two first order integral linear sub-systems, then the decoupling of the system is realized, as shown in Figure 2.

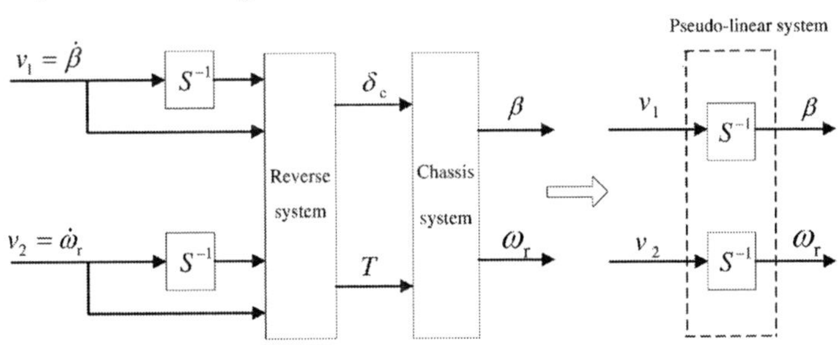

Figure 2. Equivalent structural diagram after decoupling

3.2. Construction and training of the neutral network reverse system

The static neutral network and the integrator are used to construct the system (7) and the structure of the neutral network reverse system of the integrated chassis system is obtained as 4-10-2, as shown in Figure 3.

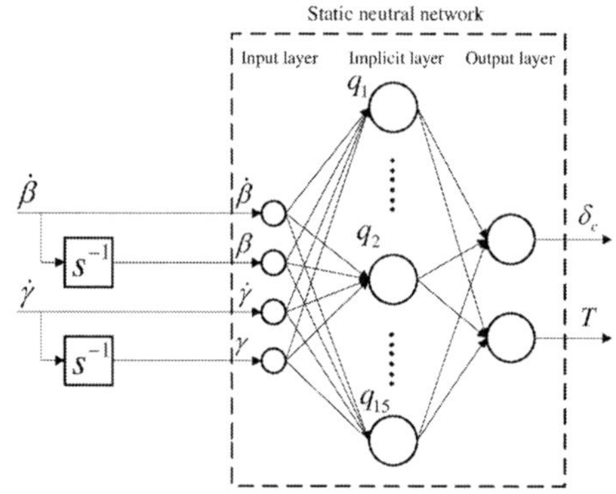

Figure 3. Structure of the BP neutral network reverse system

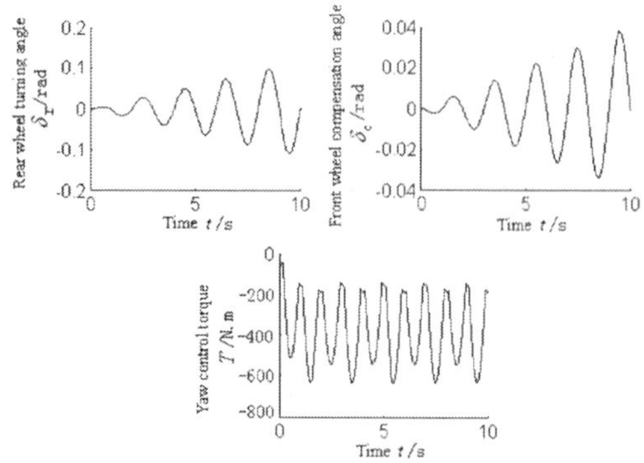

Figure 4. Excitation signal for training of the neutral network reverse system

Supposed that the velocity of the vehicle is 10 km/h and the turning angle of the rear wheels applied by the driver is variable-amplitude sine curve, the turning compensation angel δ_c of the rear wheels and the excitation signal input by the yaw control torque T are as shown in Figure 4. Set the times of training to 500, the learning efficiency to 0.05 and the network target error to 10-4, the precision required is obtained after 61 times of training.

3.3. Design of the closed loop controller

To improve the response quality of the integrated system, a composite controller comprised of a closed loop controller and a neutral network reverse system is designed to control the integrated system, as shown in Figure 5.

The PD closed loop controller:

$$v = \boldsymbol{K}_p diag\{\boldsymbol{e}\} + \boldsymbol{K}_d diag\{\boldsymbol{de}\} \tag{8}$$

wherein:

e, ***de*** - Deviation and deviation change rate of the system output signals

$\boldsymbol{K}_p$, $\boldsymbol{K}_d$ - Proportional and differential coefficients

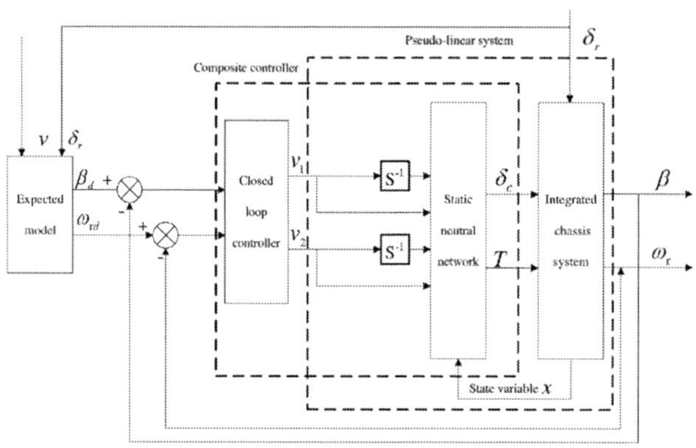

Figure 5. Structural diagram of the decoupling control on the integrated chassis system

4. Simulation and Result Analysis

The simulation to the designed decoupling control system of the chassis and relevant analysis are performed based on Matlab/Simulink. The simulation parameters are: m=4639kg, a=1.00m, b=0.97m, Iz=8428kg.m2, Caf=154.86kN.rad-1, and Car=150.45kN.rad-1. The initial velocity is set to v=10m/h and the factor of adhesion is $\mu = 0.85$.

The simulation to the single lane change is performed for the PD control without system decoupling (PD control for short) and the decoupling PD control based on the neutral network reverse system method (decoupling PD control for short), and the simulation results are compared with the expected values. The amplitude of the turning angle of rear wheels applied by the driver is set to 0.08rad and the frequency is set to 0.5Hz for the simulation; then the yaw velocity and slidslip angle response curves are obtained under the condition of single lane change, as shown in Figures 6 and 7.

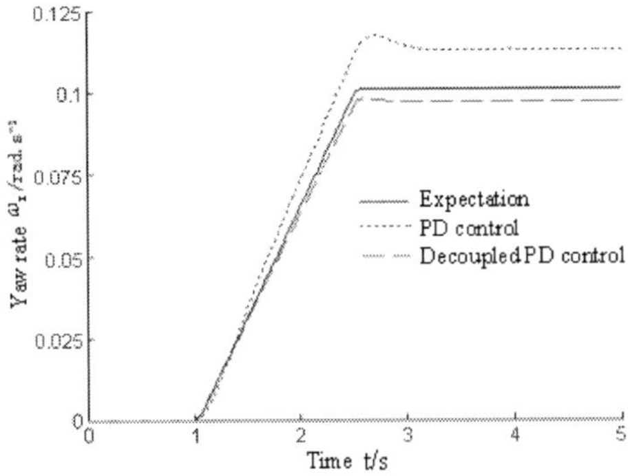

Figure 6. Yaw velocity response curve

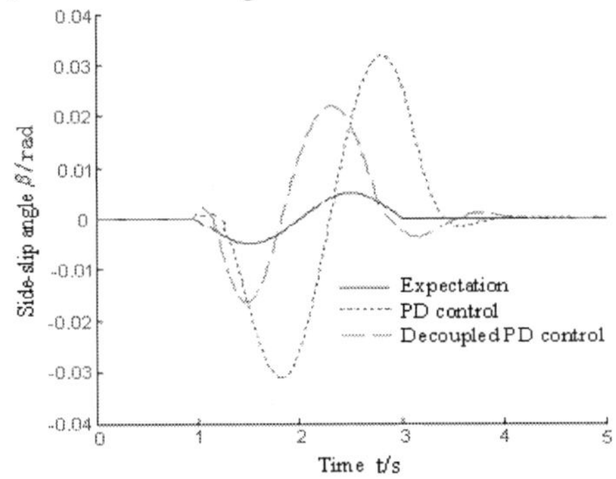

Figure 7. Slidslip angle response curve

According to Figures 6 and 7, compared with pure PD control, the yaw velocity under decoupling PD control can better follow up the expected values, the overshoot is small and the slidslip angle is controlled within a small range for stable driving.

Conclusion

This paper discusses the decoupling control on the ARS and the DYC of forklift trucks with the neutral network reverse system method, analyzes the reversibility of the integrated system of the chassis, and constructs a BP neutral network reverse system. Then a pseudo- linear system corresponding to the inputs and outputs of the original system is obtained by connecting the reverse system in series with the integrated system to realize the decoupling among control loops, thus effectively enhancing the state tracking and stability of forklift trucks and improving the steering stability of forklift trucks under extreme working conditions.

Acknowledgements

This paper is supported by National Natural Science Foundation of China (Grant No. 51205101).

References

[1] Trachtler A 2004 Integrated vehicle dynamics control using active brake, steering and suspension system. *J. International Journal of Vehicle Design* **36** 1-12

[2] [2] Karbalaei R, Ghaffari A and Kazemi R 2007 A new intelligent strategy to integrated control of AFS/DYC based on fuzzy logic. *J. International Journal of Mathematical Physical and Engineering Sciences* **1** 47-52

[3] Si Jun-de and Wang Guo-qiang 2010 Design and Performance Simulation of ROPS and Energy-Device. *J. Transactions of the Chinese Society for Agricultural Machinery* **8** 20-24

[4] Yu Zhuo-ping, Zhao Zhi-guo and Chen Hui 2001 Influences of active front wheel steering on vehicle maneuver and stability performance. *J. China Mechanical Engineer* **7** 652-657

[5] Zhang Si-qi, Zhang Tian-xia and Zhou Shu-wen 2012 Study on Vehicle Stability Control Based on Distribution of Yaw Torque. *J. China Mechanical Engineer* **6** 751-754.

[6] Zhu Bing, Li You-de and ZHAO Jian 2010 Integrated Chassis Control Based on Multivariable Frequency Domain Control Methods. *J. Transactions of Chinese Society for Agricultural Machinery* **1** 14-17

[7] Chen Wu-wei, XU Juan and Hu Fang 2007 The proportion of the input and output decoupling and decoupling of nonlinear systems of differential control vehicle. *J. Chinese Journal of Mechanical Engineerin* **3** 64-70

Analysis of Wear for a Rocket Sled Slipper

Jun Liu[1], Weihua Wang[1], Feng Zhao[1]

[1] Testing Department, Aerospace Life-support Industries Ltd., Xiangyang, China

Abstract. This study analyzed the thermal and stress field of a rocket sled slipper by establishing a thermal-mechanics coupling model. The temperature field and thermal stress field were obtained. The results are compared with real rocket sled run test. It shows that there is an agreement of wear distance between the test and simulation. However, the real test local temperature which is above 1470 ℃, in evidence of detected Cristobalite-high SiO2, is much higher than simulation result which is 818 ℃, due to high vibration impact load in real run test.

1. Introduction

Rocket sled tests are ground test platforms which enables equipment to be tested under high acceleration or high airspeed conditions. Such working condition can cause slipper wear[1].

In the study of material wear characteristic, temperature and stress are the two most important coupling parameters. Temperature rise caused by friction is a complicated nonlinear physical process. During the test, the work done by friction is converted to heat and passes into the contact area of the slippers, which increases the temperature of the slipper and produces thermal stress. The temperature rises as the rocket sled flies, resulting in change of material performance and thermal stress, influencing on boundary conditions of temperature field and stress field distribution. At present, numerical methods are generally used for the study of plane sliding friction[2].

This study aims to establish a finite element structure and thermal coupling model to obtain the temperature and stress distribution of the slipper, and compare the results with real run test..

2. Theory

2.1. Basics

In the process of friction, the frictional energy caused by deformation or destruction in contact surface material convert into thermal energy, and thus the surface temperature of the friction material is raised. When under high flying velocity, strong friction heat generated in the contact surface changing the state of the contact surface, leading to the material performance change and thereby failure in meeting the demand of actual working condition. For example,thermal stress can change the contact surface conditions; thermal expansion can change the initial parts fit etc. At the same time, the thermal stress produced by friction heat has an effect on the overall stress distribution. Therefore, it is necessary to conduct temperature analysis first. In the run test, the maximum working speed of the rocket sled can reach 2.8 Ma, therefore the slippers will cause detached shock wave, which will heat the windward side, and other surfaces are also heated due to aerodynamic heating effect. But since the overall operation time is not greater than 8s, the aerodynamic heating effect is not obvious. To facilitate the calculation, the simulation model is based on the following assumptions.

a) The heat removed by grinding and peeling is ignored;

b) The aerodynamic heat effect caused by the detached shock wave is ignored.

c) The aerodynamic heat effect of the slipper surface is ignored, and the surrounding fluid temperature is considered as the local temperature.

d) The materials are isotropic.

2.2. Heat conduction equation in finite element

In the process of plane sliding friction, the main heat transfer is the heat conduction between the contact surfaces and the heat convection between the object and the surrounding air. In the calculation of temperature, the basic equation is equation 1,

$$\rho c \frac{\partial T}{\partial \tau} = \left[\frac{\partial}{\partial x}\left(\lambda \frac{\partial T}{\partial x}\right) + \frac{\partial}{\partial y}\left(\lambda \frac{\partial T}{\partial y}\right) + \frac{\partial}{\partial z}\left(\lambda \frac{\partial T}{\partial z}\right) \right] + q \tag{1}$$

where q is the heat generates inside particle dxdydz, τ is the temperature rise in particle.

The equation can be simplified as equation 2 when there is no interior heat source and the coefficient of thermal conductivity of material is not changed with position.

$$\frac{\rho c}{\lambda} \frac{\partial T}{\partial \tau} = \frac{\partial^2 \mathrm{T}}{\partial x^2} + \frac{\partial^2 \mathrm{T}}{\partial y^2} + \frac{\partial^2 \mathrm{T}}{\partial z^2} \tag{2}$$

2.3. Calculation of heat flux density

According to the heat transfer theory, the friction heat generated by plane friction is distributed between two contact objects following the following principles:

Friction heat between two different materials is not evenly distributed to two contact surfaces, and the distribution relationship is always related to the thermal properties of materials. The generation and distribution of frictional heat can be seen as such there is a continuous exothermic plane transient heat source in between, and the heat transfer can be seen as the superposition of the transient heat source. For rocket sled slippers, the friction heat entering the contact surface of the slipper is equation 3,

$$q_1 = \frac{k}{1+k} \mu \cdot N \cdot v \tag{3}$$

where q is total heat flux, N is load on slipper, v is velocity, k is coefficient of heat partition.

2.4. 2.4 Stress analysis

In rocket sled run, the friction heat continuously enters slippers from the contact surface, making the slipper temperature rise continuously. The thermal stress can be calculated according to equation 4,

$$[\sigma] = \Delta T \cdot [\alpha(T)] \tag{4}$$

In finite element analysis, the thermal load can be considered as equivalent to force applied on the corresponding node. According to the theory of thermal stress, the stress analysis constitutive equation under thermal stress is equation 5,

$$[\sigma] = [D] \cdot ([\varepsilon] - \Delta T \cdot [\alpha(T)]) \tag{5}$$

where $[\sigma]$ is stress matrix, [D] is stiffness matrix, $[\varepsilon]$ is strain matrix, ΔT is temperature variable, $[\alpha(T)]$ is coefficient of linear expansion.

3. Simulation

3.1. Geomrtry and Model

The 3D geometry was built in ANSYS shown in Figure 1. 3mm hexahedral elements are applied in mesh shown in Figure 2. The material properties and conditions are listed in Table 1and Table 2.

IOP Conf. Series: Materials Science and Engineering **398** (2018) 012019 doi:10.1088/1757-899X/398/1/012019

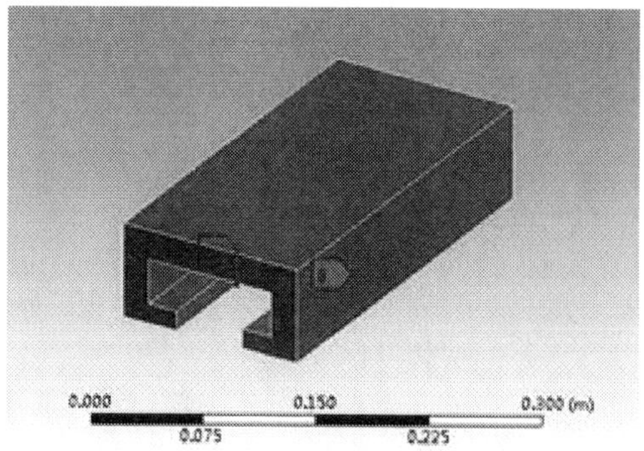

Figure 1. Geometry model

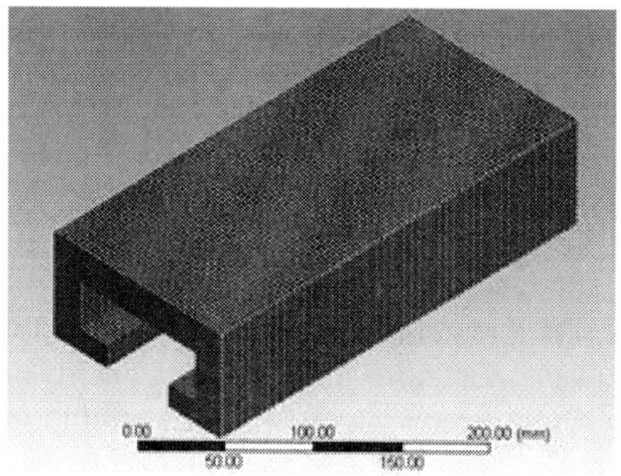

Figure 2. View of mesh

Table 1. Ial properties and condition parameters.

Parameters	Value
Load	2000N
Contact size	94mm×284mm
Maximum speed	750m/s
Slipper thickness	21mm
Acceleration	93.75m·s^{-2}
Operating time	8s

Table 2. Al parameters of slipper and track.

Object	**Temperature**（℃）	**Special heat** J/(g·°C))	**Thermal conductivity** J/(mm·s·°C)	**Density** ρ (g/cm3)
Slipper 0Cr18Ni9Ti	20	0.462	0.0146	7.75
	200	0.512	0.0161	
	400	0.540	0.0180	
	800	0.604	0.0239	
Track U71Mn	20	0.472	0.147	7.92
	200	0.480	0.143	
	400	0.524	0.138	
	800	0.615	0.132	

IOP Conf. Series: Materials Science and Engineering **398** (2018) 012019 doi:10.1088/1757-899X/398/1/012019

3.2. Boundary Conditions

The boundary conditions for the model to solve finite element heat transfer equation and stress equation are stated as following.

a)The friction contact surface of slipper applies the second boundary condition. The heat flux density is calcuated according to the equation 3 and thus yields the second boundary condition;

b)All surfaces except the contact surfaces applies the first boundary condition, and the surface temperature is the ambient temperature;

c)Normal pressure load is exerted on the surface of slipper.

d)Fixed support is applied on slipper contact surface. A, B, C, and D points apply constraints on z-axis and freedom on X and Y axis.

3.3. Simulation result

The temperature distribution is shown in Figure 3. The curves of contact surface temperature and maximum stress varying with time respectively are illustrated in Figure 4.

Figure 3. Temperature distribution

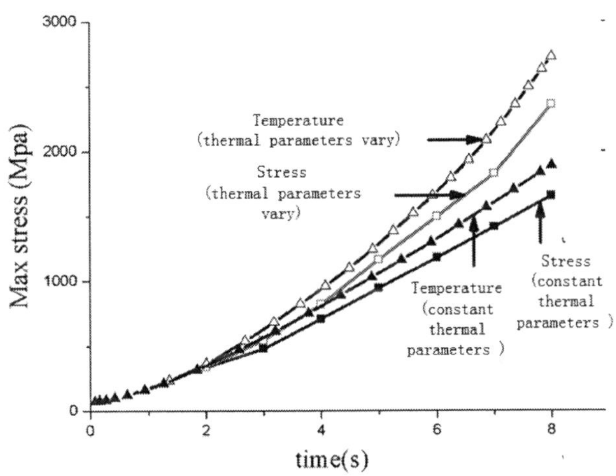

Figure 4. Contact surface temperature varies with time ; Maximum stress varies with time

From the contact surface temperature, it can be seen that, over time, friction heat flow continuously enters into the contact surface, making contact surface temperature rise[3]. The solution of the maximum temperature can reach 818 ℃, the maximum temperature located on friction contact surface. The maximum stress is 2350 MPa which can be read from Figure 4.

The solution of temperature distribution was transfer into load and solved in stress field. The result observed from Figure 5 shows that the maximum stress is also located on the contact surface and the stress and strain in friction heat response area are much higher than other areas.

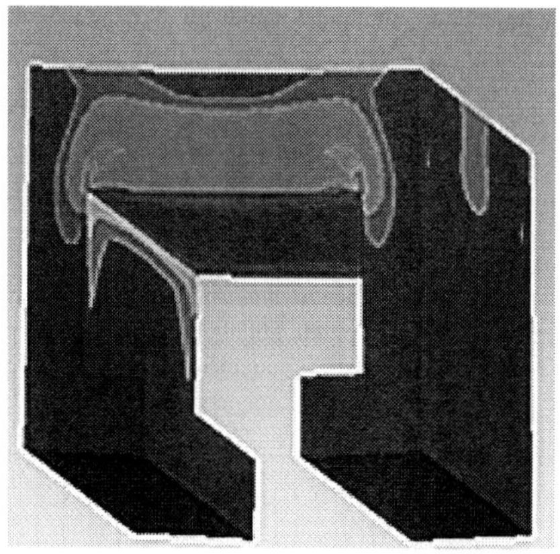

Figure 5. Stress field distribution

The variations of the total strain and thermal stress along a certain direction were mapped in Figure 6. Figure 7 shows the variations of the total strain and thermal strain along a certain direction. From the graphs, it can be seen that the temperature is high and changes dramatically within 6 mm distance from the contact interface, and the thermal strain is the dominate strain source. In the range of 6mm to 14mm distance from the interface, the difference between the strain due to thermal and the strain due to contact load are small. The thermal strain varies slowly and the load stress is the dominant influencing factor. From 14mm to the top surface of the slipper, the thermal strain does not change evidently. The strain due to load is the dominant factor and the total strain increases. The area within 6 mm from the interface is the thermal concentrated zone and has the highest strain.

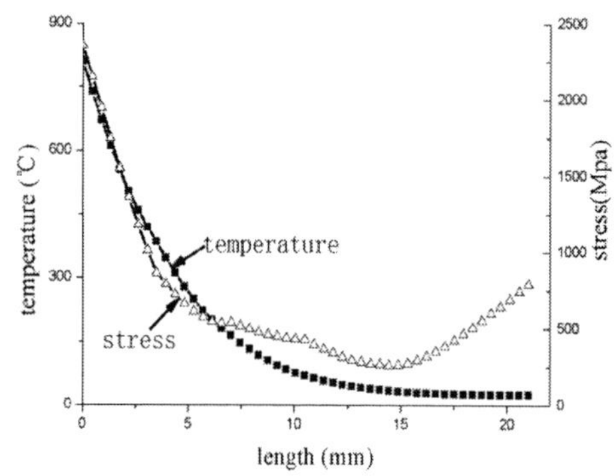

Figure 6. The variations of the temperature and stress along a certain direction

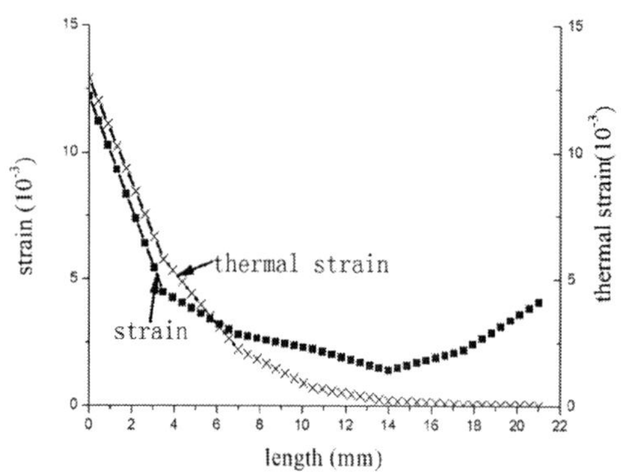

Figure 7. The variations of the total strain and thermal strain along a certain direction

IOP Conf. Series: Materials Science and Engineering **398** (2018) 012019 doi:10.1088/1757-899X/398/1/012019

4. Experiment analysis

4.1. Obversation of slippers

In real test, 6 mm plates were installed on the corresponding locations on the slipper surface. The positions relative to the track are shown in Figure 8 a. An analysis was conducted on the macro morphology of the slipper. The slipper plate samples were taken from a rocket sled after run test. The overview of worn samples is shown in Figure 8 b-f.

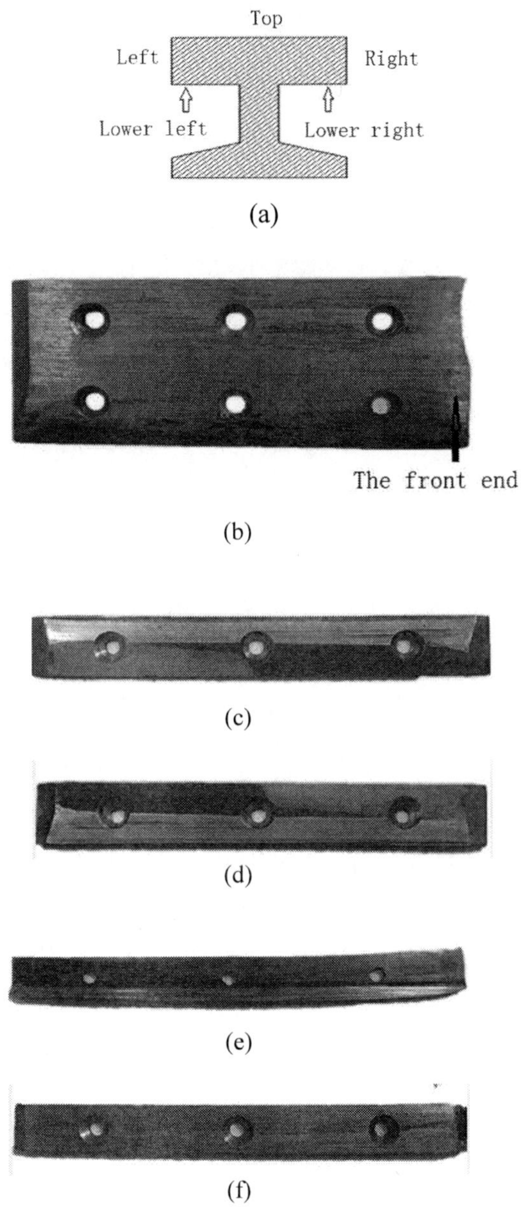

Figure 8. Sample photos

(a.Schematic of slipper plate positions b. Top slipper plate c. Left slipper plate d. Right slipper plate e. Lower left slipper plate f. Lower right slipper plate)

The changes of slipper dimensions were measured in Table 3.

Table 3. Al parameters of slipper and track.

Position	Front /mm	Middle /mm	Rear /mm	Original /mm
Top	2.7	5.6	5.8	6
Left	5.1	6	6	
Right	4.9	5.8	5.6	
Lower left	1.2	1.2	1.9	
Lower right	5.5	5.5	5.6	

From Table III it can be seen that the front and lower left slipper were seriously worn. The front end of top slipper was worn into sharp shape. The lower left slipper was highly deformed compared to original shape. The total thickness changed obviously. No obvious thickness changes were observed from other slippers.

4.2. Hardness test on slippers

The samples were analyzed by hardness tester TY10MH-5. The samples were cut into 20mm×20mm×3mm specimens. Hardness was tested every 0.2mm in thickness direction. The results are illustrated in Figure 8.

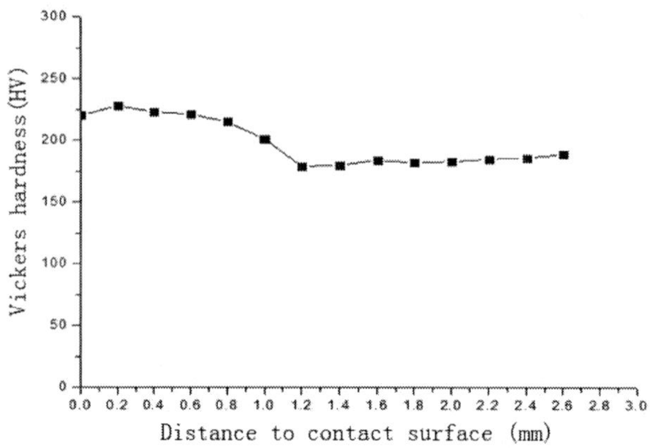

Figure 9. Hardness test results on vertical cross section

It is obvious that the surface hardness is obviously higher than the interior base. It confirms that the wear behavior resulted in plastic deformation and thereby work hardening.

4.3. Wear surface product analysis

The specimens were analyzed by X-ray diffractometer. The results are shown in Figure 10.

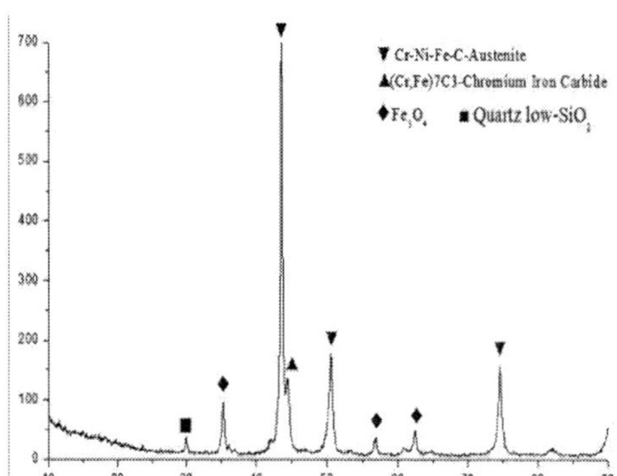

IOP Conf. Series: Materials Science and Engineering **398** (2018) 012019 doi:10.1088/1757-899X/398/1/012019

a)Top slipper plate

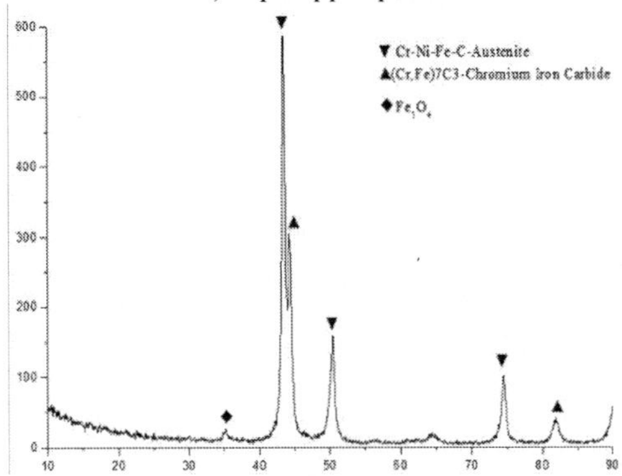

b)Side slipper plate

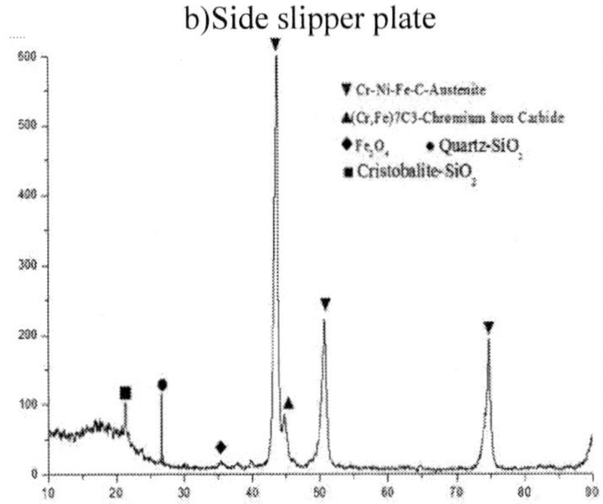

c)Lower side slipper plate

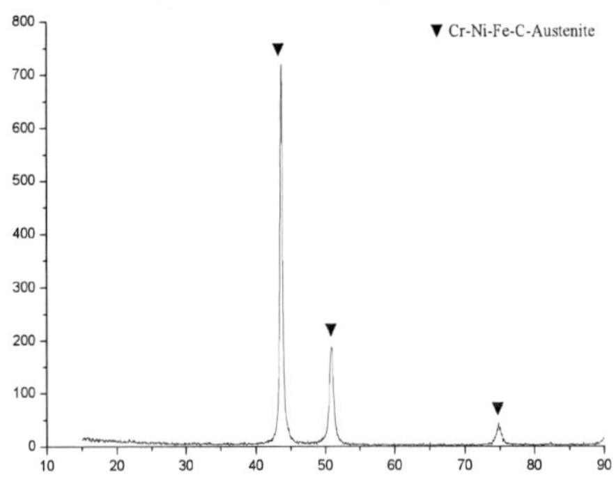

d)Original unworn slipper plate

Figure 10. XRD analysis on failure and original samples

From Figure 10 d) we know that there is only austenite microstructure in original slipper. After run, Fe3O4 and (Cr, Fe)7C3 can be observed on the surfaces. It can be inferred that the temperature reaches 650 ℃ between the contact pairs. Furthermore, Cristobalite-high SiO2 was detected on the lower left slipper, which indicates that the temperature of surface area was above 1470 ℃. Since Si was not existed in the slipper material, the compound is believed to be generated due to the mix of impurities in the air. In addition, the rocket sled is believed to be under an abnormal attitude leading a high loading condition and high friction heat was generated on the lower left side slipper.
Because of severe vibration load in real test causing high impact load, the surface temperature is higher than the simulation result.

4.4. Observation of microstructure

IOP Conf. Series: Materials Science and Engineering **398** (2018) 012019 doi:10.1088/1757-899X/398/1/012019

Figure 11 shows typical microtopography of top and side slipper plates. Obvious difference can be seen along the vertical direction starting from the contact surface[4]. Heat-stress coupled tissue can be observed in the area of contact surface and contiguous zone, where there can't see obvious grain, and is uneven. The surface shows a stairway shape, and there are cracks in the dense tissue of the surface layer. The thermal-stress coupled issue is gradually thinning as the distance from surface increases.

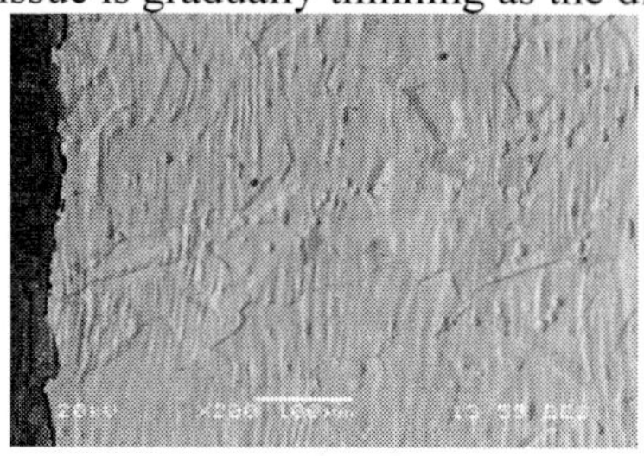

(a)

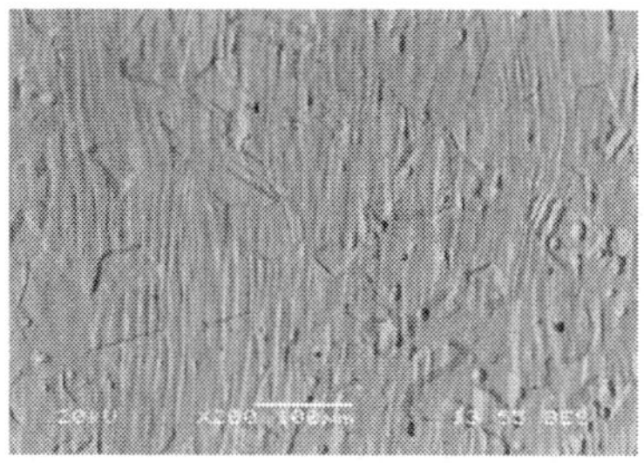

(b)

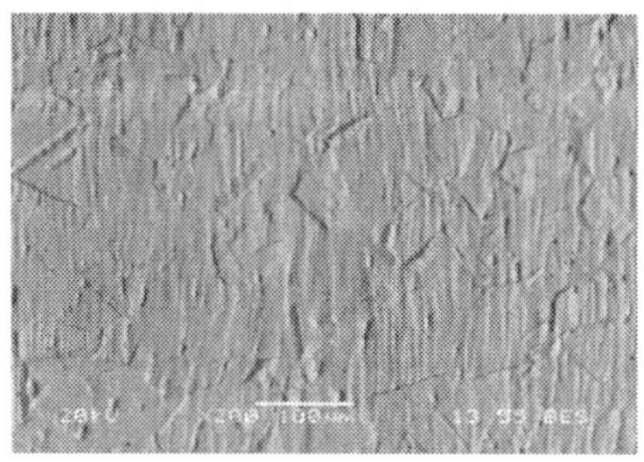

(c)

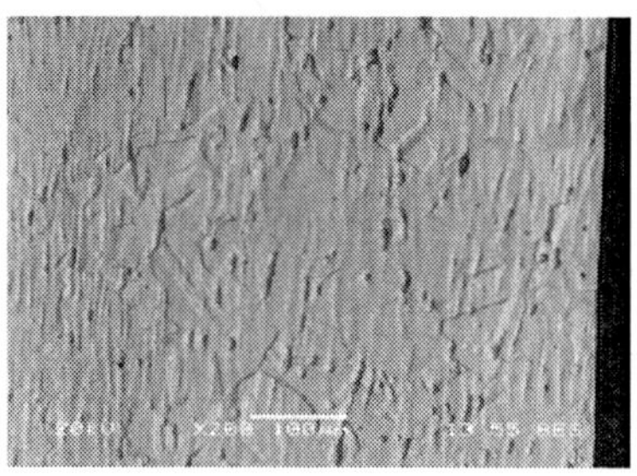

(d)

Figure 11. Typical Microscopic microstructure of samples

5. Conclusion

Simulation and Experiment results conducted in this work allowed the following conclusion:

For a rocket sled run of speed 750m/s and time of 8 seconds simulation, the maximum temperature reaches 818℃, the maximum stress is 2350 MPa. The friction heat and stress are concentrated within 6 mm distance from contact surface. In contrast, the analysis of macro morphology of the slipper taken from real run test shows that local temperature reaches 1470 ℃ under this speed and time condition in

IOP Conf. Series: Materials Science and Engineering **398** (2018) 012019 doi:10.1088/1757-899X/398/1/012019

evidence of Cristobalite-high SiO2 generated. The worn distance is in good agreement with simulation result. Under the double action of vertical impact load and frictional heat, thermal-stress coupling tissues in the material are produced proved by microstructure observation.
The temperature difference is mainly because that the simulation applies an idealistic condition that not taking account of the vibration load, and impurities in the air environment combining the materials generate Cristobalite-high SiO2 in high temperature.

6. References

[1] Cameron G. J 2007 An Evaluation of High Velocity Wear, AFIT/GAE/ENY/07-M06. Master's thesis, Air Force Institute of Technology *J. Wright Patterson AFB, OH* **2** 07.
[2] Cameron G and Palazotto A 2008 An Evaluation of High Velocity Wear *J. Elissa, "Title of paper if known, " unpublished* **7** 1066-1075
[3] Stephen P 2010 Consideration of Wear at High Velocities, AFIT/GAE/ENY/10-M16. Master's thesis *J. Air Force Institute of Technology, Wright Patterson AFB, OH* **24** 18
[4] Tachau R D M, Yew C H and Trucano T G 1995 Gouge initiation in high-velocity rocket sled testing *J. International journal of impact engineering* **4** 825-836

IOP Conf. Series: Materials Science and Engineering **398** (2018) 012018 doi:10.1088/1757-899X/398/1/012018

Numerical Simulation of the Stress, Temperature & Wear Behaviours of the Drum Brake

Ehtisham Shahid[1], Xiaoying Wang[1], Zijie Fan[1], Liangjin Gui[1]*

[1]State Key Laboratory of Automotive Safety and Energy, Department of Automotive Engineering Tsinghua University, Beijing 100084, China.

*E-mail: gui@tsinghua.edu.cn

Abstract. Thermo-mechanical coupling phenomena exist in the drum brake, which produces high contact stress field and transient temperature. Kinetic and Potential energies are applied and transformed into thermal energy and that thermal energy couples the different attributes such as stress, temperature, and wear. In this paper, the Archard abrasive wear model and Finite Element Method (FEM) are combined to simulate the normal braking condition of drum brake by means of the stress and wear coupled analysis method. Then based on stress, temperature and wear coupling method, the brake is simulated under emergency braking condition by using ABAQUS. UMESHMOTION subroutine is coded to realize the material loss. The process shows the contact pressure area, temperature field and wear abruption and their coupling process. Stress, temperature and wear are associated with expansion properties of a substance which are discussed in undeviating simulation results. Keywords-Wear; Drum brake; FEM; Heat flux; FEA.

1. Introduction

Drum brake has a strong, complex and intense level of thermal stress and coupled wear while braking. Due to pressure, friction plates and drum crushes while coming into contact with each other and cause severe friction. Work done by friction produces a huge amount of heat and temperature. The friction force is the cause of heat so it directly affects the heat transfer and conduction. The sudden rise of temperature changes the stiffness of friction plates, the hardness of brake drum and the interfacial friction factors. Meanwhile, Uneven thermal deformation affects the surface of the drum in brake which brings the change in the distribution of friction and contact pressure along with the wear attribute. In turn, wear behaviour changes the contact surface's state which often affects the distribution of contact pressure. Hence, braking process is fully related to coupled stress, temperature and wear attributes [1,2].

Coupling problem of stress, temperature and wear involve complex structural geometry and nonlinear behaviours in contact so it is hard to obtain accurate analytical solutions. In the present time, many of the researchers and scholars have been using FEM to solve the thermal-stress coupling process problems in braking systems [3,4]. However, the research on thermal boundary conditions was a breakthrough in its own and made things lot simple in the field of tribology but wear and tear factor was totally ignored on the temperature and displacement impact on the coupled surface.

At the end of last century, the research on wear began by using FEM [5,6]. Based on Adaptive Lagrangian-Eulerian (ALE) method the adaptive wear model was studied for radial bearing with the rotation process at the time of shaft contact and the bondage of wear was smaller than the surface height [7].

However, the solution is simply suitable for simple structural problems and for real life practical engineering problems but there is still a great distance to cover [8,9]. The effect of wear and influence of thermo-elastic instabilities in mechanical coupling and the seriously deviated stress results from the actual situation were studies under developed FEM methods on the bases of elastic-mechanical equation etc., just as a 2D model with Hertzian pressure distribution [10,11].

Starting with the stress and wear coupling algorithm for the simulation of the friction in a drum brake, along with that the 3D model of drum brake is developed for transient braking and temperature reduction analysis. The process includes an interaction of stress, temperature and wear depend on the thermal parameters along with the temperature of materials. The simulation results are verified in terms of stress, temperature, and wear.

2. Method and mathematical model of stress temperature and wear

2.1. Thermal-stress coupling and boundary conditions

- All materials used in the model, including brake drum and friction plates are fully isotropic, homogeneous and uniform.
- There is no source of heat inside the material because the whole deformation doesn't get into plastic deformation and no damping factor is found.
- In the whole braking process, the inertial forces are completely ignored and considered main forces are the contact force and the frictional force so that the mechanical analysis can be done including thermodynamic properties because following model's contact conditions are very strong and deformation or mesh distortion is possible.

Three Dimensional heat conduction is required to be considered in a model in which the temperature and are could be expressed with regard to spherical coordinates and cylindrical co-ordinates (Cartesian co-ordinates). Temperature field and the material's density are expressed as well along with thermal conductivity and specific heat capacity [12].

The boundary condition for convective heat transfer has to be declared along with heat convection and a decrease in temperature coefficients. The boundary condition for friction heat flux has to be declared along with the relative velocity of sliding surface or relative sliding rate [13,14].

By using FEM, deformation needs to be considered. The displacement has to be declared by following the elastic model and need a supposition for minor deformation.

$$\begin{bmatrix} 0 & 0 \\ 0 & C_\emptyset \end{bmatrix} \begin{Bmatrix} \dot{u} \\ \Delta\dot{\emptyset} \end{Bmatrix} + \begin{bmatrix} K_u & -K_{u\emptyset} \\ 0 & K_\emptyset \end{bmatrix} \begin{Bmatrix} u \\ \Delta\emptyset \end{Bmatrix} = \begin{Bmatrix} fu \\ \Delta P_{q(p)} + \Delta P_H \end{Bmatrix} \tag{1}$$

In the equation above $C_\emptyset$ is the matrix for the heat capacity, $K_\emptyset$ is the matrix for the heat conduction or thermal conductivity, $\emptyset$ is the vector of temperature at nodes, P is the load vector of temperature at nodes, P_q is the head-induced load vector caused by the heat generation and it is a function of contact pressure p. The load vector caused by convective heat dissipation is P_H. K_u is the displacement on the nodes for node matrix, $K_{u\emptyset}$ is the node temperature's contribution towards the nodal force matrix, u and fu both are the displacement vector and force vector of the node, respectively.

Coupled FEM involves stress and temperature. By considering equation (1) the whole drum brake can be heated all over. Heat convection equation is [15, 16].

$$q = h_c A \Delta T \tag{2}$$

In the equation above q is the heat transfer per unit time, A is the heat transfer area of the surface, h_c is the convective heat transfer coefficient of the outer surface of the brake drum and the scrounging and ΔT is the temperature difference between the brake drum and friction plates.

$$h_c = \frac{k_l Nu}{l} \tag{3}$$

Here, Nu is the Nusselt number, k_l is the thermal conductivity and l is the outer diameter of the brake drum with the characteristic length of the Reynold number.

2.2. Contact properties

The Suppose if there is a contact between two different objects taken as points are A and B then momentum equilibrium condition can be considered [17], which includes the component of normal force, and properties.

When objects come into contact then sometimes there is a penetration and sometimes not so it can be calculated by using mutual penetration rate of objects, their normal velocity, and normal contact conditions.

2.3. Wear modeling

The numerical integration of wear is based on the Euler method. The incremental form of generalized Archard wear model is [18].

$$\Delta h = kp\Delta s \tag{4}$$

In the above formation Δh is the increment for wear thickness, k is the coefficient for wear factor, Δs is the increment for relative displacement.

By assuming the contact nodes of the brake drum and the friction plates in finite element mesh is i and the integration step is j then the wear increment is as followed.

$$\Delta h(i,j) = k(i,j)\ p(i,j)\ \Delta s(i,j) \tag{5}$$

Then the cumulative wear is appeared as,

$$h(i,j) = \sum_{l=1}^{j} \Delta h(i,j) = \sum_{l=1}^{j} k(i,l)\, p(i,l)\, \Delta s(l) \tag{6}$$

3. Numerical simulation of normal and emergency braking process

In order to keep simulation results of stress and wear more accurate, it is necessary to simulate the brake's normal braking process to achieve smooth contact after executing finite element model because there are some errors left behind as uneven friction surface due to installation errors or scientific errors while manufacturing etc., which may cause poor contact while start-up if brake is not in normal braking process.

3.1. FEM model

The finite element model of the drum brake from the rear axle of heavy duty vehicle is created by using the Hypermesh software after illumination of some unnecessary components like springs, pins, cables, holders etc., 'figure 1'. The number of elements are 10466 (15841 NODES). The whole drum is simulated by using first-order element type C3D8RT because this is one of the units which can be simulated better under complex surfaces and improve contact conditions during analysis. The both friction plates have adopted first-order element type C3D8RT as well. Meanwhile, the brake shoes containing shell element type S4RT and the rollers are replaced, because the cam forces on the brake shoes have no direct impact on the wear and tear. Instead of rollers, the analytical rigid beams connected to control points are used to receive force and to impose a corresponding constant concentration.

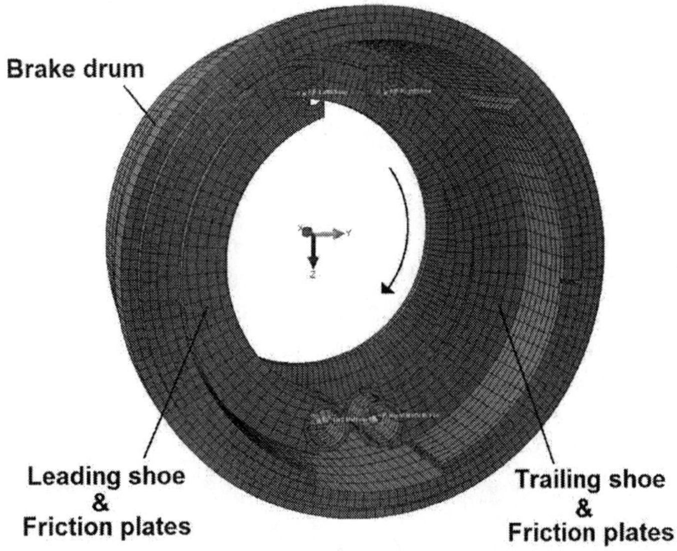

Figure 1. Finite Element Model of drum brake.

The brake drum is made up of material tagged as HT250 as per Chinese standard and regulations. Brake shoes are made up of Steel. Friction plates are made up of composite material. All the parameters and properties for brake drum, friction plates and shoes are shown in table 1.

Table 1. Material properties of drum brake.

	Temp. (°C)	Specific heat $(J/(kg.°C))$	Conductivity $(W/(m.°C))$	Thermal expansion $(10^{-6}/°C)$	Young's modulus (GPa)	Density (kg/m^3)	Poisson's ratio
	20	503	42.38	04.386	100.000	7220	0.3
	100	530	43.06	11.653	-	7220	0.3
	200	563	44.23	12.836	099.693	7220	0.3
Drum	300	611	43.55	13.580	-	7220	0.3
	400	641	40.67	13.578	096.273	7220	0.3
	500	701	39.72	13.804	-	7220	0.3

IOP Conf. Series: Materials Science and Engineering **398** (2018) 012018 doi:10.1088/1757-899X/398/1/012018

Friction plates	-	1178	01.98	07.550	007.650	2150	0.3
Shoes	-	452	48.00	11.000	205.000	7800	0.3

While the wear coefficient for the friction plates is $k = 15 \times 10^{-15} Pa$. Attributes of the model include absolute zero temperature 1×10^{-9}°C and Stefan-Boltzmann constant $5.669 \times 10^{-8} J/sm^2k^{-4}$. The value used for heat convection and surface film condition is $50\ W/(m°\text{C})$, as film coefficient depends on circumstances and varies accordingly.
The reference points (RP) are shown in the 'figure 2'. During simulation, the inner part of the cylinder is coupled to flange surface through multi-point constraint (MPC) from its surface to control point at drum's axis of rotation denoted by RP-Drum in 'figure 2a'.
In the brake shoes, the bottom end is hinged with the brake plate and during the simulation, and the nodes are coupled to the inner surface of pin holes and its contact point on the axes of nodes by using MPC denoted by RP-Left Bottom Fix and RP-Right Bottom Fix in 'figure 2b'. In an actual braking system, the upper end of shoes is connected with a roller at the top so the upper hole surface section is required and the circular nodes are coupled with the control points on their control axis point. Rollers are replaced and controlled by the rotation point on the axis denoted by RP-LeftShoe and RP-RightShoe. These two control points are linked with the control points on the axis by using beam type elements through MPC.

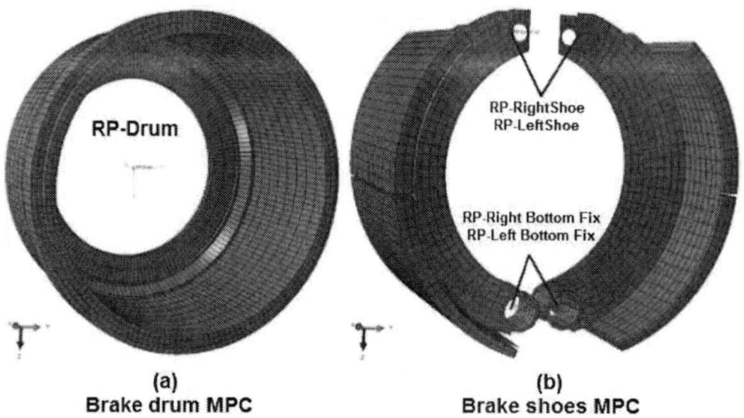

Figure 2. Reference points and multi-point constraints.

The contact between the brake drum and the friction plates is mainly defined between the inner surface of the brake drum and the outer layer of friction plates. The contacting algorithm for friction plates is the Penalty friction formulation and the friction coefficient is 0.35. The constraints between friction plates and brake shoes are tie constraints. In the normal braking process, the brake torque is about 17550.5Nm.

3.2. Solution
As it was mentioned earlier that in the whole normal braking process the inertial forces are completely ignored. Neglecting the inertial forces effect belongs in a quasi-static process (2).
The UMESHMOTION subroutine has been used to calculate the adjacent incremental steps of the stress analyzation. Then ALE is used to update the finite element mesh to realize the wear simulation by incremental stress in the incrimination step and then update the wear amount.

In the start, the displacement loads are pre-applied on the contacted surfaces so that the contact calculation of each contact pair converges properly. Then displacement loads are removed and a load is applied at RP-LeftShoe and RP-RightShoe. Then the constant rotation speed of brake drum is set at the brake drum control point denoted as RP-Drum, ω=26.455rad/s. Parameters are set for normal braking simulation process.

During the simulation, the wear of the brake drum is ignored and the friction material is set as the ALE domain. Then the wear on the surface of friction plates is calculated by UMESHMOTION subroutine.

Process using thermal-stress-wear coupling method is shown in the 'figure 3'. Processing method is split into parts as frictional contact process is combined with the material removed from surface and thermo-elastic distortion. On the bases of their FEM substructure, they have been solved in ABAQUS where mechanical and thermal results can be attained together. Adaptive mesh tool and the UMESHMOTION subroutine are used together in ABAQUS to adjust the position of on surface nodes according to material loss induced by wear.

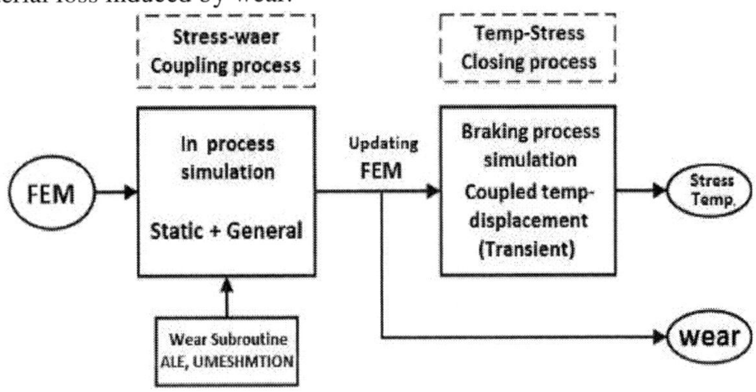

Figure 3. Thermo-mechanical & wear coupling.

3.3. Solution result and analysis

Evolution of the contact pressure of friction plates can be seen in 'figure 4'. First part represents the initial pressure distribution on leading and trailing shoe at the initial time. Second half represents the contact pressure distribution at the time of completion of the running process on leading and trailing shoe. It can be seen from the 'figure 4' that most of the friction plates area has come into contact with the inner surface of the brake drum.

At the initial moment, the contact pressure distribution is very much uneven. In the leading shoe, the contact pressure is greater than the trailing shoe due to the lifting force.

Right after the start of the running the area where the contact pressure is observed, shows that the contact is improved in the area where contact pressure is large and contact pressure increased gradually. At the end of the normal braking process, the maximum contact pressure is distributed across most of the area of friction plates. It shows that the contact pressure distribution is more uniform and graphically from most of the area it is accessible to differentiate by colours.

While in the normal braking process the contact pressure and stress distribution along the brake drum axil can't be neglected and has certain characteristics. In 'figure 4' the left side of the left friction plate is in the direction of brake drum flange. The right side of the right friction plate is in the direction of brake drum flange. The part of friction plates close to the brake drum flange has greater contact pressure than the part of friction plates at the opening of a brake drum. The flange side of brake drum has fix constraints so stiffness is relatively larger than the opening side of a brake drum.

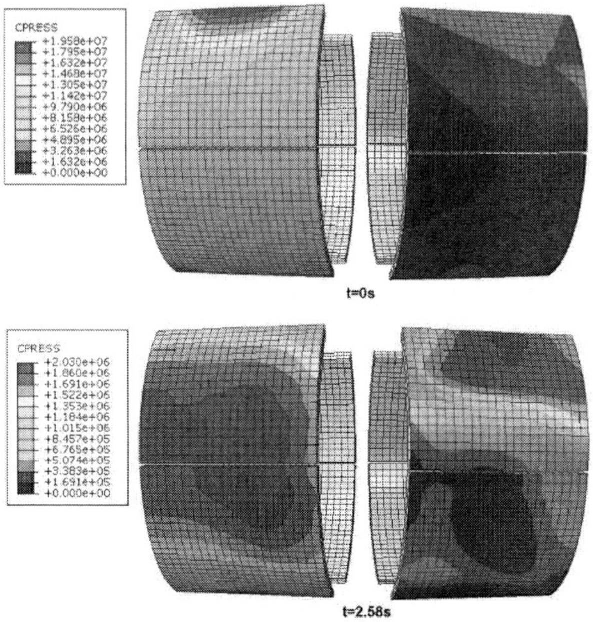

Figure 4. Contact pressure distribution in normal braking condition.

After the normal braking process, the friction plates are in the good contact form with the brake drum. Now in the emergency braking condition, the effect of temperature is considered. In the step when loads are applied the brake time is 2.58s and the rotation speed of brake drum is 26.455rad/s which constantly decelerated up to 0.

The concentrated forces added on the Left and Right friction plates are calculated by finite element software. Concentrated forces can be seen in table 2.

Table 2. Applied load on brake shoes.

	Load Point	CF1(N)	CF2(N)	CF3(N)
Leading shoe	RP-LeftShoe	000	-36037.3	12041
Trailing shoe	RP-RightShoe	000	50622.4	-18137.6

The contact and heat flux between the brake drum and friction plates are mentioned earlier in [1]. When brake drum is in contact with a friction plate, the set coefficient for heat flux is large to ensure that temperature at contact points are the same.

Distribution of contact pressure during emergency braking is shown in the 'figure 5'. In leading shoe, the contact status is improved because of the role of the entire friction plate so the distribution of contact pressure is relatively uniform and the left side of friction plate is beside flange of the brake drum, so the contact pressure is a bit more than other portions of the friction plate. Some of the visible effects can be observed in trailing shoe.

IOP Conf. Series: Materials Science and Engineering **398** (2018) 012018 doi:10.1088/1757-899X/398/1/012018

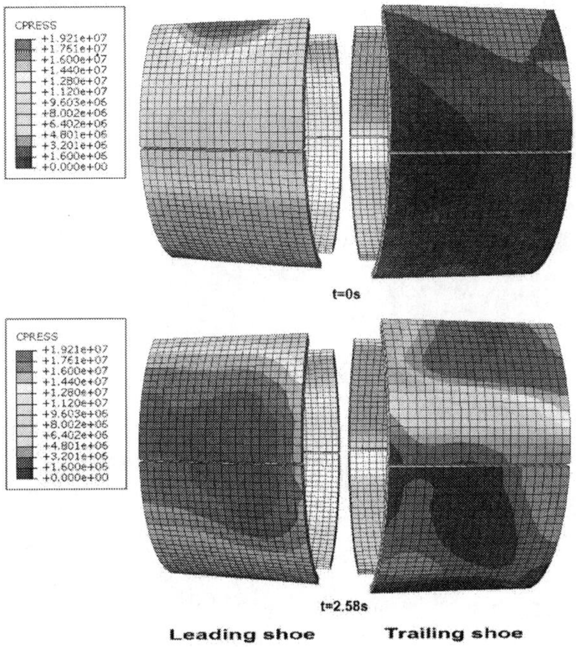

Figure 5. Contact pressure distribution in emergency braking condition.

The amount of wear of the friction plates is shown in the 'figure 6'. The sliding surface of drum brake is nearly equal to each other in the braking process so the amount of wear and contact pressure distribution is nearly close. At the end of braking process where contact pressure was maximum, the observed intensity of wear depth has a minute value of 0.425µm.

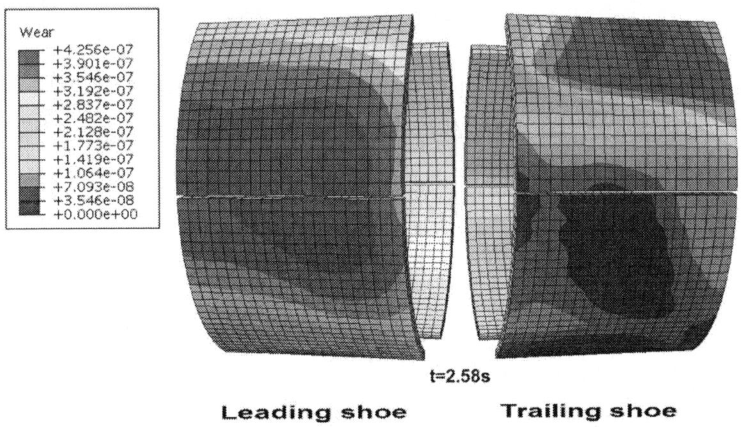

Figure 6. Wear depth - Emergency braking condition.

The changes in temperature of the brake drum and friction plates and its distribution during the whole braking process is shown in the 'figure 7'. At an initial time when t=0.10s the temperature of friction plates and brake drum is not that large but temperature distribution is so uneven.

Contact pressure and the frictional thermal flux is directly proportional to each other so because of this relation the frictional temperature of leading shoe rises higher than the frictional temperature of trailing shoe.

Along with that, the area of friction plates which are near to brake drum flange has the high frictional temperature and pressure than the other areas of friction plates on the open end of a brake drum.

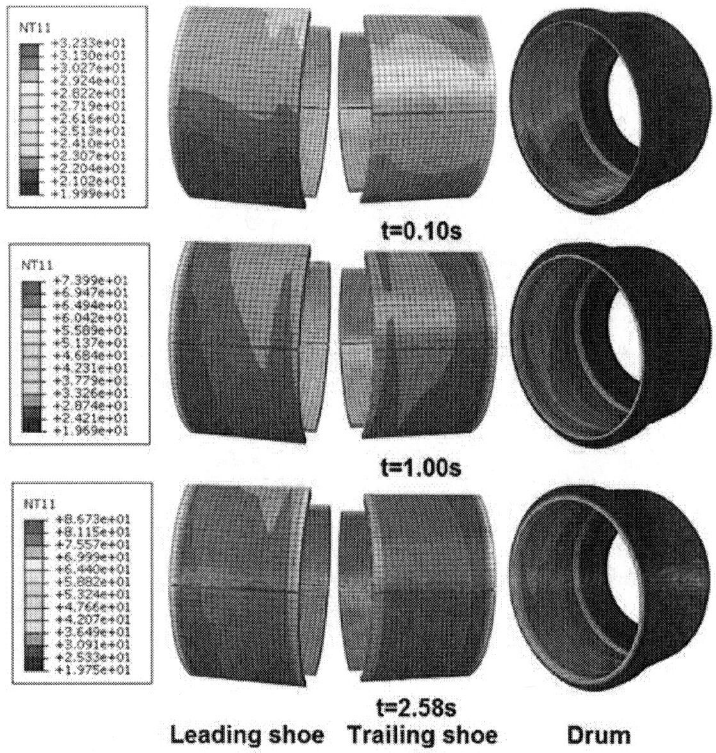

Figure 7. Temperature evaluation at different times in emergency braking condition.

The contact between friction plates and the brake drum are good enough so the temperature and pressure distributions are almost axisymmetric. Because of good contact between surfaces the inner surface of the brake drum and the outer surfaces of friction plates are not with much difference, nearly the same.

4. Conclusion

In this paper, an algorithm has been formed to consider thermo-mechanical coupling and tribological properties including stress, temperature and wear for drum brake based on FEA. The normal braking process of drum brake is numerically simulated to compare the effect and distribution of wear with contact pressure, stresses and temperature simultaneously.

- The stress-temperature-wear coupling simulation algorithm is used in normal braking process, then results along with the evaluation of the contact pressure is shown. From the results, the contact pressure can be evaluated while interacting with temperature filed.
- From the whole research, the results indicate that in normal braking process the contact pressure of leading shoe is larger than the trailing shoe. The contact pressure at the both ends of the both shoes are higher than the middle area. The contact pressure of the friction plates is

greater on the sides which are near to the flange of brake drum than the sides which are on the opening end. The contact pressure distribution tends to be uniform.

- The temperature field distribution of brake drum is approximately axisymmetric. The friction plates have the lower temperature on the sides which are near to the flange and higher temperature on the sides which are on the opening end.
- Friction plates wear is far larger than the brake drum so only friction plates wear has been considered in the simulation. ALE method sets a load and friction for materials as a whole and UMESHMOTION subroutine calculates joint wear for frictional surfaces.

5. Acknowledgement

The authors would like to acknowledge the National Natural Science Foundation of China (No. 51475255) for their financial support of this research.

6. References

[1] Belhocine A, Bouchetara M, "Transient analysis of Thermoelastic contact problem of disk brakes," *Frontiers of Mechanical Engineering*, 2013. **8**(2): p. 150-159.

[2] Majcherczak D, Dufrenoy P, Berthier Y, "Tribological, thermal and mechanical coupling aspects of the dry sliding contact," *Tribol Int.,* 2007. **40**: p. 834-43.

[3] Simo JC, Miehe C, "Associative coupled thermoplasticity at finite strains: formulation, numerical analysis and implementation," *Comput Methods Appl Mech Eng.,* 1992. **98**: p. 41-104.

[4] Heckmann A, "A brake model with Thermoelastic disc for the analysis of vehicle judder vibrations". *Vehicle system dynamics,* 2006. **44**: p. 360-367.

[5] Podra P, Andersson S, "Simulating sliding wear with finite element method," *Tribology International,* 1999. **32**(2): p. 71-81.

[6] Chantrenne P, Raynaud M, "A microscopic thermal model for dry sliding contact," *Int. J Heat Mass Transfer,* 1997. **40**: p. 1083-94.

[7] Rezaei A, Van Paepegem W, De Baets P, "Adaptive finite element simulation of wear evolution in radial sliding bearings," *Wear,* 2012. **296**: p. 660-671.

[8] Cai P, Wang Y, Wang T, Wang Q, "Effect of resins on thermal, mechanical and tribological properties of friction materials," *Tribol Int.,* 2015. **87**: p. 1-10.

[9] Ireman P, Klarbring A, Stromberg N, "Finite element algorithms for Thermoelastic wear problems," *European Journal of Mechanics-A/Solids,* 2002. **21**(3): p. 423-440.

[10] Wu L, Wen Z, Li W, Jin X, "Thermo-elastic-plastic finite element analysis of wheel/rail sliding contact. In: Proceedings of the 8th international conference on contact mechanics and wear of rail/wheel systems," *Florence, 2009.* **271**; 2011. p. 437-43.

[11] Ostermeyer GP, Graf M, "Influence of wear on Thermoelastic instabilities in automotive brakes," *Wear,* 2013. **308**(1): p. 113-120.

[12] Liangjin Gui, Xiaoying Wang, Zijie Fan, Fangyu Zhang, "A simulation method of thermo-mechanical and tribological coupled analysis in dry sliding system," *Tribology International,* 2016. **103**: p. 121-131.

[13] Mazahery A, Shabani MO, "The accuracy of various training algorithms in tribological behaviour modelling of A356-B4 Composites," *Russ Metall,* 2011. **46**(21): p. 699-707.

[14] Zienkiewicz OC, Taylor RL, Zienkiewicz OC, Taylor RL, "The finite element method," *London:* McGraw-*hill,* 1977.

[15] *Abaqus Analysis User's Manual,* §6.5.2 Uncoupled heat transfer analysis. Adaptive manual version **6.12-1**, Dassault Systems Simulia Corp., Providence, RI; 2012.

[16] Huang H-C, Usmani AS, "Finite element analysis for heat transfer," *London: Spinger,* 1994.

[17] *Abaqus Analysis User's Manual,* §12.2ALE. Adaptive manual version **6.12-1**, Dassault Systems Simulia Corp., Providence, RI; 2012.

[18] Soderberg A, Andersson S, "Simulation of wear and contact pressure distribution at the pad-to-rotor interface in a disc brake using general purpose finite element analysis software". *Wear,* 2009. **267**: p. 2243-51.

IOP Conf. Series: Materials Science and Engineering **398** (2018) 012020 doi:10.1088/1757-899X/398/1/012020

Effective Ways of Enhancing Communications in Design Process: A Preliminary Study

Shou-Chen Lin, and Kim C K Lee*

Department and Graduate Institute of Industrial Design,
Chaoyang University of Technology, 168 Gifong E. Rd. Wufong Taichung, Taiwan

kim.cklee@msa.hinet.net

Abstract. The major task of Industrial Design Department is to accomplish the company's goal through product creations. Since this is an interdisciplinary task inherently, effective communications with various peer departments are crucial. However, different departments always have conflicting agendas and varied aspect of focuses towards the company's goal. Thus effective communication between peer departments is a requirement of building consensus, which will enhance cooperation between them to achieve the company's goal collectively. The aim of this research is to summarize effective ways of communications in product development and design process that industrial design team managers can take as the reference to enhance their design management performances. Since the research project is still going on, this paper will include the preliminary study of this research only. We investigated existing research results and developments on communications and product development comprehensively through literature review. In conclusion, we summarized existing knowledge concerning the topic of communication in product development into a questionnaire, twelve questions was developed in this questionnaire for further in-depth interviews with design managers to elicit more replies of their valuable experiences. We hope that the result of this research could help to enhance communication performance in the industrial design team, which will be beneficial for the companies that they are serving.

1. Introduction

In a speech on quality and competitiveness at a major conference among top executives in London held by Michael Smith of APV /Baker Perkins Machines in 1981, he showed his audience figures of the notorious 'swing syndrome' [1] (Figure 1) which effectively illustrated the perils of poor communication between departments during product development.

In this illustration, Smith interestingly presented the confusing results entailed by erroneous interpretations on consumer needs. It clearly demonstrated the common facts that are occurring in businesses during product development. Moreover, this illustration had also made it self-evident that effective communications are highly crucial for satisfiable performance in product development. Therefore, this research is set out to investigate the effective ways of enhancing interdepartmental communication performances for industrial design teams in Taiwan. The research objective is to conclude with some useful references for Taiwanese managers to take as the toolkit to enhance communication performance in their industrial design teams. Although the research conclusions can serve as references to managers in Taiwan specifically due to the cultural distinctions of the topic of communications, nevertheless, they will still be worthwhile for taking into considerations of benchmarking for managers who are even not in the similar cultural background as Taiwan.

IOP Conf. Series: Materials Science and Engineering **398** (2018) 012020 doi:10.1088/1757-899X/398/1/012020

Source: Peters [1] p.120

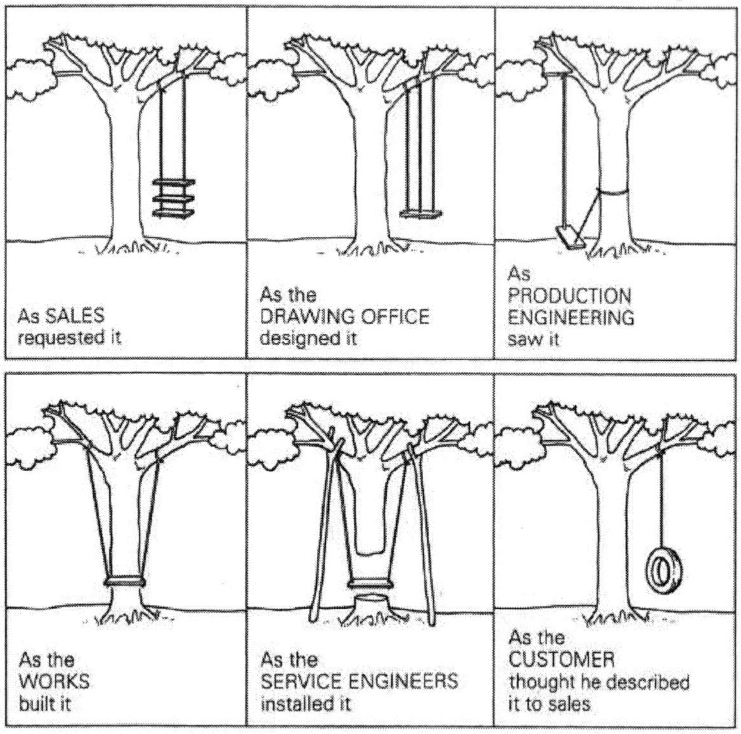

Figure 1. The swing syndrome.

2. Literature reviews

Before we began to conduct our in-depth interviews, it is obvious that a sound foundation is a prerequisite for designing an appropriate set of questions to elicit useful comments from our interviewees. There is no doubt that such a sound foundation must come from reviewing literature in the vast area of communications. Following some initial surveys, we defined the topic domains as communication theories, organizational communications, industrial design teams, and communications in the design process. The essential literature reviewed in this stage are as follows:

2.1. Communication Theories

Before we can send our signals to others effectively, others will find us hard to understand mostly, and the other way around is always also true. Therefore, communication remains a tough problem for every one of us to tackle with great efforts. Since this is also the core issue of this research, the essential literature was reviewed here briefly.

2.1.1. Defining Communication.

The definitions of the word "communication" provided by Oxford Learner's Dictionary [2] are:

- *The activity or process of expressing ideas and feelings or of giving people information.*
- *Methods of sending information, especially telephones, radio, computers, etc. or roads and railways.*
- *A message, letter or telephone call.*

The significant points in these definitions are "activity or process of expressing ideas and feelings or of giving people information," "Methods of sending information," and "A message." To summarize the definitions made by Shannon & Weaver [3], Anderson [4], Compton & Bennett [5], Barnard [6], Schramm & Roberts [7], Deaux, Dane, Wrightsman, & Sigelman [8], and Coulter & Robbins [9], communication involves three crucial conditions:

a) It must be performed mutually and is impossible to accomplish unilaterally.

b) The signals utilized in the communication must be understandable by the both sides.

c) There must be an intention to send signals.

2.1.2. The Process of Communication

Shannon and Weaver [3] were pioneers of defining the communication system as a combination of the elements including information source, message, transmitter, signal, noise source, received signal, receiver, message, and destination, which clearly delineated the components and the process of communication as shown in Figure 2.

Source: Shannon & Weaver [3] p.7

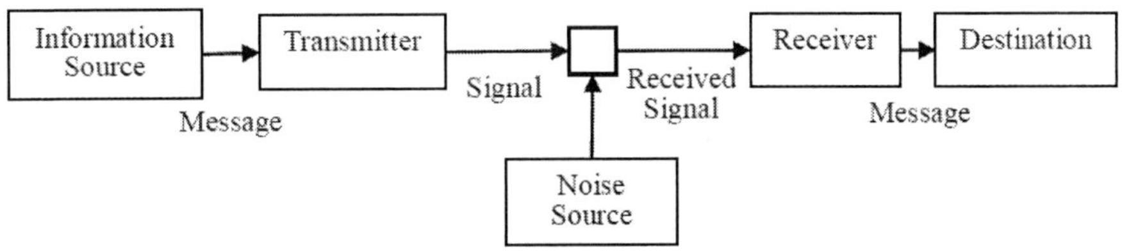

Figure 2. The communication system.

2.1.3. The Principles of Communication

Based on their research on communications, Cutlip & Center [10] pointed out "The Seven C's of Communication" as principles of effective public relations as follows:

- *Credibility:* Communication starts with a climate of belief.
- *Context:* A communication program must square with the realities of its environment.
- *Content:* The message must have meaning for the receiver and it must be compatible with his value system.
- *Clarity:* The message must be put in simple terms.
- *Continuity and Consistency:* Communication is an unending process.
- *Channel:* Established channels of communication should be used - channels which the receiver uses and respects.
- *Capability of the audience:* Communication must take into account the capability of the audience.

2.2. Organizational Communications

Organizations are complex combinations of a number of people. Stemming from the fact that even communication between two persons would be complicated enough, the complexity of organizational communications will certainly increase in proportion to the size of the organization in question. Also, the effectiveness of organizational communications will affect the organizational effectiveness heavily.

2.2.1. Defining Organizational Communications

In a survey conducted by Thomas and Schmidt [11], their research result clearly suggests that managers from middle to top levels are all holding an active and growing interest in learning more about both the prevention and management of conflict. Also, Luthans, Hodgetts, & Rosenkrantz [12] found that the constitution of managers' contribution to organizational effectiveness was as follows: interpersonal network (11%), traditional management (19%), human resource management (26%), and daily communications (44%). These figures showed clearly that daily communications are occupying a significant portion of managerial contribution, and this number must be in proportion with their workload as well.

2.2.2. Categories of Organizational Communications

There are several ways to categorize organizational communications, Rasberry and Lindsay [13] suggested the following forms of categorization:

a) Formal organizational communications

There are two ways to categorize formal organizational communications. The first way is to categorize according to the direction of communications such as upward, downward, horizontal, and diagonal communications. Another way of categorization is divided according to the styles of communication loops such as circle, chain, Y, and wheel shaped loops, as shown in Figure 3.

Source: Rasberry & Lindsay [13] p.93

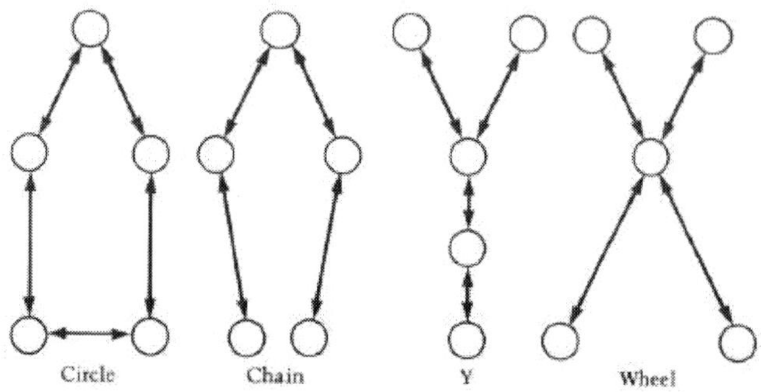

Figure 3. Four types of communication loops.

b) Informal organizational communications

Organizational communications happen in informal settings include types of path such as single strand, gossip, probability, and cluster communications, as shown in Figure 4.

Source: Rasberry & Lindsay [13] p.80

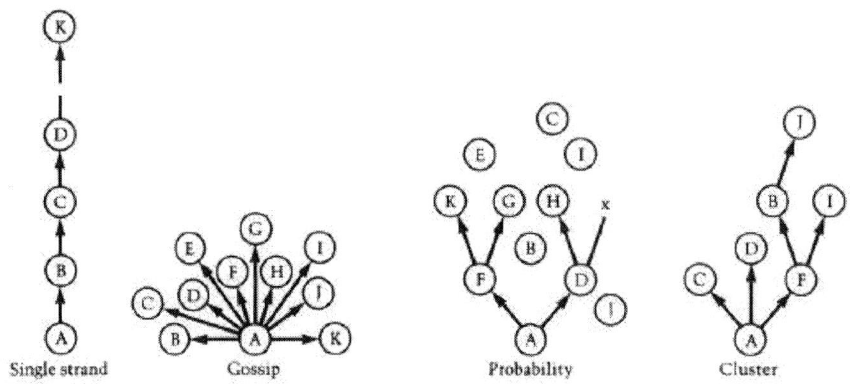

Figure 4. Four paths of informal communication.

According to the description of ways to categorize organizational communications by Rasberry and Lindsay [13], their research suggested the system of categorizing organizational communications as shown in Figure 5.

Source: This research

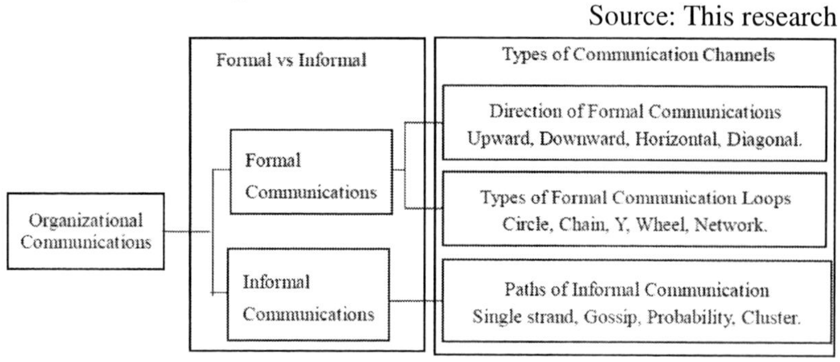

Figure 5. The system of categorizing organizational communications.

2.2.3. Roles in Organizational Communication

Besides the traditional roles of organizational communications in terms of transmitter and receiver, Rogers and Agarwala-Rogers [14] suggested many additional important roles in their arguments including gatekeepers, liaison, opinion leader, and cosmopolite. Their brief explanations are as follows:

a) Gatekeeper

An individual who controls the messages flowing through a communication channel.

b) Liaison

An individual who connects two or more groups of members without themselves involved in any one of the groups.

c) Opinion leader

An individual who is capable of influencing other members' behaviors or attitudes.

d) Cosmopolite

An individual who has a relatively high capability of communication outside of the organization.

2.3. Industrial Design Teams

In addition to acquired understandings on the theories of communication and organizational communications, this research had explored the combination of these theories with the unique characteristics of industrial design teams. There are two aspects of explorations in this regards, including the tasks and the in-house design process of industrial design teams.

2.3.1. The Tasks of Industrial Design Teams

Cooper & Press [15] argued that there is certainly no need to restrict the task of industrial design teams to merely the creation of shapes, colors, and dimensions of products. Successful designs must also take the economic and technological considerations into account as well, so that industrial designers can satisfy user needs to a good extent with appropriate design strategies. At the 29th general assembly in Gwangju, South Korea, the ICSID (International Council of Societies of Industrial Design) [16] announced a renewed definition of industrial design as:

> *Industrial Design is a strategic problem-solving process that drives innovation, builds business success, and leads to a better quality of life through innovative products, systems, services, and experiences (http://wdo.org/about/definition/).*

To summarize the above views on the task of industrial design, it clearly is a synthesized consideration of the relationship between mass produced products and their users. To better fulfill such a widely related content of tasks, it is crucial that there must be a well-orchestrated interdisciplinary collaboration among the peer departments of product development. There is no doubt that such a smooth collaboration between disciplines must come from successful communications.

2.3.2. The Design Process and Industrial Design Teams

In searching for better collaborations among peer departments of product development, Zaccai [17] found that the key to well-integration in product development after a split of tasks between departments of various disciplines is the process. He suggested that the standardized process of product development is the modulator of communication flows between disciplines, and further summarized three models of collaboration in product development.

a) Traditional product development

Traditionally, the development project are handed over from one department to another, and industrial designers are viewed as decorators and usually brought in at the last moment but isolated from the product development process, as shown in Figure 6. This type of product development process typically leads to the perils of poor coordination depicted in Figure 1.

Source: Zaccai [17]

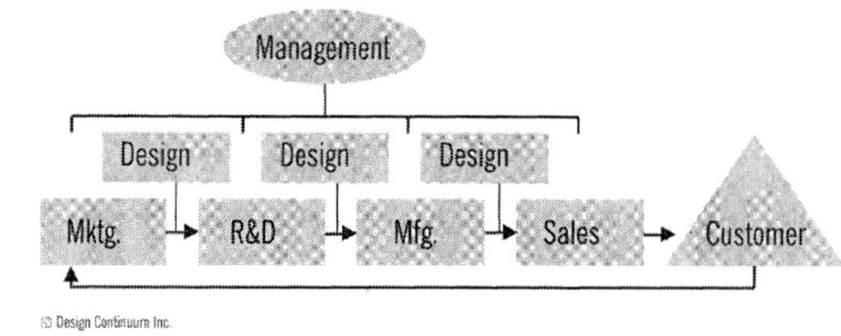

Figure 6. Traditional product development.

b) Partially integrated product development

As an improvement of the traditional process, there emerged among American manufacturers was a new model, which Zaccai [17] referred to as the Partially Integrated Product Development Process, as shown in Figure 7. In this improved model, R&D, marketing and manufacturing began to work more closely. However, top managements still remain distant from these development activities. As indicated in the figure, the word "partially integrated" was denoting the triangle consists of R&D, marketing, and manufacturing, in which they are better integrated than before, whereas industrial designers are still excluded from this integration.

Source: Zaccai [17]

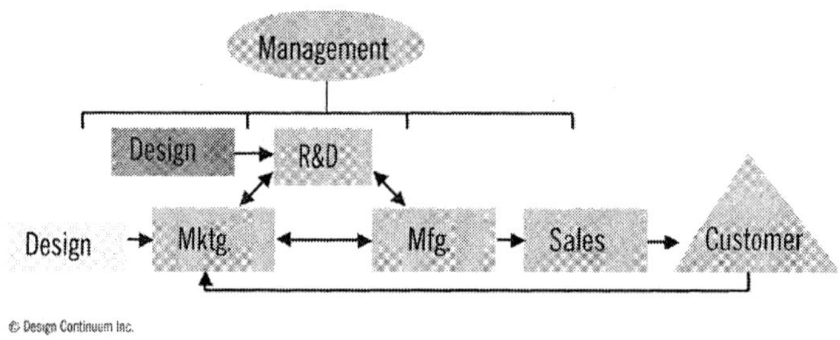

Figure 7. Partially integrated product development.

c) Fully integrated product development

Zaccai [17] found that more and more firms had come to realize that design is a precise problem-solving methodology instead of merely making things pretty. To garner the merit of designers' contribution, these firms developed a "Fully integrated product development process," as shown in Figure 8. In this model, constant cooperation and communication are encouraged, and management began to take the critical role as central coordinator. Also, designers had now become equal partners in it.

Source: Zaccai [17]

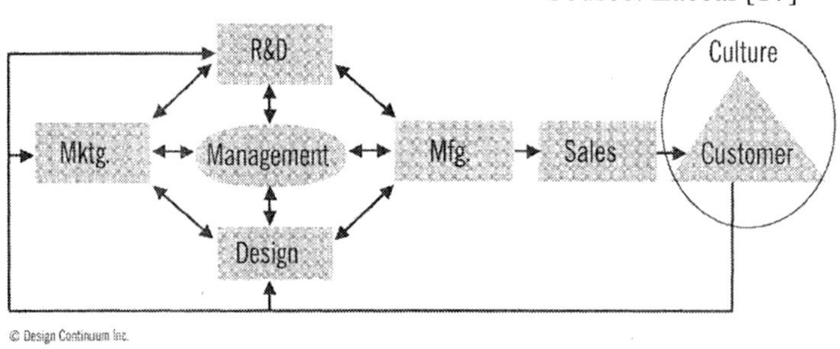

Figure 8. Partially integrated product development.

In addition, Cooper and Press [15] also concluded that most Japanese firms have design process lie at the heart of their innovation activities. They interpreted this process in two ways:

d) Task oriented internal process

This model represents the creative activities that designers are performing to tackle with specific problems. There are five phases in this model including "Defining problem", "Understanding problem", "Thinking about problem", "Developing idea", as well as "Detail design and test".

e) Action oriented external process

This model of process represents the design activities that are operating in organizations. There are four phases in this process, including "Concept", "Embodiment", "Detail", and "Production". Furthermore, Hollins & Hollins [18] held similar concept as Walker [19] to develop the model of "Total Process of Design within Management." They both see design as applying technologies and developing product concepts and meeting the needs of the market externally so that their model had included the "Action Oriented External Process" as a part of it, as shown in Figure 9.

Source: Cooper & Press [15] p.38

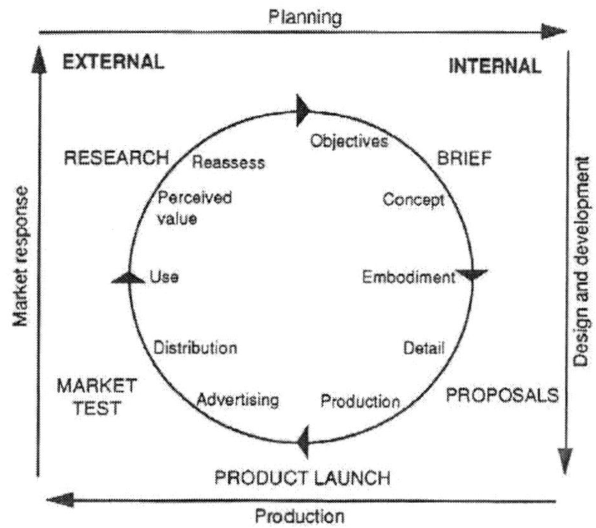

Figure 9. The total process of design within management.

In the above three models of product development process, they had revealed various routes and contact points of communication between functions involved in activities of product development. These facts are valuable information for reference of this research.

2.3.3. Collaborative Design

The intensified international trade and more globalized enterprises had long forced the process of product development to extend far beyond national borders. Also, product development among collaborative companies or between different locations of a company had become a common phenomenon. Such kind of collaboration between scattered locations had given rise to the issue of communications particularly since it has become much more difficult than those who are in the same place at the same time. In solving communication problems in this regard, Saad & Maher [20] proposed a Time-Space Matrix (as shown in Figure 10) that is useful in analyzing different situations under such kind of conditions.

Source: Saad & Maher [20] p.728

	Same Time	Different Time
Same Place	Face to Face Interaction (Meeting Room)	Asynchronous Interaction (Bulletin Board)
Different Place	Synchronous Distributed (Video Conferencing)	Asynchronous Distributed (E-mail)

Figure 10. The Time-Space Matrix.

2.4. Communications of Industrial Design Teams

The interdisciplinary nature of product development and the fact that industrial design is only a part of the product development process both made communication destined to be a key of success for managing industrial design teams. This section explores the issue of communication of industrial design teams in terms of its key skills, strategies, and roles:

2.4.1. Key Communication Skills in the Design Process

In investigating the key skills required in managing client relationships and guiding design decisions, Best [21] summarized that communication is the most crucial part of these requirements. She further analyzed these skills into the following branches:

a) Verbal skills

Individual differences in viewpoints and communication styles can create major barriers to effective communication. Therefore, effective communicator must be competent in verbal dialogue so as to get message across to the receiver persuasively and promote good mutual understandings. Also, a strong will to understand the viewpoionts of the receiver is essential to be understood seccessfully.

b) Interpersonal skills

Communication is basically an issue of dealing with other people, therefore, interpersonal skill is always crucial for effective communication. Listening instinctively while speaking clearly, diplomatically, and sensitively are deemed as good interpersonal manners mostly. More importantly, empathy is a useful manner, which can build an avenue for seeing things from others' point of view clearer and make the other interpersonal skills more effective.

2.4.2. Communication Strategies in the Design Process

In the research conducted by Williams & Cowdroy [22], they identified the strategies available for design teams as two categories and six strategies. The category of verbal strategies includes strategies of technical language and analogy, while the category of visual strategies includes strategies of gesture, sketching graphics, existing graphics and actual objects as follows:

a) Verbal Strategies

- *Technical Language:* This strategy refers to the deliberate use of specific technical terms in communication activities to represent an image of professionalism.
- *Analogy:* This strategy includes three ways of analogy, i.e. project specific analogy, domain specific analogy, and external to domain analogy. The strategy of Project Specific Analogy refers to the use of examples drawn from the specific project that both sides of the communication are working on. The strategy of Domain Specific Analogy involves the use of examples within the industry domain where both sides of the communication sharing experiences. The strategy of External to Domain means communications using examples taken from other industry domain where both sides are not belonging.

b) Visual Strategies

- *Gesture:* This strategy mainly uses hand and arm movements to illustrate aspects of design such as size, function or mechanism, the relationship between product parts, and shape.
- *Drawing:* This strategy is the use of graphics and freehand drawing in communicating specific designs. It is after the above strategies had failed to get messages across when drawings come to the fore usually, although some members may have a better capability in illustrations and thus tend to adopt this strategy earlier than others.
- *Existing Graphics:* This strategy involves the use of drawings and pictures accumulated in the progress of the project in hand or those documents collected for this project.
- *Actual Object:* There are times when communications require complements of real objects, especially prototypes or mock-ups which can be very effective in proving some points than many words.

In addition, Williams & Cowdroy [22] had analyzed more than 14,000 interactions in product development research; they found that the most frequently used communication strategies is technical language (54.8%), followed by charts (24.73), the gap with other strategies are very significant, as shown in Table 1 below:

Table 1. The frequency of communication strategies used.

Technical language	54.80%
Project analogy	2.65%
Domain analogy	2.92%
External analogy	2.45%
Gesture	8.25%
Sketch	3.45%
Chart	24.73%
Object	0.70%

Source: Williams & Cowdroy [22]

2.4.3. Communication Roles in the Design Process

Through her research on communication roles during product design process, Sonnenwald [23] concluded the result of five roles that she observed spanning across formal organizational boundaries: sponsor, interorganizational star, intraorganizational star, intergroup star, and intragroup star. She found that although traditional organizations usually do not have these roles formally, they are instrumental in multidisciplinary tasks such as product development.

a) Sponsor

This is a role which secures the support and funding for the design project in the organization, and aligns the goal and strategies of the design project and the organization.

b) Interorganizational Star

The function of this role is to interact with others in the larger organization and peer departments, such as a project leader.

c) Intraorganizational Star

This role takes charge of transmitting and filtering information on the project's goals, plans, tasks and budgets within the design project, such as group leaders in the project.

d) Intergroup Star

This is a role which represents his or her group to coordinate activities and strategies across groups of the design team.

e) Intragroup Star

This role mainly facilitates interactions among members of his or her group by providing them with socio-emotive supports and conflict resolution helps.

3. Summary

Based on the above literature review, we distilled four aspects of knowledge concerning the topic of communication in product development. They were Communication Theories, Organizational Communications, Industrial Design Teams, and Communications of Industrial Design Teams. Then we further divided these four aspects into ten issues for our further research into communications in product development. They include Principles of Communication, Defining Organizational Communications, Categories of Organizational Communications, Roles in Organizational Communications, The Tasks of Industrial Design Teams, The Design Process and Industrial Design Teams, Collaborative Design, The Key Skills of Communication in the Design Process, The Strategies of Communications in the Design Process, and Communication Roles in the Design Process.

IOP Conf. Series: Materials Science and Engineering **398** (2018) 012020 doi:10.1088/1757-899X/398/1/012020

3.1. Communication Theories
As communication is such a crucial knowledge for better interactions between people, it has long been a popular research topic widely. Therefore, the existing theories are abundant for exploration, and they will help a lot to lay the cornerstone for this research.

3.1.1. Principles of Communication
For a solid fundamental understanding of the basics of good communication, the principles of making effective communication in the existing literature must be studied first.

3.2. Organizational Communications
Even communication between two persons would be complicated enough; nevertheless, most organizations consist of three or more members. Therefore, the issue of effective organizational communications will certainly be a highly complex one to tackle. However, it will be worthwhile to pay great effort since communication holds the key to organizational success no matter how good the members are.

3.2.1. Defining Organizational Communications
Seeking definition involves digging into the essential characteristics of any issue in question. Therefore, it always is a highly important topic to explore when we are trying to be knowledgeable in any area. Thus there is no exception when we are at the outset to acquire the existing knowledge in the field of organizational communication.

3.2.2. Categories of Organizational Communications
While we studied any topic in depth enough, it eventually comes to the stage where we become capable of analyzing characteristics of the subject, and thus we may find that the subject has its characteristics vary according to certain circumstances; such a variation can divide the subject into categories. Therefore, identifying categories of a subject is also a way of deepening our understanding of it.

3.2.3. Roles in Organizational Communications s
Most organizations are consist of so many individuals that the interactions among its members become a complex web with various kind of relationships. To enhance these various relationships, it eventually requires different roles being played by specific members. Therefore, identifying the roles in organizational communications is also a useful way of understanding these relationships. Therefore, we will like to know what kind of communication roles our interviewees' companies have for their product development.

3.3. Industrial Design Teams
Since this research is going to apply our understandings on the issues of communication theories and organizational communications in industrial design teams, thus we should take a look at the unique characteristics of industrial design teams here. There are two aspects of explorations in this regards, including the tasks and the in-house design process of industrial design teams.

3.3.1. The Tasks of Industrial Design Teams
The traditional task of industrial design teams is merely the creation of shapes, colors, and dimensions of products. However, many successful in-house industrial design teams had also demonstrated their capability to take the economic and technological considerations into account, so that industrial designers can satisfy user needs to a good extent with appropriate design strategies, such as the example that we have seen in Apple. Behind this scene of success, there must be an interdisciplinary collaboration among the peer departments of product development, which requires successful communications. Therefore, we will like to know how our interviewees' companies are doing in this regard.

3.3.2. The Design Process and Industrial Design Teams

According to the collaborative and multi-stepwise nature of product development, a well-structured design process is required to govern certain flow of communications. Therefore, we will have to understand the design process of the interviewees' companies.

3.3.3. Collaborative Design

The trend of globalization had extended the process of product development far beyond national borders. This extension did not limited within one company, but also among collaborative companies. Such kind of product developing collaboration between organizations at scattered locations had made the issue of communications so crucial since it has become much more difficult than before. Therefore, we will have to know what our interviewees' companies are doing to enhance this kind of communications.

3.4. Communications of Industrial Design Teams

The interdisciplinary nature of product development and the fact that industrial design is only a part of the product development process both made communication destined to be a key of success for managing industrial design teams. This section explores the issue of communication of industrial design teams regarding its key skills, strategies, and roles:

3.4.1. Key Communication Skills in the Design Process

In this research, we set out to choose design managers of established and successful companies as our interviewees. Therefore, we can also assume that they all have an institutionalized design process in place. Although it will be difficult to make their design process revealed, however, we will have to find out the key skills that they are utilizing to keep their design projects go through the process smoothly.

3.4.2. Communication Strategies in the Design Process

As our interviewees are all design managers of successful companies, we can assume that they must have some useful strategies for dealing with the issue of communication in their design process, which are worthwhile for others to take as reference. Therefore, we will have to unveil the key strategies of this part from our interviewees.

3.4.3. Communication Roles in the Design Process

In the complex web of organizational communication in our interviewees' companies, we can assume that there are some instrumental roles which are working well enough to make product development projects keep on flowing more smoothly along their design process. Therefore, we will have to figure out what kind of communication roles do our interviewees deem as key factors of success in their product development.

4. Conclusions

Based on the summary of existing literature reviewed on the topic of communication in product design process in chapter three, we developed twelve questions to use in our interviews. They are as follows:

- What are the major ways and channels of inward communication in your company?
- What are the major ways and channels of outward communication in your company?
- What are the major ways and channels of outward communication in your company?
- What kind of product development process does your company adopt?
- What kind of positions dedicated for communication in product development does your design team have?
- How do your company hand down design tasks to your design team?
- How do your design team report the progress of their projects to your management?
- How do your design team conduct their inward communication in design?
- How do your design team conduct their outward communication in design?

- What kind of communicational problems do you usually encounter in your design tasks?
- Do you also encounter other kinds of communicational problems? What are they and how do you deal with them usually
- Do you have any more comment on our interview today?

We hope that this questionnaire will be able to help us elicit valuable experiences from successful design managers through in-depth interviews. Their precious answers will be our materials to analyze by the Grounding Theories Methodology in our subsequent research, to distill them into principles of communication in product development, which will be helpful for many other design managers to take reference for enhancing the effectiveness in communications of their design teams.

Furthermore, although the subsequent research is still going on and the final result is yet to report here. Nonetheless, this paper has made a comprehensive review and report on the literature regarding the issue of communication. Therefore, we hope that this paper can be a useful steppingstone for those who are in need of acquiring a better understanding of the knowledge of communication, especially in the field of new product development.

5. Acknowledgement

This paper is the result of a research project that was only made possible under the generous sponsorship awarded by the Ministry of Science and Technology of the Executive Yuan, Republic of China. The serial number of this research project is NSC 106-2221-E-324-014. Also, the authors had received strong support from their affiliated university, i.e. Chaoyang University of Technology, Taiwan. The authors would like to express their special appreciations here.

6. References

[1] Peters T 1990 *The design dimension: the new competitive weapon for product strategy & global marketing* (Oxford England: Basil Blackwell)

[2] *Oxford Advanced Learner's Dictionary* 2017 "Communication". Retrieved from http://www.oxfordlearnersdictionaries.com/definition/english/communication?q=communication

[3] Shannon C E and Weaver W 1949 *The Mathematical Theory of Communication* (Urbana IL: University of Illinois Press)

[4] Andersen M P 1964 *The Speaker and His Audience: Dynamic Public Speaking* (New York NY: Harper & Row)

[5] Compton H and Bennett W 1967 *Communication in supervisory management* (London England: Nelson)

[6] Barnard C I 1968 *The functions of the executive* (Cambridge MA: Harvard University Press)

[7] Schramm W and Roberts D F 1971 *The Process and Effects of Mass Communication* (Urbana IL: University of Illinois Press)

[8] Deaux K, Dane F C, Wrightsman L S, and Sigelman C K 1993 *Social psychology in the '90s* (Pacific Grove CA: Brooks/Cole Pub. Co.)

[9] Coulter M K and Robbins S P 2001 *Management* (Upper Saddle River NJ: Prentice Hall)

[10] Cutlip S M and Center A H 1964 *Effective public relations 3rd ed* (Englewood Cliffs NJ: Prentice-Hall) pp 166–167

[11] Thomas K W and Schmidt W H 1976 A Survey of Managerial Interests with Respect to Conflict *Academy of Management Journal* June 1 1976 **19** 2 pp 315–318.

[12] Luthans F, Hodgetts R M, and Rosenkrantz S A 1988 *Real managers* (Cambridge MA: Ballinger)

[13] Rasberry R W and Lindsay L L 1994 *Effective managerial communication* (Belmont CA: Wadsworth)

[14] Rogers E M and Agarwala-Rogers R 1976 *Communication in organizations* (New York NY: The Free Press)

[15] Cooper R and Press M 1995 *The design agenda: a guide to successful design management* (Chichester England: John Wiley & Sons)

[16] World Design Organization 2015 *Definition of Industrial Design* Retrieved on November 2 2016 http://wdo.org/about/definition/

[17] Zaccai G 1991 How to Make the Client / Consultant Relationship More Like a Basketball Game than a Relay Race *Design Management Journal* **2** 2 Spring 1991 pp 43–48.

[18] Hollins G and Hollins B 1991 *Total design: managing the design process in the Service Sector* (London England: Pitman)

[19] Walker D 1989 *Managing Design* (Milton Keynes: Open University Press)

[20] Saad M and Maher M L 1995 Exploring the possibilities for computer-supported collaborative designing *CAAD Futures'95* (Singapore)

[21] Best K 2006 *Design management: managing design strategy, Process and Implementation* (Lausanne Switzerland: AVA Publishing SA)

[22] Williams A and Cowdroy R 2002 How Designers Communicate Ideas To Each Other In Design Meetings *International Design Conference - Design* (Dubrovnik) May pp 14–17, pp 947–952

[23] Sonnenwald D H 1996 Communication roles that support collaboration during the design process *Design Studies* **17** 3. pp 277–301.

IOP Conf. Series: Materials Science and Engineering **398** (2018) 012021 doi:10.1088/1757-899X/398/1/012021

The Opportunities and Challenges of "Made in China" under the New Model of Intelligent Manufacturing

Lijie Feng[1], Weinan Li[1] Junju Yue[1]

[1]School of management engineering, ZhengzhouUniversity,Zhengzhou, China

18703995021@163.com

Abstract. This paper analyzes the opportunities and challenges faced by China's manufacturing industry under the situation of intelligent manufacturing mode, and puts forward two routes which should be taken in the development of intelligent manufacturing in our country, and take the new round of global technological revolution and industrial revolution as the background and combine the development of our country's manufacturing industry. Emphasizing to step-by-step, step by step to promote the construction of intelligent factories, in the process to focus on product innovation and enhance the reliability of product quality, to develop intelligent manufacturing and intelligent manufacturing services for China's intelligent manufacturing In the industry to promote the application.

1. Introduction
At present, the new round of technological revolution and industrial revolution led by information technology is deeply affecting the human production and way of life. The deep integration of informatization and industrialization has led to the development of traditional manufacturing industry in the direction of digitalization, networking, intelligence and service. The new industrial revolution marked by intelligent manufacturing has brought the world manufacturing industry into a new stage of development.

Manufacturing industry is an important pillar of China's national economy, China's manufacturing industry since 2009 go beyond the United States, becomes the world's largest manufacturing industry, but there is still a big gap between the world's power from,how to improve the level of manufacturing development is currently a strategic task[1]. After the 2008 world financial crisis, the world's manufacturing industry developed countries have developed national strategy, focusing on the development of their own manufacturing industry, to seize the future development of the world manufacturing industry, the high point of compression of China's traditional manufacturing industry has the size, cost and other advantages, China's manufacturing industry is facing the international demand shrinking, the rapid rise in costs, developed and developing countries, "before and after the attack" and many other challenges. In the global manufacturing network, intelligent, service development trend, how to clearly analyze the new situation under the opportunities and adjustments in China's manufacturing industry, China's manufacturing industry has an important reference significance for long-term development.

2. Overview of intelligent manufacturing development

2.1. Intelligent manufacturing definition and characteristics

Intelligent ,manufacturing (IM) has no internationally accepted definition[2].Intelligent manufacturing as a production mode is the general term of intelligent manufacturing technology and intelligent manufacturing system. The carrier of intelligent manufacturing is intelligent manufacturing system or intelligent factory, which relies on artificial intelligence technology to realize the intellectualization of production activities. As a unique production mode, intelligent manufacturing has its unique characteristics compared with the traditional production mode and the current advanced production mode

(1)Intelligent manufacturing is a complicated system engineering on the time dimension, intelligent manufacturing activities throughout the product design, manufacture, use, maintenance, scrap the whole life cycle, the purpose is to make the product whole life cycle of the lowest cost of production, the value of the minimum resource consumption, the product itself to achieve the highest; On the spatial dimension, intelligent manufacturing activities involving enterprise internal departments to work together on one hand, on the other hand, involves the related to enterprise synergy between upstream and downstream enterprises and related industries, makes the production activities are no longer isolated, closed, make the product between related industries and enterprises to form a stereo system.

(2)Intelligent manufacturing brings disruptive change to traditional production activities and makes industrial production process more flexible and stable than traditional production mode. Traditional process to manufacture the product as the core, rely on enterprise workers, engineers and managers work together, in a relatively closed the factory produce the required final product, in this process, people, machine, material, method, ring and other elements determine the properties and quality of products. Because the traditional discrete production process each link, each link between information cannot transmit, efficient and fast system once appear abnormal, if not found in time, will bring great loss, and a system exception if there is no technical personnel after adjust, repair, system can not run normally. In intelligent manufacturing mode, however, due to the artificial intelligence technology, network technology, a new generation of information technology such as technology advances, the whole production process of each link together through the network, a different link information can realize high speed circulation through the network, make the production process has the adaptive, since the decision, self diagnosis and self repair ability, reduce the influence of uncertain factors, the stability of the production process is higher.

(3)Intelligent manufacturing is the inevitable trend of manufacturing development in recent years, the information technology to lead the world a new round of technological revolution and industrial revolution is sweeping the globe, the depth of the integration of informatization and industrialization, prompting manufacturing industry to develop in the direction of digitization, network, intelligence and service, the representative is Germany put forward the strategy of "industrial 4.0", "advanced manufacturing country strategic plan" in the United States, is known as a smart manufacturing leading a new round of industrial revolution.

2.2. Technological development status

The development of manufacturing industry has experienced the stage of manual production, assembly line manufacturing, automatic manufacturing of flexible automation and integrated manufacturing, parallel planning design and agile manufacturing. Since the 1980 s, the advanced manufacturing technology and computer technology is widely used in modern manufacturing industry, the traditional design methods and management methods cannot effectively and quickly solve the new problem in the modern manufacturing system.

As a result, people started using modern tools and methods, using the latest research results, through the traditional Manufacturing Technology, artificial intelligence science, computer Technology and science, and other organic integration, the development of a new type of Manufacturing Technology and System, which is Intelligent Manufacturing Technology (IMT) and Intelligent Manufacturing System (IMS), they make general Intelligent Manufacturing, (IM). Since the concept of intelligent manufacturing (IM), intelligent manufacturing system (IMS) has been focused by many countries.

In China in the late 1980 s also take "artificial intelligence" into the main topics of the national science and technology development planning, has in the expert system, pattern recognition, robotics, machine understanding of Chinese has made a number of results. Since 1993, the national natural science foundation (NSF) has been supporting research projects on intelligent manufacturing in a modest amount each year; The country's "nine-five" plan also places advanced manufacturing technologies (including IMT and IMS) as one of the key areas of development. In the strategy of rejuvenating emphasizes the advanced manufacturing today, our country attaches great importance to the development of intelligent manufacturing, especially since May 2009, the equipment manufacturing industry adjustment and revitalization of planning ", the national policy support will be increasing. On March 27, 2012, China's ministry of science and technology organized the "special plan for the development of intelligent manufacturing technology".
Since the financial crisis, governments of the United States, Germany, Japan and other countries have proposed to revitalize manufacturing by developing intelligent manufacturing. Launched in June 2011by the United States, including industrial robots, "advanced manufacturing partnership", introduced in February 2012 and "advanced manufacturing national strategic plan", in 2012 set up the network for manufacturing innovation, and has set up a material manufacturing innovation research and institute of digital manufacture and design innovation. Germany, through the government, the fraunhofer institute and state governments, has invested in manufacturing research in the automation of numerical control machine tools, manufacturing and engineering automation. Japan has offered to boost the international competitiveness of its manufacturing industry by speeding up the development of collaborative robots and unmanned chemical plants. In 2013, Germany formally implemented the "industrial 4.0" strategy of intelligent manufacturing as the main body, consolidating its manufacturing years, the growth momentum of China's intelligent manufacturing and equipment industry has been growing rapidly, which has initially formed certain rules. This section must be in one column.

3. Challenges and Opportunities Faced Under China 's Manufacturing Industry

3.1. The challenge

(1)International division of labor for the intense after the financial crisis, the manufacturing industry presents the developed countries and developing countries to step in for the situation of a new round of international division of labor and global manufacturing division map may therefore refactoring, attack before and after the manufacturing industry in China is facing a situation. On the one hand, the financial crisis is the United States, the European Union and other developed countries and regions to attaches great importance to the development of the real economy, put forward development strategy of industrialization, again stressed that make full use of information technology to enhance the level of manufacturing, development of emerging industries, to consolidate manufacturing in the developed countries in technology, industry leading advantage, further widened and the gap of our country. On the other hand, the export price of low value-added products in China is weakened by the appreciation of labor costs and RMB exchange rate. In addition, India, Vietnam, Indonesia and other developing countries with lower labor costs to undertake the transfer of labor-intensive industries, China's manufacturing industry in some labor-intensive industry with developing countries battle has begun.
(2)China's manufacturing industry is still large and weak. Manufacturing is the main body of the national economy. The strength of manufacturing determines the international status of a country. After decades of rapid development, China has become a real manufacturing power, yet it is still not a manufacturing power. Although the "made in China" brand has been known all over the world, but our country manufacturing industry as a whole big but not strong, independent innovation ability, product quality, resource utilization, industrial structure, informatization level and industrial powers in the world, there is still a large gap in the core technology, additional value, product quality, production efficiency, energy resource utilization and environmental protection problems, sharp contradictions.
(3)Manufacturing traditional advantage gradually lost after the financial crisis in 2008, the world's developed countries and developing countries to step in a new round of international division of labor, the global manufacturing division of labor is thus refactoring, makes our country manufacturing industry on the one hand, by the advanced industrial countries in the world of the extrusion of

IOP Conf. Series: Materials Science and Engineering 398 (2018) 012021 doi:10.1088/1757-899X/398/1/012021

"industrialization" development strategy, the other party facing India, Vietnam, Indonesia and other countries of low labor costs on the strong competition of manufacturing industry in our country, to our country formed "attack" before and after the serious situation of [3]. Along with the rising cost of human resources, environment and other aspects of the increasingly prominent, the traditional dimensions, resource consumption mode of production has come to an end, manufacturing advantages gradually lost.

(4)Technology innovation ability is insufficient for over 30 years of reform and opening-up, through the introduction of foreign advanced technology, digestion and absorption, the overall level of manufacturing industry in our country has greatly promoted, formed a relatively complete industrial system. However, in the process of rapid development of manufacturing industry, there are still outstanding problems,[4] which are embodied in the following aspects: first, the overall manufacturing industry is large and not strong. Although our country has been truly manufacturing power, but is still not a manufacturing powerhouse, manufacturing power depends on the element, and the manufacturing powerhouse rely on innovation, knowledge, and our country manufacturing industry in the international industrial division of labor in the chain, at the stage of low technology content and added value of "processing, assembly" link, in the high value of research and development, design, engineering contracting, marketing and after-sales service, lack of competitiveness; Second, core technologies are still under control. It is shown that the core technology of basic manufacturing equipment is dependent on people, core equipment and parts, key raw materials and devices. Third, the basic research and development investment that supports the product core technology is insufficient, the product core competence lacks. Compared with developed countries in foreign countries, China still has a significant gap in investment in basic technology research and development. It also leads directly to the gap between the core technology and developed countries in the world.

(5)Product quality reliability issues highlighted by the thought of extensive economic growth, the speed of product quality reliability level for a long time lag behind the growth of economy of scale, product quality and relevant technical standards of the overall level is not high reliability. In numerical control machine tool, for example, although China's machine tool use and manufacturing power, but in our country, more than 50% of the use of CNC machine is imported machine tools, among them, automotive, electronics, aerospace and other fields used by the machine almost dependent on imports, the main or even the only reason of this situation is the quality of the machine tool reliability problems. Although the product can be close to or even reach the international advanced level in the function of precision index, this function can't be maintained effectively. Compared with foreign products, the reliability difference is obvious. In the use of domestic machine tools, the fault frequency is frequent, which brings high maintenance cost and huge downtime loss to the user enterprises. Meanwhile, it also has a great impact on the reputation of machine tool manufacturers. At the same time, in the related technical standard field, our country is almost empty, compared with the developed countries, the gap is obvious.

(6) the development of resource utilization is low in the previous process, make the enterprise rely on large-scale inputs to obtain the economic efficiency and growth rate, however, low utilization rate of resources and environment pollution caused by serious [5]. Take the 2013 statistics as an example. In 2013, China's GDP accounted for 12.1 percent of the world's gross domestic product, while China's energy consumption accounted for 22 percent of global energy consumption in the same year. It can be seen that China's energy resources utilization efficiency is still low, contradiction is very prominent.

3.2. Development Opportunities

After more than 30 years of development, our economy has become a huge economic giant, and the capacity of resources and ecological environment has been saturated. Already say goodbye to the era of rapid development, economic development in our country into a new orbit, towards more advanced form, the stage of evolution of division of labor is more complex, the structure more reasonable, the manufacturing industry development mode in addition to the urgent demand, manufacturing shift from elements drive toward innovation is needed.however, a new round of scientific and technological revolution led to the development of new manufacturing models in the development of smart manufacturing provides a rare opportunity for China's manufacturing industry.

(1)Manufacturing technology and information fusion technology brings the new industrial revolution in recent years, by leading the global information technology a new round of technological revolution and industrial revolution is sweeping the globe, the depth of the integration of informatization and industrialization, prompting manufacturing industry to develop in the direction of digitization, network, intelligence and service, is a representative put forward by the German "industrial 4. 0" strategy, is known as intelligent manufacturing as the leading factor, the fourth industrial revolution for the transformation and upgrading of China's manufacturing industry, and promote the depth of the fusion of industrialization and information, provide a new direction. A new round of technological revolution and industrial revolution, will bring the world within the scope of the manufacturing of disruptive change, main performance for the fusion of the digital technology and artificial intelligence technology to the product, apply digital technology and artificial intelligence technology and product design, manufacturing, production process, and based on the digital technology, the Internet, the Internet of things, cloud computing, big data, under the support of emerging technologies such as manufacturing from tradition to open, the service way transformation.
(2)Four modernizations synchronous development huge space in accordance with the "industrialization, informatization, urbanization and agricultural modernization" synchronous development, urbanization and agricultural modernization provides demand for manufacturing industry. Urbanization will drive a large number of rural people into towns and cities, bringing huge demand for consumption, thus driving the growth of domestic demand. The agricultural modernization demand will drive rural infrastructure construction, agricultural equipment and other market demand; The integration of informatization and industrialization, in line with the current trend of manufacturing industry in the world, is in line with the actual demand of China's manufacturing industry, which will effectively promote the manufacturing industry in China. At the same time, the expansion of demand and the rapid growth of the consumption level brought by the development of China's "four modernizations" also provide a rare opportunity for the development of China's manufacturing industry.
(3)High-end manufacturing equipment in our country go abroad in the golden age for a long time, our country high-end manufacturing equipment an awkward situation, on the one hand, domestic manufacturing of high-end manufacturing equipment are almost used by foreign monopoly, users would rather spend high price to buy imported brands choose domestic brands, domestic high-end manufacturing equipment in the domestic market in a dilemma; On the other hand, due to the impact of the global financial crisis, demand in emerging developed countries has shrunk, and the foreign market of high-end manufacturing equipment in China has been squeezed. Under the impetus of the intelligent manufacturing model, the global manufacturing industry will be more open, will form a highly organic network connectivity, it will be for our country, including high-end manufacturing equipment, products to the international market provides an excellent opportunity.

4. The development of intelligent manufacturing recommendations

Intelligent manufacturing is the deep integration of artificial intelligence technology and manufacturing technology, and the development of artificial intelligence leads to the emergence of intelligent manufacturing. As a production mode, intelligent manufacturing has its unique presentation carrier, and the carrier of intelligent manufacturing performance is intelligent factory and intelligent production. Therefore, the goal of developing intelligent manufacturing in our country is to cultivate and build intelligent factories in different industries in manufacturing, and gradually realize intelligent production.
(1)Clear main line combined with current situation of the development of manufacturing industry in our country, the development of intelligent manufacturing can be around two main line, the first line, development of intelligent manufacturing can make industry, through the development of intelligent manufacturing promote the development and perfection of related enabling industry, such as industrial robots, high-end CNC machine tools, intelligent 3 d printing equipment, complete sets of equipment such as intelligent manufacturing equipment industry, and industrial control system, information security system represented by high-end industrial software industry, provide software and hardware foundation for the development of intelligent manufacturing. Second main line, to promote the

application of intelligent manufacturing mode in manufacturing, intelligent upgrade in the traditional manufacturing enterprise, the transformation as the breakthrough point, the construction of digital, intelligent and factory production lines, workshop, finally realizes the intelligent production.

(2)It is a long-term process to promote intelligent factory construction in a hierarchical and orderly manner, and should plan and step forward in an orderly way. First of all, should be developed intelligent manufacturing medium and long-term development planning, orderly advance in stages, step by step to choose the bibcock enterprise demonstration pilot, to point with surface, eventually to promote the all-round development of intelligent manufacturing and application; Secondly, in the early development of intelligent manufacturing should pay attention to the leading of the government policy, from the aspects of land, taxation, financial, personnel and so on enterprise support, and play their own intelligent upgrade internal demand, effectively promote the application of intelligent manufacturing model in enterprises; Third, Combine the actual situation of enterprises and promote intelligent manufacturing development in a hierarchical manner. It is a gradual process for the production of automatic production from industrial 1.0, industrial production and industrial production, and industrial 3.0 automation production to the current industrial 4.0. However, due to the development of China's manufacturing process of special process, the different levels of development of manufacturing enterprises in China, both in the industrial 1. 0 of the era of enterprise, also has in the industry 3. 0 and towards industrial 4. 0 of enterprise, to combining the actual development level of different industry, different enterprise orderly, pay attention to the industrialization and intelligent.

(3)Pay attention to product manufacturing activities of technological innovation is the core of the product, the purpose of popularization and application of intelligent manufacturing is to improve the intelligent level of manufacturing activity, make intelligent, efficient, green manufacturing, personalization, as a service, the core is still around the product. Therefore, in order to promote the application of intelligent manufacturing, the government should pay more attention to the improvement of product innovation ability and enhance the added value of the product itself. Enterprises should pay attention to the establishment of independent innovation system, and change traditional passive innovation to active innovation. The government should play a positive role in guiding enterprises to enhance their independent innovation ability and cultivate the good atmosphere of independent innovation.

(4)The competition of product quality and reliability level is the primary way to win the domestic and foreign market. Enterprises shall establish perfect quality reliability work system, to carry out the system reliability design, test and verify the quality of the work, continuously enhance the level of components, parts, product quality reliability; The government should increase the investment in the development of the common basic technology of quality reliability, conquer the key technology of quality reliability, and provide support for the improvement of product quality and reliability. Pay attention to the quality reliability of intelligent equipment design, test, test, evaluation and analysis of the basic theory and application of intelligent equipment and process reliability of the whole machine, functional components, parts, quality consistency, such as technology research, enhance the level of high-end equipment machine reliability.

(5)Attaches great importance to the development of intelligent products and intelligent manufacturing services along with the advance of artificial intelligence technology, intelligent of the products is the development trend of the future, in advance of intelligent manufacturing activity at the same time, should pay attention to product intelligent level of ascension, make through intelligent manufacturing to produce a product with high intelligent level. At the same time, developing smart manufacturing services vigorously around the main line of "manufacturing services" and "service manufacturing" is necessary. First, to develop manufacturing services, both for producer and producer services, as well as for consumer and consumer processes. The former includes the product design and development of related services industry, transportation and logistics, financial services, information transmission and computer software, e-commerce and other services; The latter involves product recycling, disposal and remanufacturing services. Second, develop the manufacturing information services, such as the Internet industry, industry of big data, cloud computing and other new service forms, build innovation service system, improve the next generation Internet, a new generation of mobile communication, triple play, Internet of things, cloud computing and other emerging industries in the field of independent innovation ability, strengthen the integration of information technology and manufacturing technology development. Third, the government to guide, the development of the third party inspection, certification, metrology, transformation of scientific and technological achievements, talents cultivation, the standard system revision related services, such as forms, which guarantees our province development of intelligent manufacturing.

Conclusions

As intelligent manufacturing as the core of a new round of technological revolution and industrial revolution of the world's arrival, our country manufacturing industry ushered in the unprecedented development opportunity, to develop intelligent manufacturing as the breakthrough point, to make our country by manufacturing power into a manufacturing powerhouse, is a strategic task facing us. , however, There are still many challenges and difficulties to realize the change of manufacturing industry from large to strong won't happen overnight, the situation of development of China's manufacturing industry, hierarchical, popularization and application of intelligent production step, improve enterprise's capacity for independent innovation, the development of intelligent manufacturing can make industries, improve product quality reliability level and intelligent level, development intelligence services, eventually to promote intelligent manufacturing model is widely used in industry.

References

[1] Jiansheng Pan, Jing wang and Jianfeng Gu 2015 Research on the development strategy of high-sex intelligent manufacturing in China. *J. Metal heat treatment* **1** 1-6

[2] Dongbiao Zhao and Jianying Zhu 1999 Development and research of intelligent manufacturing technology and system. *J. Chinese mechanical engineering* **8** 95-99

[3] Yongxiang lu 2010 Towards green and intelligent manufacturing-China manufacturing development road. *J. China mechanical engineering* **4** 379-386

[4] Bolin Sun 2013 Future trend of intelligent equipment manufacturing in the future. *J. Automation instrument* **1** 1-5

[5] Jianying Zhu 2007 Reflections on manufacturing technology and manufacturing development strategy. *J. Mechanical manufacturing and automation* **1** 1-9.

IOP Conf. Series: Materials Science and Engineering **398** (2018) 012022 doi:10.1088/1757-899X/398/1/012022

Research on Promoting the International Competitiveness of China's Manufacturing Industry under the Background of TPP

Ke lu[1], Yimeng Chen[2] and Xiaocheng Zhai[3]

[1]School of Literature, Law and Economics,Wuhan University of Science and Technology, Qingling Street, Wuhan, China
[2]Hubei SME Research Center, Wuhan University of Science and Technology, Wuhan, China
[3]School of Literature, Law and Economics,Wuhan University of Science and Technology, Wuhan, China

1261157657@qq.com,271547685@qq.com,519308485@qq.com

Abstract. As the United States, Japan, Australia and other 9 countries joined trans-pacific partnership (TPP), it has become the highest standard and the largest free trade agreement in the history. But China is excluded, and with the development of the negotiation, contracting states gradually formed a sort of circle around China, which will no doubt have a significant impact on China's foreign trade development, especially on China's manufacturing industry. No matter whether China will join TPP or not, it will never take the easy way out. So, exploring how to improve China's international competitiveness of manufacturing has been an important topic.

1. Introduction

In 1980, China's total exports amounted to $18 billion 120 million, of that manufacturing sector, accounting for only 50%. Since the reform and opening up, the export scale of China's manufacturing industry is expanding day by day, of that was inevitably affected by the global financial crisis in 2009. Although the total export volume has fluctuated, while the proportion of manufacturing exports has risen, without declining, this has led successfully to a rapid recovery of China's economy. China has succeeded in exporting more than $2 trillion and 50 billion of manufactured goods to the United States, making it the world's largest manufacturing country in 2011. In 2015, it reached $2 trillion and 274 billion 949 million, accounting for 18% of global manufacturing exports, second only to 38.5% of EU economies, compared with 0.8% in the early period of reform and opening up, it has a qualitative upgrade. The average annual growth rate of China's manufacturing exports is 11.8% during the period of 2005-2010; China still maintains its first annual growth rate of 11% at that time. It can be seen that the manufacturing industry is not only a pillar industry in China, but also a real "the workshop of the world"[1].

1.1. Competitive Advantage Enhancement

Since joining WTO in 2001, the international competitiveness of China's manufacturing industry has been continuously improved, and the share of international market has steadily increased. Both comparative advantage and competitive advantage play a major role, but the decisive role is the enhancement of competitive advantage. This result is inseparable from China's efforts to integrate into the global economic system, make use of two markets and resources at home and abroad, conform to comparative advantages and focus on developing labor-intensive industries since the reform and opening up. These measures not only greatly reduce the cost of manufactured goods in China, improve the product structure, but also promote the quality of intangible factors such as management, technology, human resources and so on, and the advantages are constantly enhanced, resulting in a substantial increase in the manufacturing industry's capital and international competitiveness[2].

The export growth rate of China's manufacturing industry fluctuates greatly, and even negative growth occurs in some previous years; In recent years, the export growth rate has obvious weakening trend, and the export growth rate advantage is almost zero in 2013 and 2014; From 2009, the international competitiveness index of China's manufacturing industry has been maintained at around 0.22, and there has been no obvious upward trend in recent years, China's exports of manufactured goods facing more and more competitive pressure. For a long time, the profits and international competitiveness of China's manufacturing industry mainly rely on extensive expansion to obtain, which allows manufacturing industries and enterprises to maintain international competitiveness by reducing the cost of tangible elements of labor and resources. China's resource endowment advantage began to show weakness, labor-intensive industries began to face the pressure of transformation and upgrading since 2010.

1.2. High Technology Becomes a New Growth Point

With the growing scale of China's manufactured products and the development of science and technology, China's manufacturing industry began to focus on innovation and development of science and technology, product structure and industrial structure adjustment have achieved good results, high technology become a new growth point.

First of all, since the reform and opening up, the proportion of China's manufactured goods in total exports has steadily risen, it has been smoothly transformed from the initial exporter of manufactured goods to a big producer of finished products. The structure of China's export commodities has been further optimized.

In addition, according to customs data statistics, in 1992, China's high-tech products exports accounted for only 5.9% of finished goods, and began to improve continuously since its accession to the WTO in 2001; In 2002, the export of high-tech products accounted for 20.8% of the total exports of manufactured goods, and has been stable over 2003-2015 for 29% years, and reached the highest 33.11% in 2009 when the financial crisis spread to the world; In 2015, the export of high-tech products reached 655 billion 211 million US dollars, accounting for 30.18%, which has obviously become the main force to promote the export of China's manufacturing industry[3].

1.3. Internationally Renowned Brands in Small Quantities

According to the 2015 BrandZ Top 100 Global Brands, in the list of the most valuable top 100 brands in the world, the United States occupies 59 seats, Japan 7 seats, as the emerging economy of South Korea also has 3, although our country has 14 enterprises to be listed in the top 100, but only HUAWEI is a manufacturing enterprise, ranking in the seventieth place, which is still the first time to be listed in this list for China. According to the statistics of 2015 Machine Top 500 Global Brands, the United States occupies 26 of the top 100 enterprises, Japan ranks second as its 18, and China has 16 enterprises to enter the top 100. Although China has become the world's third largest manufacturing power seen by it's the total number of brands, based on the huge population of 1 billion 300 million, compared with other developed countries, China's well-known brands are numbered, only one company into the top ten. After all, it is still the market situation that Chinese enterprises won by quantity rather than quality. The existing brands have a low proportion of independent intellectual property rights and lack of well-known brands with extensive international influence.

2. Factors Restricting the Upgrading of International Competitiveness of China's Manufacturing Industry

2.1. Factor Cost Increase

At the beginning of reform and opening up, China has been developing labor-intensive and resource intensive industries to maintain export oriented economy with the advantages of labor and resource endowments, up to now; the advantage of China's manufacturing industry in the global market competition is still dominated by factor cost advantage. However, as the scale of manufacturing industry continues to expand in an extensive mode, China's resources, energy and labor costs are changing dynamically, the rise of labor cost and the deterioration of human resources make the original comparative advantage of China's manufacturing industry become weaker. On the one hand, from the cost of resources and environment, China has begun to show the trend of relative shortage of resources, environmental carrying capacity weakened, depending on the demand for natural resources is difficult to survive in the industry policy and long-term development, and with the national attention on the environment, the costs of resource exploitation in improving constantly. On the other hand, from the perspective of labor cost, with the development of China's economy and the change of the age structure of the population, the labor cost of our country is rising rapidly, and the labor advantage is constantly being weakened. From 2000 to 2014, a total of 15 years, China's manufacturing average wage increased by nearly five times from 8750 Yuan to 50620 Yuan. In addition, there is still a big gap between China's educational level and technical level and developed countries, which affect labor productivity and makes the market demand for high-quality labor constantly rising[4].Industries have to offset the impact of low cost advantage by improving the quality of labor.

In addition, the tariff, environmental protection, labor and the high standards of origin in the TPP negotiation terms will directly or indirectly enhance the manufacturing cost of China's manufacturing industry. The customs union between the TPP Contracting States that "98% of the product tariff will be exempted" will undoubtedly have a huge impact on the competitiveness of manufacturing industries in the international market of our country. China's manufacturing costs will be relatively high, while labor, environmental protection and high standards of origin will also directly increase the cost of factor inputs in manufacturing. As non-contracting countries, in the international market competition with the contracting parties, the related enterprises can maintain their competitiveness only by reducing costs in a short period of time. The first is affected by the interests of workers, in the long run, it will cause tense labor relations in China, the formation of labor costs continue to be forced to reduce and international competitiveness can't enhance the vicious circle.

2.2. Lack of Technological Innovation Ability

Technology is the key element to constitute the competitive advantage of industry, the lack of technological innovation in China has been one of the important factors that restrict China's manufacturing industry to improve economic efficiency, development level and international status[5]. First, Vlow intensity of R & D investment.R & D activity is the core of scientific and technological activities, is an important mean to achieve independent innovation. According to the data of China Statistical Yearbook 2015, in 2014, the R&D expenditure of China's manufacturing industry was 1 trillion and 301 billion 560 million Yuan, accounting for only 2.05% of GDP, however, the share of R & D funds in GDP in developed countries is generally around 3% or even higher. Moreover, the experimental development accounts for a large proportion in R&D investment in China, while the proportion of basic research and application research is very small, which indicates that the original innovation ability of China's manufacturing industry is insufficient.

Second, the core technology is controlled by People.From the overall industry point of view, the manufacturing technology in China, especially the core technology, mainly relies on imports from developed countries such as the United States, Japan and Germany. The core and key technology dependence of China's manufacturing industry is 50%~60%, while that of advanced countries is 30%. 70% of new product development depends on foreign technology. Such as high-speed rail bearings, construction machinery, hydraulic system seals, automotive engines and other important components and key materials, more than 80% rely on imports; Since 2009, China's automobiles have ranked No. 1

in the world with a production volume of 13,791,000 units and sales volume of 13,644,800 units, but the core technologies mainly rely on imports from Germany and Japan; According to the data released by the Ministry of Commerce, in 2015, the total output of air conditioners in China was as high as 103.8462 million sets with a total export value of 9.382 billion U.S. dollars, ranking first in the same field. But even the Haier and GREE, which focus on product R & D, and the core technologies of refrigerants and compressors, are still in the hands of developed countries. On the whole, although some industries in the manufacturing industry in China have led the world in output, they still do not possess the absolute technical superiority through the massive labor input and policy support.
Third, patent application structure is not balanced.According to the statistics of China Statistical Yearbook 2015, in 2014, the number of patent applications accepted and authorized number ranked first in the world, However, judging from the types of patents, the proportion of the number of invention patents accepted and the number of authorized patents is at a relatively low level, which are respectively 39.3% and 17.9%, the proportion of applications and authorized amount of foreign invention patents are 86% and 80.1% respectively. This has directly led to the lack of independent property rights and well-known brands in our manufacturing industry and hindered the improvement of the overall level of our manufactured products in the international market[5].

2.3. *Irrational Industrial Structure*

First, based on low value-added products.Since the reform and opening up, the country has been improving the international competitiveness of China's manufacturing industry for a long time with the continuous development of labor intensive industries and a large number of tangible elements such as resources, and has expanded the scale of manufacturing industry in a flat, lack of depth and depth, which leads to manufacturing technology for labor market development model of heteronomy, eventually decide that China's industrial added value of manufactured goods is much lower than the developed countries to master the core technology. According to the World Bank WDI database data, the results show that the added value of China's manufacturing industry has maintained a steady upward state since joining WTO, even in the financial crisis, the growth rate in 2010 is still more than 16%, and the proportion of manufacturing value added in GDP has remained above 30%. But from the point of view of value-added sources, the situation is not optimistic. In 2014, the proportion of domestic indirect industrial added value of the manufacturing industry as a whole was higher, which was 37.33%, the proportion of domestic direct added value of nine manufacturing enterprises is generally low, and the proportion of domestic direct added value is 29.76% and 25.96% in addition to transport equipment and mechanical equipment, and other industries are below 25%, especially the industry electronic and optical equipment that needs strong technical support, only 17.09%. Therefore, the value-added of China's manufacturing exports still remain in the low value-added stage and mainly rely on the production of low-tech production processes such as component production and product assembly to obtain added value[6].
Second, low proportion of high-end manufacturing.The proportion of low-end industries in China is far higher than that of high-tech industries, although our country enjoys a higher position in the clothing, textile, chemical, home appliances and other manufacturing technology field, such as aircraft manufacturing factory, while in the automatic control system, scientific instrument and precision measuring instruments and other high value-added effect the development of high-tech fields, basically by developed countries monopoly, or even blocked. The status of China's manufacturing industry in the global value chain division is still very low, which has led to the phenomenon that the gross domestic product and the manufacturing industry's international competitiveness in the past 30 years of reform and opening up have become extremely high, exposing the awkward situation of "big but not strong" in our manufacturing industry.
Third, unreasonable industrial agglomeration.Generally speaking, the countries or regions that realize industrial agglomeration will occupy the leading advantage in the international market, such as Silicon Valley and Germany's tool manufacturing industry cluster. On the one hand, when industrial agglomeration occurs, enterprises can understand each other more quickly in competition and game, at the same time, the survival of the fittest, the free and cruel way of competition make them constantly

improve in technology and management, and ultimately achieve the overall level of competition in the industry. On the other hand, in addition to the industrial agglomeration, the cooperation, exchange and complementarity of each enterprise can promote their own technology and product quality, previous horizontal cooperation between firms in different parts of the value chain can also play a big role in reducing manufacturing costs in all sectors. Furthermore, enterprises in the cluster can quickly perceive changes in the market, seize market opportunities ahead of time, and create leading products.
And so far, China's industrial agglomeration has mainly occurred in the labor-intensive industries of traditional manufacturing industry in Jiangsu, Zhejiang and Shanghai coastal areas, Beijing, Shanghai, Guangzhou and Shenzhen Internet industry, agglomeration within the industry has reached a relatively saturated degree, but the inter industry division of labor to reduce costs has not yet achieved remarkable results, new high technology manufacturing industry agglomeration is in the initial state. China's geographical reasons make the industry agglomeration in different regions relatively difficult, which limits the efficiency of agglomeration. In addition, the immaturity of technology and the backwardness of innovation make the structure and product upgrade of various industries slow, and can't occupy the international market technically[7].
Fourth, low status of global value Chain division of labor.In recent years, with the decrease of communication and transportation costs, the level of international division of labor gradually goes from product to specific production link, The cutting of the value chain between different places makes the global value chain come into being. With rich natural resource elements and low-cost labor force participating in low-tech and low-profit manufacturing processes such as processing and assembly in the global value chain, China has obtained extremely limited processing fees. This passive participation in the international division of labor has made China's manufacturing industry in the international division of labor in the low position has been determined for a long time, which is not conducive to the construction of independent brands and the improvement of innovation and R & D capabilities, thus restricting the development space of manufacturing industry and its pulling effect on the overall economy.
Taking into account the industrial structure and technological level of China's manufacturing industry at this stage, although the continuous investment and improvement of high-tech has already achieved initial success in promoting the division of global value chain in China's manufacturing industry, the results are still difficult to compare with the developed countries, the reason behind the manufacturing technology of our country makes the ability to create high value-added sectors to create accounts for a low level, in addition, in the global value chain, R & D, design, manufacturing, marketing and other key parts of the complex high-end part is difficult to imitate innovation, a short time is difficult to achieve the effect of rapid, which requires the support of the national innovation system, but also needs long-term exploration of enterprise transformation.
Fifth, imperfect quality supervision system.In recent years, "made in China" has become a hot word all over the world, but at the same time of being crowned "the workshop of the world" and "manufacturing power", "made in China" is also faced with many problems. The unqualified rate of the products of the national supervision and inspection is as high as 10%, and the export products are the first to recall the products in foreign countries for the first time, the annual direct manufacturing industry loss of more than 200 billion Yuan. As labor costs continue to rise, prices rise in the international market, and many major components and core systems quality assurance is not high, the quality image of product and brand did not achieve synchronous improvement, which directly led to the decline of the international competitiveness of China's manufacturing industry. On the other hand, China's department products are shoddy, and the product standards and quality supervision system are far less than developed countries, leading to "made in China" become synonymous with "poor quality", consumers of China products do not trust the China manufacturing industry is facing a serious quality problem of credibility, directly affect the product export volume, indirectly restricts our country to improve the international competitiveness of the manufacturing industry.

3. Suggestions on promoting the international competitiveness of China's manufacturing industry under the background of TPP

First, strengthening education and training.Strong industrial workers are the strong cornerstone of "made in China", China has entered an aging society, demographic dividend is disappearing, and various labor costs and labor utilization issues have begun to gradually highlight. Good education and training can not only improve the quality of labor, but also provide a good human foundation for the R & D and production of high-tech products.

The strategy of education structure should be adjusted. First of all, on the basis of continuing to improve the basic education, vocational education should be placed in a higher position, strengthen vocational training and education fit for industrial needs, and promote the deep integration of industry and education, form the new force of government, enterprise and social capital to promote the development of advanced vocational education, transport more efficient workers for the manufacturing industry. Secondly, the cultivation of technical research talents should be strengthened simultaneously; the government should implement more favorable policies for innovative education, and provide a solid human capital to improve the international competitiveness of China's manufacturing industry.

Second, increase technology investment.In order to promote the international competitiveness of China's manufacturing industry, the most important thing is to further accelerate the construction and improvement of technological innovation capability system and support system. First of all, improve the technological innovation capability system of manufacturing industry. Strengthen the innovation investment, make the innovation investment structure rationalization, on the basis of the introduction of advanced production technology from developed countries, improve the innovation output efficiency, improve the ability and level of re innovation, perfect the market competition system, and stimulate the two innovation ability in the industry. Strengthen the manufacturing sector overall technical level and the ability of independent research, increase the added value of the products, the ability to adapt to the international competition in the market has improved technology. Secondly, perfect the supporting system of technological innovation in manufacturing industry, the government should strengthen the awareness of manufacturing industry as the main body of technological innovation through a series of measures, increase the subsidy and support of the government to the corresponding industries, and strengthen the financial support for the manufacturing technology innovation activities, at the same time, it is necessary to take various measures for commercial banks and other financial institutions to improve the financing environment for the manufacturing industry to participate in technology research and development.

Third, speed up efforts to upgrade its industrial Structure.we should strengthen personnel training, increase the policy and investment of technology introduction and re innovation, improve the proportion of high value-added industries, optimize the product structure of enterprises and industries by increasing the added value of products, and transform labor-intensive industries by high and new technology. Then we cultivate outstanding enterprises in each industry, encourage independent innovation, play industrial agglomeration effect from the technology, make it become the supporting point of industrial agglomeration; The government supports the policy, promotes the cooperation among industries, optimizes the agglomeration structure of China's manufacturing industry, and expands the economies of scale. In addition, we should further improve the investment environment and enhance the ability of enterprises to attract foreign investment, further expand the channels to obtain investment, diversification of investment methods, so as to enhance the effect of attracting foreign investment, guide foreign investment in technology intensive industries, so as to promote the overall upgrading of China's manufacturing industry, enhance international competitiveness.

Fourth, strengthen quality management. Facts have proved that the closer to the end of the consumer market, the higher the profits of enterprises, if we want to firm the heels of the terminal market, we must vigorously develop their own brands. China's manufacturing industry to create can affect the global international brands, in addition to invest heavily in marketing, fundamentally need more strict quality management system, only by ensuring the quality of the product quality can we talk about the reputation and credibility of the brand. Enhance the quality of management, we must first upgrade production technology, promote the brand construction with the scientific management system, and constantly improve the quality standard system, certification system, inspection system and quality

IOP Conf. Series: Materials Science and Engineering **398** (2018) 012022 doi:10.1088/1757-899X/398/1/012022

safety warning system, the whole process of research and development, production, testing, sales and service of high standard and make the complete system throughout the product. To achieve this, we need to start from two aspects of enterprise and government: On the one hand, enterprises need to strengthen the quality management of the production process, to be in the industry model, establish quality benchmarking; On the other hand, the government needs strict supervision according to the standard system, from the national level to establish a sense of quality and promote "the spirit of craftsman". At the same time, we should strengthen the construction and maintenance of brands, establish a good image of "made in China", and offer preferential financing and tax benefits to international brand enterprises with high added value and high reputation.

Fifth, promote China's regional economic.Integration.From this point of view, many topics in TPP will have a negative impact on China's manufacturing industry, the development of high standards and benefits of treaty between the parties on a lot of technologies are directly or indirectly suppress the manufacturing industry in China. But in the long time, whether or not China joins the TPP, the agreement can reverse the manufacturing industry in China. Because in the context of economic globalization, the overall situation based on regional economic integration, it is impossible to completely isolate the non-contracting states from the terms and standards of the contracting states. Facing this inevitable trend, China should pay close attention to the positive response and actively promote regional economic integration with China as the center, bypassing the free trade agreement between TPP and more countries. It includes signing bilateral free trade agreements with TPP member countries and promoting East Asian Strategy of China, Japan, South Korea, FTA and "ASEAN 10 + 3", speed up the Regional Comprehensive Economic Partnership (RCEP) negotiations, complete the ongoing Chinese - Australia and China - GCC FTA negotiations as soon as possible, hedge TPP adverse impact to our country's, and combined with the development advantage of the The Belt and Road "further accelerate the deepening of the reform of the RMB, continue to expand the market advantage, so as to make our country occupy a leading position in the international market.

Conclusion

Through the analysis of the current situation of China's manufacturing industry and the factors that restrict the international competitiveness of China's manufacturing industry, we can see that since the reform and opening up, China's manufacturing industry has developed rapidly, the international competitiveness has been greatly improved, but compared with the developed countries there is still a big gap, the main reasons are that Factor cost is rising, technological innovation ability is insufficient, industrial structure is irrational, global value chain division of labor is low, and quality supervision system is imperfect. At present, China is a "manufacturing country" rather than "manufacturing power", the manufacturing industry in the world has not yet get rid of the low value-added, low quality status quo. And facing the big trend of TPP, China's manufacturing industry must fully upgrade, in addition to strengthen education and training, increase investment in technological innovation, accelerate the upgrading of industrial structure, strengthen the quality management of outside, we should pay close attention to TPP development progress, and actively take response measures to promote regional economic integration in China as the center, combined with the deepening reform of The Belt and Road, grab seize power at, and occupy the initiative.

References

[1] Ding 2017 Spatial temporal changes of FDI in China's manufacturing industry after China's entry into WTO *J. Theoretical Research* **5** 81-87

[2] Liu 2015 The development of Chinese manufacturing industry: current situation, problems and trend. *J. Yuejiang Academic Journal.***4** 15-21

[3] Lv Li 2015 Changes of International Economic Environment in the Thirteenth Five - year Plan and Their Impact on China 's Industrial Development *J. International Trade.* **6** 21-31

[4] Huang 2015 Core competence, functional orientation and development strategy of China's manufacturing industry *J. Made in China 2025.* **6** 5-17

[5] Peng Zhang 2013 Economic Effect of TPP and APFTA and China 's Countermeasures *J. International Trade Issues.* 4 83-95

[6] Qu 2016 A study on the employment dynamics of China's manufacturing industry *J. China Industrial Economics.* **2** 83-97
[7] Zhang 2017 Development trend of China's manufacturing industry from the perspective of value added decomposition *J. China's prices.* **3** 73-76

Simulation Tool for Assessing the Performance of a Flexible Manufacturing System

Adriana Florescu*, Sorin Adrian Barabas

Engineering and Industrial Management, Transilvania University of Brasov, Brasov, Romania

e-mail: fota.a@unitbv.ro; sorin.barabas@unitbv.ro

Abstract. The flexible manufacturing system (FMS) is as complex whole, raises concerns regarding the configuration and dimensioning, as well as on the conditions for implementation. The scientific researches in this paper, are directed towards the study and analysis of flexible manufacturing systems, in order to know their behaviour and performances as well, and if possible, even before their physical realization, and to establish the scientific basis for sizing, configuration and simulation patterns of flexile manufacturing systems. The ARENA simulation software can describe the complex interactions between resources and processes that take place within a FMS, a system of discrete events. In the simulation phase, evaluation of performance criteria allows the introduction of planning strategies in the use of system resources. After developing these models the verification and validation of the models, confirms their validity. The researches will be also extended in the domain of flexible manufacturing systems management and control in real and virtual environment.

Keywords. manufacturing systems; modeling; simulation, Arena software

1. Introduction

Development of advanced manufacturing systems involve profound changes in manufacturing processes by increasing automation, robotics and information technology, and the application of methods and techniques of management, organization and quality assurance.

These systems, which ensure continuous production, respond to the current need for competitiveness, the main advantage being the almost total flexibility and adaptability to changes in the economic environment, in terms of efficiency and minimum response time, [1]. In this context, flexible production systems are superior to the systems they replace, especially in terms of technical and economic performance. Optimal use of system capacity, staff, and high technical availability and flexibility of replacement and change, results in significant increases in productivity over long periods of system use (8-10 times the classic system). The Flexible Manufacturing System (FMS) is defined as an integrated complex, composed of a set of computer-controlled machining centers and containing automatic handling, material transfer / transfer devices. FMS is designed to achieve the cost-efficient production key as it is a good combination of diversity and productivity [2].

The introduction of an FMS into an integrated manufacturing system is an important result, which leads to the optimization of the entire material and information flow of the enterprise. Given the complexity of these systems, the difficulties in designing and implementing them require intensive design, planning and management. Flexible Manufacturing Systems thus require a synthesis approach and, by implication, the appropriate specialist, designing these systems with deep knowledge of the system components and the interdependencies between them, as well as the relationships between the system and the environment. The theoretical and practical knowledge base, the newest approaches in the automation of production systems [1, 3], and the practical examples [2, 4, 5] refer to the application of modern concepts of integrated management of production systems, as well as to the analysis and evaluation of their performance.

Implementation and entry into service of a flexible manufacturing system involves a lot of expenses and the expected economic effect is hard to determine. For this reason, both during the design and the operation of the FMS, modeling and simulation technique has been used [6, 7]. With regard to Flexible Manufacturing Systems modeling operation, there are still several attempts, some recent, [8] to define and analyze the performance indicators of the system.

In many of current simulation approaches, despite the existence of well-developed tools, it is often difficult to transform system requirements into the simulation program and the whole implementation process is still carried in an intuitive manner [3].

The objective of this paper is to study and evaluate the flexible manufacturing system through modeling and simulation. Thus, one can develop a decision support tool for assessing the performance of a flexible production system that is available to companies interested in the production. Using simulation we intend to clarify some issues regarding the description of system operation, development of theories and hypotheses based on previous operation, using the model to predict future operation or establish the effects of changes in the system.

Modeling and simulation technique allows, through a computer model to make predictions on various aspects of the operation of dynamic Flexible Manufacturing System, existing or future, in order to respond best to the problems involved in its operation.

The aim of the paper is to increase the technical and economic performance of these systems by optimizing the processes based on the new knowledge in the field and the authors' previous research [10, 11].

2. Modeling FMS. case study

2.1 Presentation of the analyzed system

The study was conducted on a flexible manufacturing of cylindrical parts, designed and developed physically, according to stand in Figure 1, existing in research laboratory and the applicative character of the research theme is demonstrated.

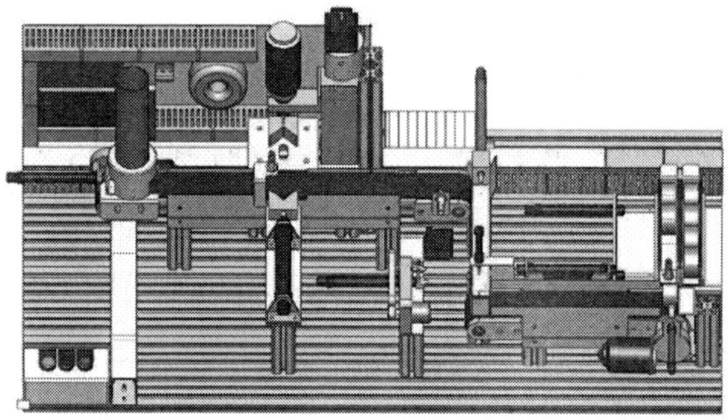

Figure 1. Sequence of Flexible Manufacturing System – FMS01

The production flow consists of the operations required for the machining of the cylindrical parts on drilling / boring-milling, turning and control stations. The system is managed by a PLC and includes stations supply / exhaust parts and equipment logistics conveyor type transport.

In a first step, a simulation program was developed and implemented using SCADA's Cx-Supervisor Developer programming environment to monitor and run the process in real time (Figure 2).

IOP Conf. Series: Materials Science and Engineering **398** (2018) 012023 doi:10.1088/1757-899X/398/1/012023

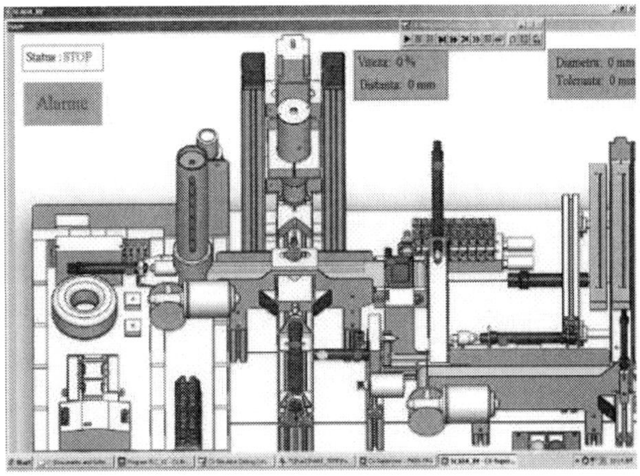

Figure 2. The sequence of the application - Simulation software

Operational parameters for the existing flow were also estimated using the ARENA software, the results being presented in paper [10].

The results have led to the need to optimize and extend the system by introducing industrial logistics elements.

The results have led to the need to optimize and extend the system by introducing some industrial robot industrial logistics elements. In the present simulation study, we aimed at balancing the production line, minimizing the asynchronous occurring in the system and ensuring a continuous flow, considering the increase of the production capacity.

The evolution in time of the system state was described by algorithms or complex procedures that better express its actual functioning. Verification and validation of models will be performed on the existing system.

2.2 *Technical and economic parameters of the model*

To achieve a mathematical model to be used in the design and performance analysis of the manufacturing system there are a number of necessary input data, shown in Figure 3: MF - material flow, IF - informational flow, EF - energy flow, time and space. The models must contain a number of parameters and optimization criteria, objective function, performance indicators, optimal synthesis in real time and environment.

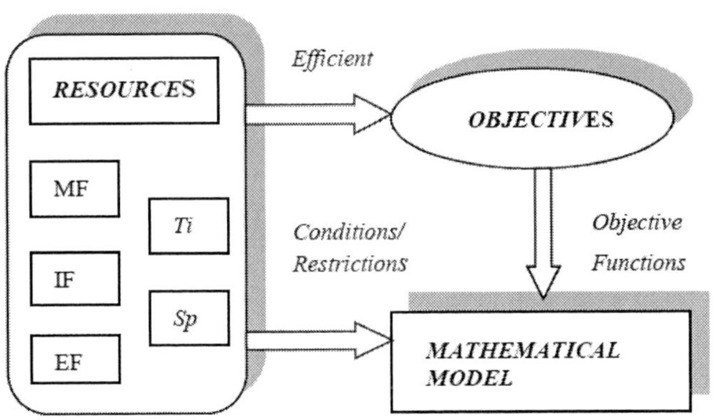

Figure 3. Input data in modeling FMS

In modeling a FMS it must respond to major issues concerning: determining resources and their characteristics that most affect system performance, formulating a model or a description, representing

IOP Conf. Series: Materials Science and Engineering **398** (2018) 012023 doi:10.1088/1757-899X/398/1/012023

these resources and the relationships between them and determining values for the performance in conditions of known scenarios.

To study and analyse the performance of a cylindrical parts processing FMS, [11] will consider the following initial data of entry into model:

Current part in manufacturing (P_k, $1 \leq k \leq p$), p - number of types parts from typological diversified manufacturing task;

The processing of parts: P_r (k, $k = ct$).

The crowd parts involving crowd of the manufacturing processes:

$$\{P_k\} \rightarrow \{P_r(P_k), k = var\} \quad (1)$$

Manufacturing task:

$$MT \rightarrow \{P_k,\ P_r(P_k), 1 \leq k \leq p\} \quad (2)$$

Manufacturing task is obtained on output from the system and is carried through the manufacturing system, as a means, defined as the set of workstations:

$$MS \rightarrow \{WS_{ij}, 1 \leq i \leq q, 1 \leq i \leq n\} \quad (3)$$

where: WS_{ij} represents workstations, machinery from the system that directly contribute to the transformation of material flow in the finished piece.

It considers serial production, manufacture executing on the size batches of the size of $n_b\,[pcs/batch]$, for $\forall P_k \in FP_k, k \in \aleph, k = \overline{[1,r]}$ with the frequency of occurrence of the lot $v_{bk}\,[no.batch/\tau]$. Any model is developed on a time interval τ (week, month, and year).

Performance analysis simulation system will be made in relation to a number of parameters listed below. The duration of manufacturing (T_m) is the total time that is needed to completely process all identical parts (P_k) from one batch and is expressed by equation (4), in which (i) is the index of the technological operation and (q) - the number of machines, workstations through which the semi-finished goes successively, ordered for the purpose of processing the finished piece:

$$T_{mk} = \sum_{i=1}^{q}(t_{opk} \cdot n_b + t_a + t_{ad})\ [t.u] \quad (4)$$

Knowing the average operative time ($\bar{t}_{op}$), auxiliaries times (t_a) and adaptation times (t_{ad}) and handling times (t_h) it can be calculated the average manufacturing duration (T_{mk}) of a part on the whole system of machines, according to relationship (5), where t_h is the handling time (transport/transfer) of the part along the entire process of processing parts, on the entire manufacturing system:

$$\bar{T}_{mk} = q_k \cdot n_b \cdot \bar{t}_{op} + \frac{t_h + t_{ad}}{n_b}\ [t.u./batch,\ syst.] \quad (5)$$

The production rate (R_p) on the part, on the machine (m_i) is calculated using equation (6) and expressed in [pcs / $t.u$, m_i], in which (T_c) is the period of the machine cycle:

$$R_p = \frac{1}{T_c} \quad (6)$$

Production capacity is available or necessary means of production required for production plan and expresses the relationship (7), where n_p is number of productive units, n_c - shifts number / u.t., h- hours number / shift, R_p production rate:

$$Q_p = n_{up} \cdot n_c \cdot h \cdot R_p \quad (7)$$

Average unit through production time - $\bar{T}_{mu}$ [h/pcs]:

$$\bar{T}_{mu} = \frac{T_m}{n_{PLk} \cdot n_{LK}} \quad (8)$$

The period of the machine cycle - T_c [*min/pcs*]:

$$T_c = t_{op} + t_{aux} + t_h \tag{9}$$

The degrees of utilization in time for workstation (η_{tu}) and in days (η_{du}):

$$\eta_{tu} = \frac{T_m}{F_{rt}} \tag{10}$$

$$\eta_{du} = \frac{n_b \cdot v_{bk}}{R_p \cdot T_m} \tag{11}$$

Production in processing (P_p), the number of semifinished or their percentage of the total number of parts to be processed, where z_p is machined number of the parts, z_w- waiting number of the parts and z_t - total number of the parts:

$$P_p = \frac{z_p + z_w}{z_t} [\%] \tag{12}$$

Optimization of these indicators aims to reduce the auxiliary time that can be achieved by automating activities and transferring human resources to intellectual activities.

The parameters and performance indicators will be set in order to optimize the synthesis of the system in real-time environment. The developed model aims to know the system behaviour as well as it performance, and if possible, even before their physical realization.

3. Simulation of FMS

3.1 Simulation tools

The use of simulation techniques in large systems is now widespread. Thus, the Flexible Manufacturing System is decomposed, its internal structure detailed, so that the processes of operation of all its component subsystems can be algorithmized and implemented step-by-step into the computer, while respecting the accuracy and taking into account the interaction with the external environment.

Simulation is one of the most powerful means of analysis used to design complex systems for planning / scheduling and control. The current trend is, simulation to become a next-generation tool for analyzing computer systems. It is now possible to provide for the complete integration of simulation with other software packages that collect, store and analyze system data [4].

For the case study analysed, ARENA Simulation Software version 15.00.00001 (Arena Trial version) built on SIMAN language was used [12]. This software can be used to describe complex interactions between resources and processes occurring within FMS, considered a discrete event system [4, 13]. In Arena Software, the basic process offers the highest level of modeling to allow rapid, easy and flexible creation of models in most systems (Kelton et al., 2010). SIMAN, simulation language from ARENA allows the dynamic deployment in time of the system. It describes and reproduces the dynamic structure of the system, the interdependencies within the system.

ARENA offers high capacity simulation of advanced manufacturing systems, relatively easy to use. With this software can be described complex interactions between resources and processes occurring within FMS considered discrete event systems. Arena is an easy-to-use, powerful modeling and simulation software tool that allows the user to construct a simulation model and run experiments on the model.

ARENA is a software for dynamic simulation of discrete technical systems that includes FMS and uses terms like: *entity* that ia a part of the system which constitutes a subsystem; *attributes* that are characteristics of the entities; *activities* than are dynamic processes that lead to modification of attribute values; *event* that changes the value of an attribute. ARENA requires knowledge of random distributions, probability, and statistics.

3.2 Developing the simulation model. Results

IOP Conf. Series: Materials Science and Engineering **398** (2018) 012023 doi:10.1088/1757-899X/398/1/012023

After collecting all necessary information to implement the study ARENA simulation model using modular flow (flowchart) and data modules will be build. With the graphics animation module that allows you to view the simulation behaviour on the display screen, errors can be determined and the model can be redesigned. After creating the complete logic model, it will be checked whether the model resulting from the computer simulation corresponds to the conceptual model. Validation will show that the model behaves just like the real system.

A manufacturing model has been developed using ARENA simulation software, a simulation sequence being shown in Figure 4. In Arena, modules are basic building blocks basic (or objects) that consist of Arena models. There are tow types of modules: flowchart modules and data modules. The flowchart modules are used to describe the flow of entities and placed in the flowchart view of the model window, [12]

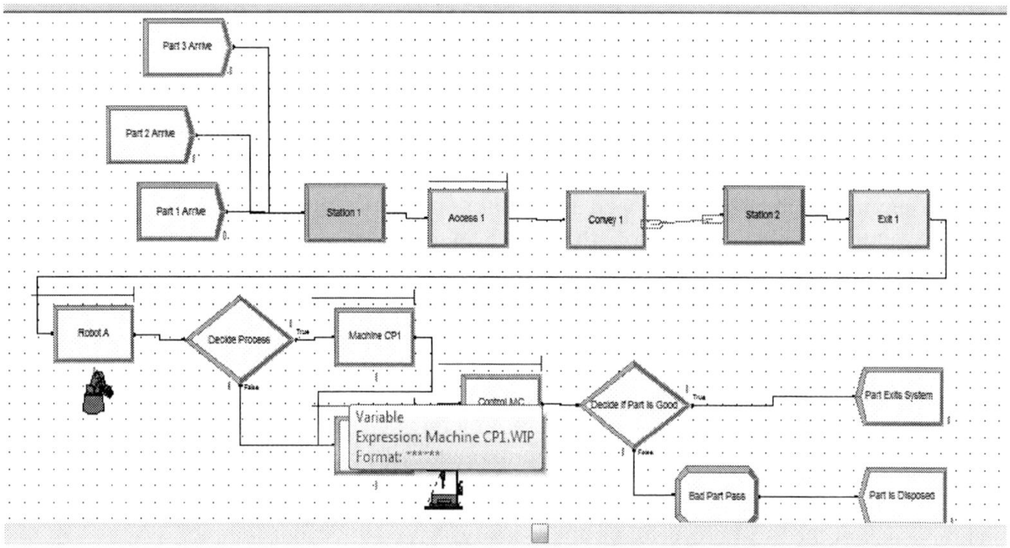

Figure 4. The flowchart of the FMS model. Simulation sequence.

Measurements, analytical models and simulation models are alternative performance analysis techniques. The models based on the simulation are preferable because they allow to obtain accurate results for complex and larger calculations, with fewer assumptions [10].

The model has been designed according to the events and processes taking place in the system. The model created evaluated the dynamic process flows, including technology manufacturing operations, but also the handling and the transferring of the blanks. It performed a simulation of the simultaneous processing of three parts, which typically occurs in the real environments. In the system the three processed pieces each have their own technological route, with different phases.

Complete description of system operation is achieved by specifying the various interactions between processes. In Flexible Manufacturing Systems, the parts interact, in general, all other processes, which express their movement into the system. The simulation is done by describing the movement of the parts through different processes, corresponding to the machine tools, the means of transport and the means of handling.

As a first result obtained by evaluating the flow of parts on machine and processing stations and examining conflicts regarding the application of limited resources, it could be evaluated and optimized the system layout, the selection of equipment and industrial logistic subsystem, as well as the operating procedures. It will explore in real environment and real-time, without interruption of the running operations new operating procedures.

IOP Conf. Series: Materials Science and Engineering **398** (2018) 012023 doi:10.1088/1757-899X/398/1/012023

The conducted simulations allowed to define some aspects of production management, to detect "narrow places" that slow the production flow and to study the impact of failures on the manufacturing process (Figure 5).

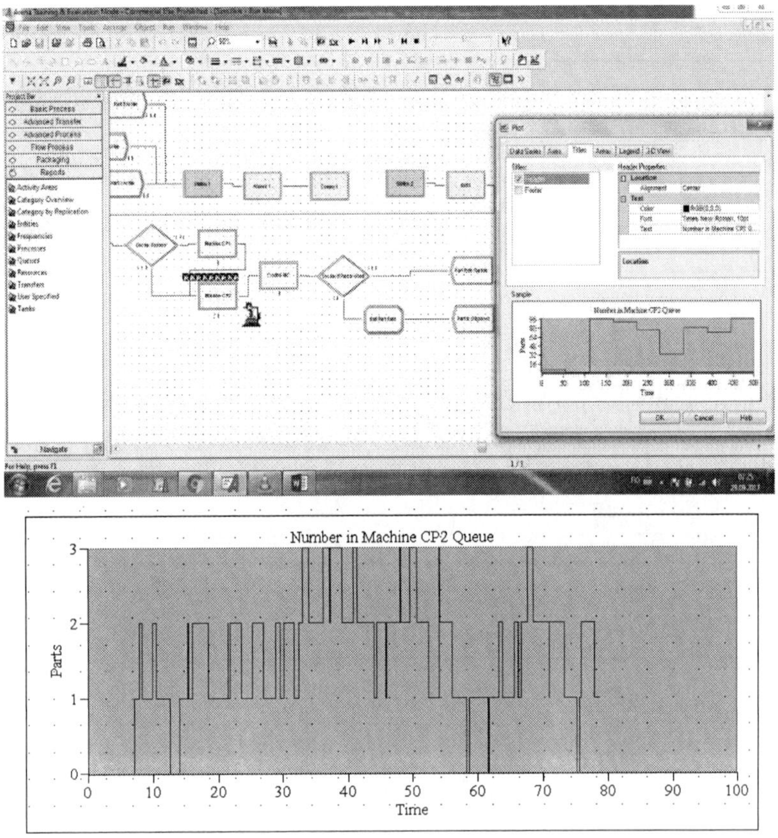

Figure 5. The simulation sequence and bottlenecks in the production process

After running of the simulation model, the result of throughput analysis such as average, minimum and maximum times for production, work in process – WIP are shown (Fig.6 and Fig.7).

Entity

Time

VA Time	Average	Half Width	Minimum Value	Maximum Value
Part	8.2585	(Insufficient)	5.1106	12.2576

NVA Time	Average	Half Width	Minimum Value	Maximum Value

Figure 6. Category overview time in FMS

Part	190.79	(Insufficient)	10.3327	337.59

Other

Number In	Value
Part	230.00

Number Out	Value
Part	210.00

WIP	Average	Half Width	Minimum Value	Maximum Value
Part	86.9864	(Correlated)	0.00	134.00

Figure 7. Work in process (WIP)

Another important indicator for performance evaluation is the degree of loading of workstations in the system. Figure 8 shows the resources report of the FMS model, after the simulation run. The resource (Machine1, Machine 2, for example) utilization was 86,96% and 97,70% (average) and 100% (maximum).

Resource

Usage

Instantaneous Utilization	Average	Half Width	Minimum Value	Maximum Value
Machine 1	0.8696	(Insufficient)	0.00	1.0000
Machine 2	0.9770	(Insufficient)	0.00	1.0000
Machine control	0.8417	(Insufficient)	0.00	1.0000
Robot	0.9083	(Insufficient)	0.00	1.0000

Number Busy	Average	Half Width	Minimum Value	Maximum Value
Machine 1	0.8696	(Insufficient)	0.00	1.0000
Machine 2	0.9770	(Insufficient)	0.00	1.0000
Machine control	0.8417	(Insufficient)	0.00	1.0000
Robot	0.9083	(Insufficient)	0.00	1.0000

Resource

Usage

Total Number Seized	Value
Machine 1	111.00
Machine 2	211.00
Machine control	210.00
Robot	230.00

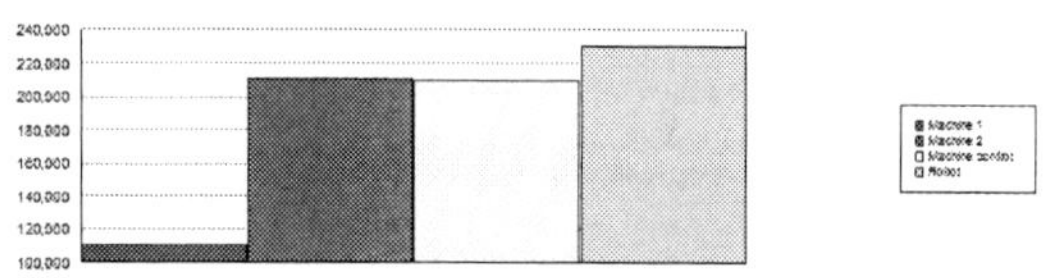

Figure 8. The Resources report of the FMS model

4. Conclusions

The purpose of this simulation study is to provide a decision support tool in the evaluation of FMS. The simulation will help assess the impact of production rates, resource use, operating time, and buffer stock levels in different scenarios. After comparing the results, we suggest the best solution for the newly designed manufacturing system. It was developed a production model for the study of cylindrical parts processing, using simulation software Arena. The following indicators of performance evaluation of the analyzed system have been optimized: real time manufacturing, the pace of manufacturing, the workload of the machines, and the blocking time in the production process. Increasing the performance of the system was achieved by introducing an industrial robot that reduces handling times, operational time, manufacturing time and waiting time. This work identified opportunities to shape the research and the application in industry.

4.1 Acknowledgments
This work was supported by a mobility grant of the Romanian Ministry of Research and Innovation, CNCS-UEFISCDI, Project number 253/8.12.2017, code PN-III-P1-1.1-MC- 2017-1155.

5. References

[1] Groover, M. P. 2010. Fundamentals of modern manufacturing, materials, processes, and systems. 4th Ed. Publisher: John Wiley & Sons.
[2] Nguyen, Nha. 2015. Simulation Modeling for Designing and Evaluating a New Flexible Manufacturing System. Master's thesis, Aalto University, School of Business.
[3] Anglani, A., Grieco, A., Pacella, M. and Tolio, T. 2002. Object-oriented modeling and simulation of FMS a rule-based procedure.Simulation Modelling Practice and Theory 10, 209-234.DOI=http://doi.org/10.1016/S1569190X(02)00100-4.
[4] Kelton, W. D., Sadowski, R. P. and Swets, N.B. 2010. Simulation with ARENA, 5th Ed. New York: McGrawHill.
[5] Reddy, B., and Rao, C. 2011. FMS Modelling and Performance Evaluation Using Automod, International Journal of Simulation Modelling, 10, 78-90.
[6] Koren, Y. and Shpitalni, M. 2011. Design of reconifigurable manufacturing system. Jurnal of Manufacturing Systems, 29 (4), 130-141. http://doi.org/10.1016/j.jmsy.2011.01.001
[7] Chan, F., 2004. A comprehensive survey and future trend of simulation study on FMS, J. of Intell. Manuf., 15 (1), 87-102.
[8] Seebacher, G. and Herwig, W. 2014. Evaluating flexibility in discrete manufacturing based on performance and efficiency Int. J. Prod. Ec., 153, 340–351.
[9] Khaleel, A. 2016., Modeling and Optimization in a new Machining Production Line by using Manufacturing System Simulation. Int. Conf. Ind. Eng. Op. Manag. (Detroit, Michigan, USA), September 23-25.
[10] Florescu, A., Barabas, S. and Sarbu, F. 2017. Operational parameters estimation for a flexible manufacturing system. In Proceedings of International Conference IManE&E, Iasi, DOI= https://doi..org/10.1051/mateconf/201711205008.
[11] Fota (Florescu) A. 2004. Designing of the machines systems. Transilvania University of Brasov Publish House, Romania.
[12] Arena User's Guide, 2014. Rockwell Automation Inc., USA.
[13] Garridi, J.M. 2009. Object oriented simulation. A modeling and programming perspective, Springer.

IOP Conf. Series: Materials Science and Engineering **398** (2018) 012024 doi:10.1088/1757-899X/398/1/012024

Numerical Simulation of the Airflow Characteristics of Drying by Ground Air Film

Zheng Yu-lan[1], **Peng Xiao-yong**[1]

[1] School of civil engineering, University of South China, Hengyang Hunan China

1170218114@qq.com,pengxiaoyong@126.com,huwei@126.com

Abstract. According to the interference and coupling characteristics among sub-systems of the chassis of the forklift truck, the the neutral network inverse system method of the non-linear system is applied for the decoupling control of the active rear steering (ARS) and the direct yaw moment control (DYC) of the forklift truck. Based on the analysis of the reversibility of the chassis system, a back propagation neutral network reverse system model is set to decouple the chassis system into two independent pseudo-linear system; A PD closed loop controller is designed to comprise a composite controller with the neutral network reverse system and the simulation is performed for verification. The simulation result shows that the decoupling control strategy with the neutral network reverse system can eliminate the interference and coupling among sub-systems of the chassis and thus enhances the status tracking and operating stability of forklift trucks.

1. Introduction

Indoor water easily cause personnel fall and conducive to breeding mold[1]. It seriously affects the indoor environment and the health of workers. The ground dryer is a device used to dry the surface of the water and improve the indoor wet environment. The traditional ground drying machine uses large air jet to dry the ground. Only a small part of the air flow is used to dry the ground, and most of the air flow is wasted. This kind of ground drying machine has the characteristics of quick drying ground and convenient use. However, with the greater noise, and the utilization rate of air is very low, it is not conducive to save energy .

Coanda effect is also called attaching effect. It points out that the fluid (water or airflow) has a tendency to move away from the original flow direction to the surface of the protruding object[2]. At present, the effect of Coanda is widely used in the field of aviation, building ventilation and industry. Zhang Shiyu et al. studied a supercritical circulation control airfoil of dual-radius, and optimized its shape, and obtained the optimal model of lift resistance performance. This circulation control technology is derived from the Coanda effect, it can solve the traditional high lift system for aerodynamic noise problems. Bladeless fan is a typical application of the Coanda effect, and the air volume can be increased by 15 times of the air intake rate of the initial air inlet, which solves the problem of strong wind blowing and noise. In a similar study, Ye Yaling used the numerical simulation method to study the airflow characteristics of the airfoil air inducer, and verified the numerical simulation results from the experimental point of view. In view of the fact that the traditional range hood is not easy to be cleaned, the noise is big and the energy consumption is big, Huang Jun you proposed a new range hood based on the effect of Coanda, which has the advantage of energy saving and noise reduction, and its theoretical displacement can reach 24.3m3/h, much higher than the national standard (7m3/h).

The Coanda effect was adopted by this paper in the application of ground dryer. The surrounding air was induced by small air volume to flow closely on the Coanda curved surface, resulting in the formation of an air film attached on the ground. The numerical simulation technique was used to simulate the segment length and inclination angle of Coanda surface, as well as the air film formation and airflow characteristics when the curvature of Coanda surface changed.

2. Governing equations and numerical methods

2.1. Control Equations and Turbulence Models

The air flow of the ground dryer is low-speed incompressible turbulent flow, satisfying the N-S equation, and the control equation is as follows:

continuity equation

$$\frac{\partial u_j}{\partial x_j} = 0 \tag{1}$$

momentum equation

$$\frac{\partial u_i u_j}{\partial x_j} = -\frac{1}{\rho}\frac{\partial p}{\partial x_i} + \frac{\mu}{\rho}\frac{\partial}{\partial x_i}\left(\frac{\partial u_i}{x_j}\right) + f_i \tag{2}$$

Among them, ρ is the air density, fi for the volume force, u for the air velocity, p for the gas pressure, μ for the dynamic viscosity coefficient, two-dimensional simulation, i,j,k value of 1 and 2. Three-dimensional simulation, i,j,k value of 1,2 and3, ui for the average wind speed.

The turbulence model uses the standard k-Ɛ model, and the turbulent kinetic energy k and turbulent dissipation rate ε equation are as follows:

$$\frac{\partial(\rho u_i k)}{\partial x_i} = \frac{\partial}{\partial x_j}\left[\left(\mu + \frac{\mu_t}{\sigma_k}\right)\frac{\partial k}{\partial x_j}\right] + G_k + G_b - \rho\varepsilon \tag{3}$$

$$\frac{\partial(\rho u_i \varepsilon)}{\partial x_i} = \frac{\partial}{\partial x_j}\left[\left(\mu + \frac{\mu_t}{\sigma_\varepsilon}\right)\frac{\partial \varepsilon}{\partial x_j}\right] + C_{1\varepsilon}\frac{\varepsilon}{k}(G_k + C_{3\varepsilon}G_b) - C_{2\eta}\rho\frac{\varepsilon^2}{k} \tag{4}$$

Among them, the turbulent viscosity $\mu_t = \rho c_u \kappa^2 / \varepsilon$. The universal constants of the model are $C_{1\varepsilon} = 1.44$, $C_{2\varepsilon} = 1.92$, and $C_{3\varepsilon} = 0.09$ respectively. The turbulence of the turbulent kinetic energy and the dissipation rate Ɛ is $\sigma_k = 1.0$, $\sigma_\varepsilon = 1.3$, respectively.

2.2. Numerical Methods

The numerical simulation was conducted based on the SIMPLE algorithm and the standard pressure compensation equation [3]. The discretization was performed by finite volume method. The first-order upwind spatial difference discrete scheme was applied in the continuity equation, the momentum equation, the turbulent kinetic energy equation and the dissipation rate equation. Ye Yaling et al. used the same numerical method as described above to numerically simulate the airflow characteristics of the airfoil-shape air induction unit based on Coanda effect. The relative deviation between the numerical simulation result and the experimental result of the airfoil-shape unit's air flow velocity was 11.4%. This has proved the reliability of numerical method.

3. Numerical simulation of air film and air flow characteristics of ground dryer

3.1. Physical Model, Model Parameters and Mesh Generation

The physical model of the ground dryer is shown in Figure 1, consisting of an arc-shaped cover plate, a slit outlet, an air inlet and a Coanda surface. Among them, the angle between the Coanda surface and the floor is Coanda surface inclination angle (α) , The air inlet is v1,The air outlet is v1 and Coanda curvature is k. Specific parameters of physical model are shown in table 1.The computational domain, shown in Figure 2, is 4000mm * 1030mm.FLUENT software is used for numerical simulation, in order to achieve the accuracy of simulation calculation, mesh encryption is carried out in the inlet and outlet and solid wall regions, The other parts of the grid are distributed gradually, and the surface mesh is divided by Quad (quadrilateral), which ensures the accuracy of calculation and meets the requirements of computational efficiency. The grid number is about 120 thousand. The surface and ground boundary conditions are set as wall, the narrow slot outlet and inlet are set as velocity inlet, and the flow field boundary is set as pressure outlet[4].

Table 1. The parameters of 1,2,3,4,5,6 model

model	Size (length mm × width mm × height mm)	α	V_1 (m/s)	V_2 (m/s)	k
1	666×330×85	11°	5.56	20	0.076
2	516×330×85	17°	5.56	20	0.076
3	446×330×85	23°	5.56	20	0.076
4	400×330×85	30°	5.56	20	0.076
5	446×330×85	23°	5.56	20	0.059
6	446×330×85	23°	5.56	20	0.091

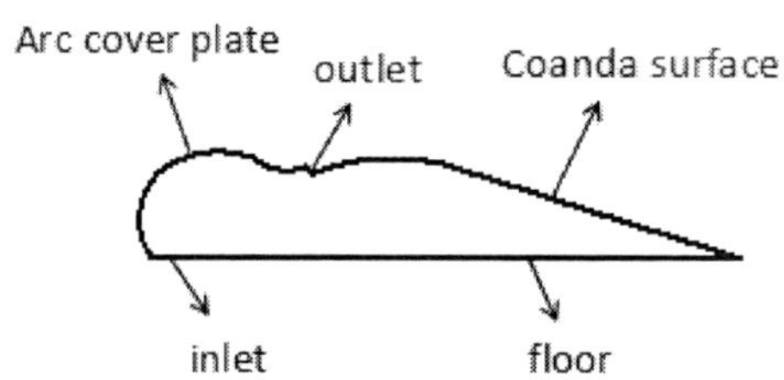

Figure 1. Physical model of ground dryer

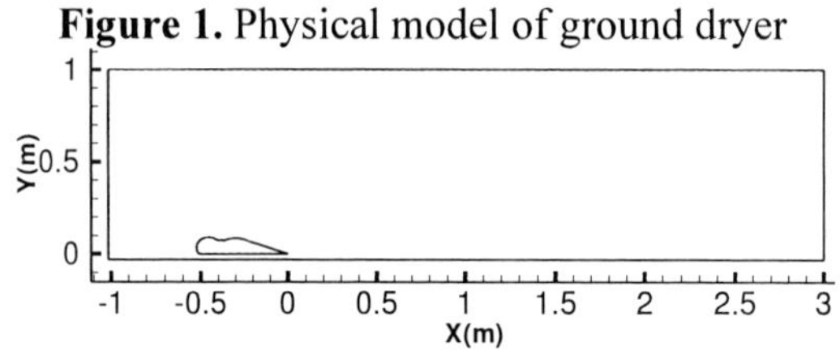

Figure 2. Computational domain

3.2. Simulation Results and Analysis

3.2.1 Effects of The Coanda's Surface 'S Inclination Angle and the Length on the Formation of Air Film and Airflow Characteristics

The selection of the Coanda's surface's inclination angle and model length has an important effect on the performance of the ground dryer[5]. In this paper, the model 1, 2, 3 and 4 in Table 1 were simulated with the outlet size being 8.4 mm. The influences of the model change on the airflow field and the formation of ground air film were observed and investigated in this paper.

(1) Analysis of velocity field

Figure 3 is the flow chart for model 1, which is similar to Models 2, 3 and 4. The air flow goes into the ground dryer through the air inlet, then sprays out from the narrow outlet, flows on the Coanda surface, induces the surrounding air to flow to the ground area, and thus continues to drive the surrounding air to flow forward. This is because the air flows on the Coanda surface flow, forms a negative pressure in the vicinity of the surface, then the surrounding air continues to supply the negative pressure area, in order to achieve the induction of small air volume.

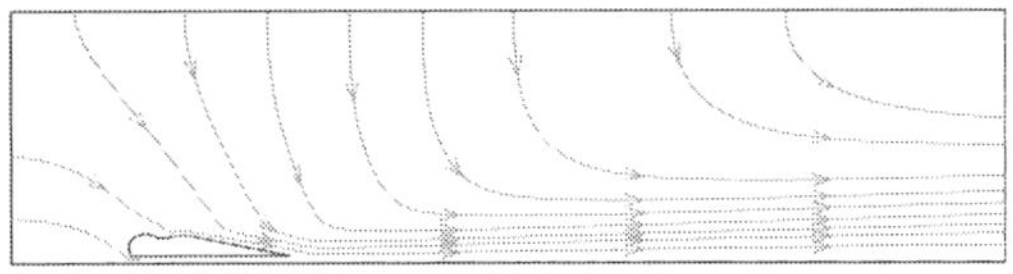

Figure 3. Model 1 streamline diagram

(2) Velocity analysis of equal cross section

The airflow that can effectively dehumidify the ground is the airflow near the ground. So the velocity distribution along the Y axis from the ground to the height of 0.1 m was analyzed. The distance in the positive direction along the X-axis from the ground to the tail of the dryer is set as d. The distance from the ground to the height of 0.1 m in the positive direction along the Y axis at d=0.33 m, 1.33 m and 2.33 m is shown in Figure 4. It can be seen from Figure 4 that when d=0.33 m, the velocity near the ground of model 3 is the largest, followed by model 4, model 2 and model 1. With the height increasing 0.06 m, the velocity of model 1 becomes the largest while the velocity of model 4 becomes the smallest. With the distance d increasing, the difference among the velocities of different models get more obvious, and the best velocity is ordered by model 3, 2, 1 and 4. Therefore, the ground air film of model 3 has the best effect.

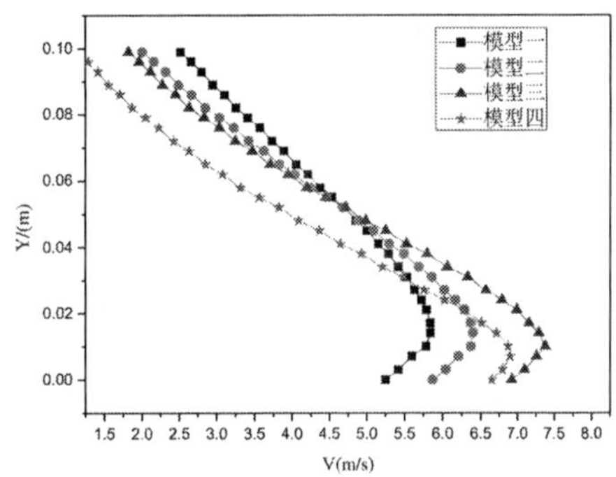

(a) d=0.33m

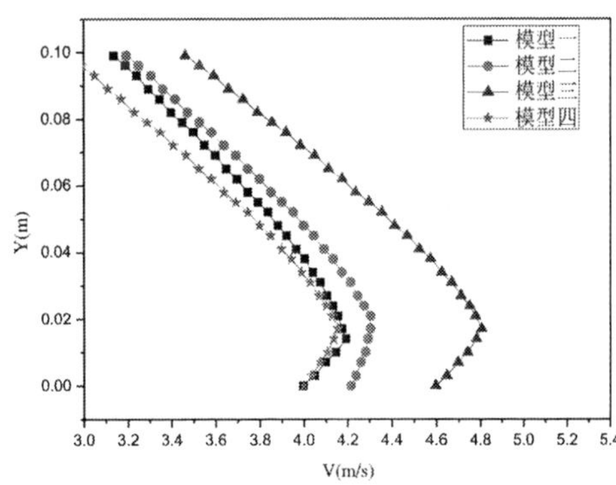

(b) d=1.33m

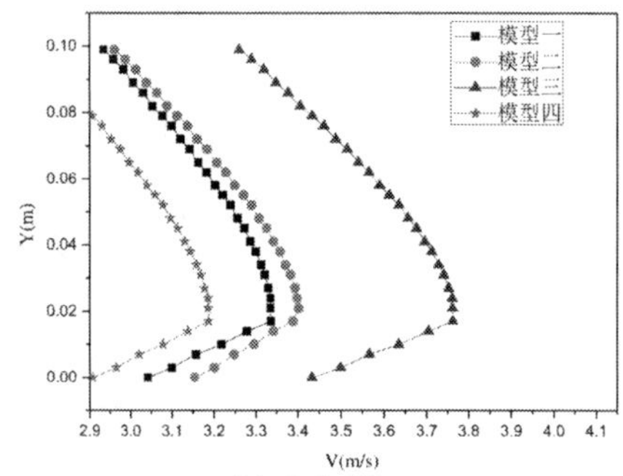

(c) d=2.33m

Figure 4. The velocity distribution of different models in different cross-sections along the positive Y-axis

3.2.2 Effects of Different Coanda Surface'S Curvatures on the Formation of Air Film and the Characteristics of Airflow

In order to observe the influence of Coanda's surface curvature on the flow field characteristics of the ground dryer, the air film formation and air flow characteristics of model 5 (k=0.059), model 6 (k=0.091) and model 7 (k=0.13) were calculated based on model 3 with the curvature being 0.076.

(1) Analysis of velocity field

Figure 5 shows the near-ground velocity skew of the ground dryer with the Coanda surface's curvature being k=0.076, and models 5, 6, 7 are similar to that. Figure 6 is the local airflow field map of the ground dryer under different Coanda surface curvatures. From Figure 5 and Figure 6, it can be seen that the velocity attached to Coanda surface is relatively higher; that the velocity becomes larger when getting closer to the Coanda surface; and that the stratification of the velocity region is obvious, with no irregularity in the high and low velocity regions. It can be seen from Figure 6 that the velocity distribution of the Coanda surface near the ground is related to the curvature of the Coanda surface. The velocity region of 14-18 m/s becomes wider when the curvature increases from k=0.059 to k=0.091, and then the region is narrowed when increased to k=0.13.

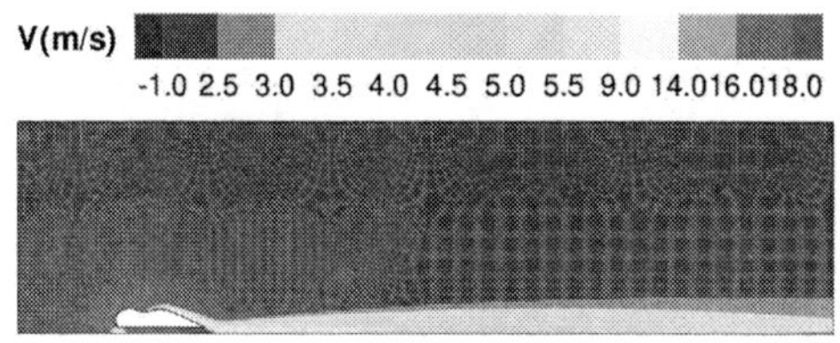

Figure 5. Near-ground velocity cloud map with different curvatures

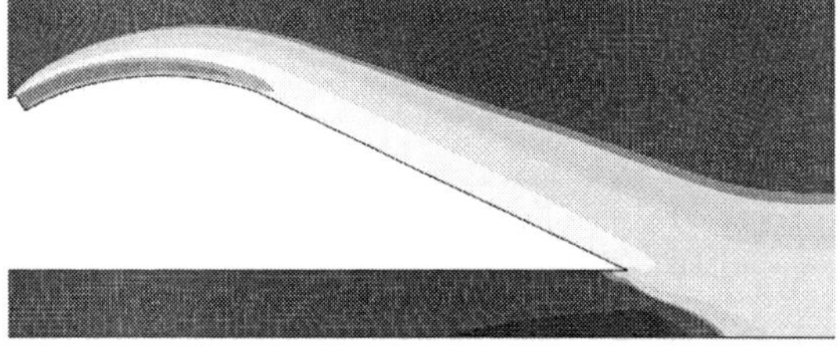

(a) k=0.059

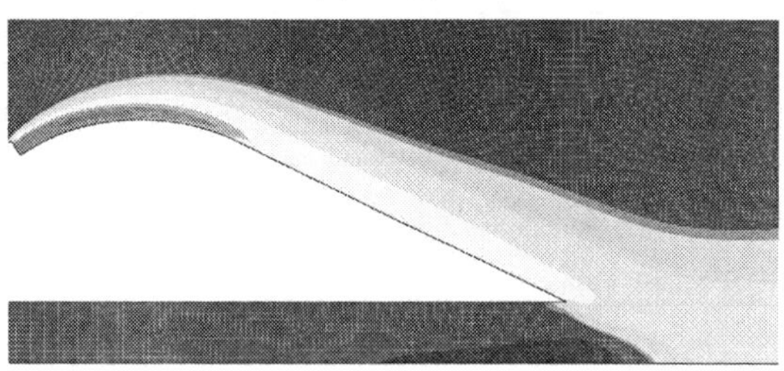

IOP Conf. Series: Materials Science and Engineering **398** (2018) 012024 doi:10.1088/1757-899X/398/1/012024

(b) k=0.076

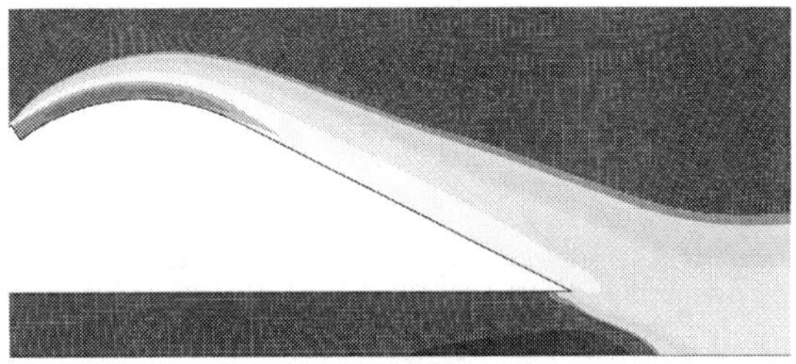

(c) k=0.091

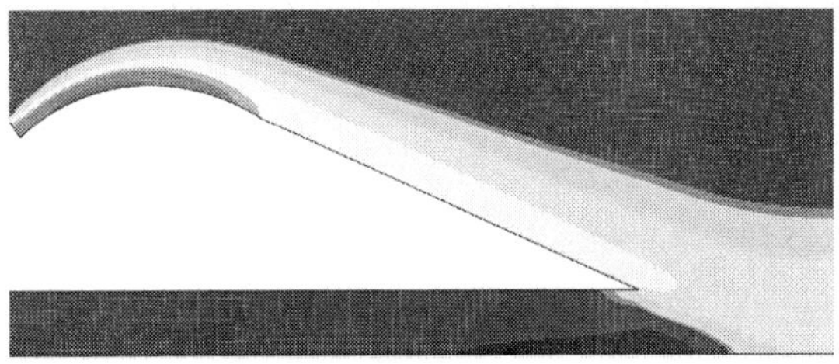

(d) k=0.13

Figure 6. Enlarged local velocity cloud map with different surface curvatures

(2) Analysis of cross-section velocity

The airflow that can effectively dehumidify the ground is the airflow near the ground. So the velocity distribution along the positive Y axis from the ground to the height of 0.1 m was analyzed. Figure 7 shows the velocity distribution of the cross sections at d=0.33 m, 1.33 m and 2.33 m from the ground to the height of 0.1 m along the positive Y axis. It can be seen from Figure 7, no matter how the height or distance increases, the near-ground velocity of model 6 has the maximum velocity, and the ground air film of model 6 has the best effect.

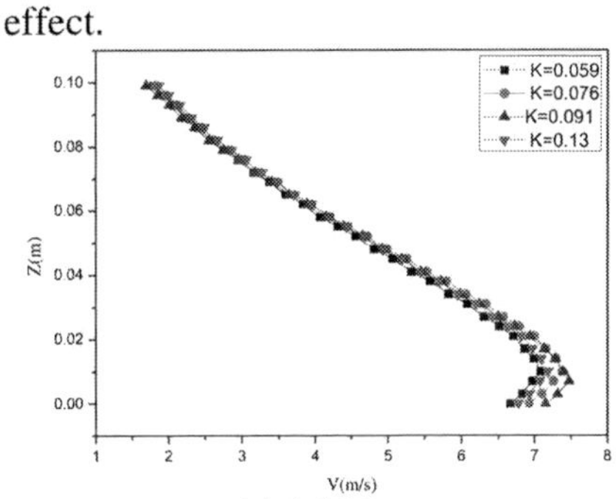

(a) d=0.33m

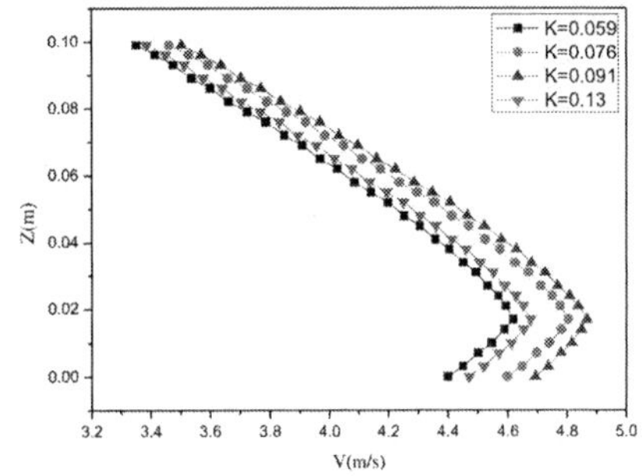

(c) d=2.33m

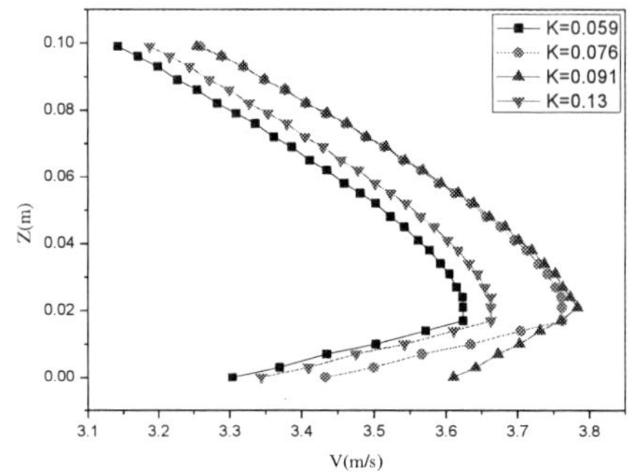

(b) d=1.33m

Figure 7. The velocity distribution of different cross sections along the positive Y-axis under different Coanda surface curvatures

4. Conclusion

The CFD technique was adopted to simulate the airflow characteristics of ground air film in a two-dimensional pattern. The air film was formed by ground dryer where the small air volume was used to induce the airflow to flow closely on the Coanda surface.

The two-dimensional numerical simulation results show that the change of the segment length, the inclination angle and the surface curvature of Coanda surface have great influence on the air film effect. As the segment length of the Coanda surface decreases and the inclination angle increases, the maximum velocity in the air film increases first and then decreases. In the four models, model 3 exhibits the best performance. With the increase of Coanda surface's curvature, the film effect becomes stronger and then weaker, and model 6 shows the best performance among the four models..

5. Reference

[1] Charpin D 2007 Moulds in indoor environments and respiratory diseases. *J. Rev mal respire* **24** 246-247

[2] Coanda H 1936 Device for deflecting a stream of elastic fluid projected into an elastic fluid. *J. U.S. Patent* **1936** 205-286

[3] Guoq Li 2014 Influence of coanda surface curvature on performance of bladeless fan. *J. Journal of thermal science* **23** 422-431

[4] Junyou Huang and Huaming Yang 2013 Research of lampblack machine optimization. *J. Journal of mechanical & electrical engineering* **11** 1388-1392

[5] Shiyu Zhang, Yicheng Zhong and Pengzhe Fu 2011 Numerical investigation of dual-radius in circulation control airfoil. *J. Aircraft design* **4** 1-6

The Application of GA-BP Algorithm in Prediction of Tool Wear State

J TANG[1], W X Li[1] and B ZHAO[2]

[1]Department of mechanical and electrical engineering, Xinxiang University, Xinxiang 453003, China,
[2]School of Mechanical and Power Engineering, Henan Polytechnic University, Jiaozuo, 45400, China

E-mail:TangJun13915929095@163.com

Abstract. In CNC shaping milling machine, the prediction of the state of tool wear has important application significance to improve productivity, reduce scrap rate and avoid security risks. In this paper, the detection and control system of disk milling cutter is set up by the current monitoring method, the input characteristic quantity and target characteristic quantity of BP neural network for tool wear diagnosis are measured, and the disk milling cutter wear condition prediction neural network is established based on the GA-BP algorithm. At last, the online prediction of milling cutter wear state is realized. The network test results show that the prediction rate of tool wear condition is more than 92.78%.

1. Introduction

At present, the domestic and foreign scholars have done a lot of research on monitoring tool wear. According to the principle of cutting tool wear detection, it can be divided into direct and indirect monitoring [1]. The indirect monitoring method mainly includes cutting force method, vibration signal method, acoustic emission method and power method [2].Compared with direct detection method, the indirect monitoring is easier to realize on-line monitoring, so it is more and more widely used in engineering practice. Among them, the current monitoring is a kind of power detection method, which is mainly based on the relationship between tool wear, cutting force, motor torque, motor power and motor current.

The cutting tool state signal is the only carrier to reflect the cutting state [3]. The domestic and foreign scholars mainly focus on the tool wear signal processing methods which include wavelet analysis, neural network, and threshold judgment and information fusion technology [4]. The neural network mainly through some basic functions of simulation of biological neurons (i.e., neurons), a neural network according to different links organized, so as to realize the fault detection and diagnosis, data feature extraction and information prediction function. Because of the influence of many factors on the wear state of the disc shaped milling cutter, and these factors are nonlinear and fuzzy, the neural network technology has a unique advantage. Based on B-spline function, the B-spline fuzzy neural network is used to calculate the tool wear, and the problems of slow learning speed, low accuracy and low recognition accuracy were solved [5]. In 2006, Onder Yumak put forward a kind of online monitoring method of tool wear which based on the adaptive fuzzy inference. The fuzzy rules are used to study the structure of neural network, in order to determine the optimal value [6]. Chen Jie got the input variables of BP neural network, established the tool wear diagnosis model of the disc shaped

IOP Conf. Series: Materials Science and Engineering **398** (2018) 012025 doi:10.1088/1757-899X/398/1/012025

milling cutter, and realized the online prediction function of disc forming cutter [7].BP neural network tool wear prediction model of which prediction error is not more than 5.4% is established, and the cutting parameters of cutting tool wear were optimized with the minimum tool wear, based on the genetic algorithm. The optimization results show that the cutting parameters are reduced 6.734% [8].

2. Design of control system for disk shaped milling cutter

In order to monitor the wear condition of the cutter, the hardware platform of the monitoring system is needed to be constructed. The hardware platform includes the processing platform, the current sensor, the data acquisition device, the computer and so on. The main function of the hardware platform of the system is to collect the spindle motor current signal through the sensor, and then the analog current signal is converted into a digital signal which can be identified by the data acquisition card. At last, the signal is transmitted to the computer through the data bus. Current sensor type is YDF-I-A4-P1-04. The data acquisition card is Advantech USB471. Data bus is USB. The detection system software is Lab VIEW. The overall structure of the hardware is shown in Figure 1.

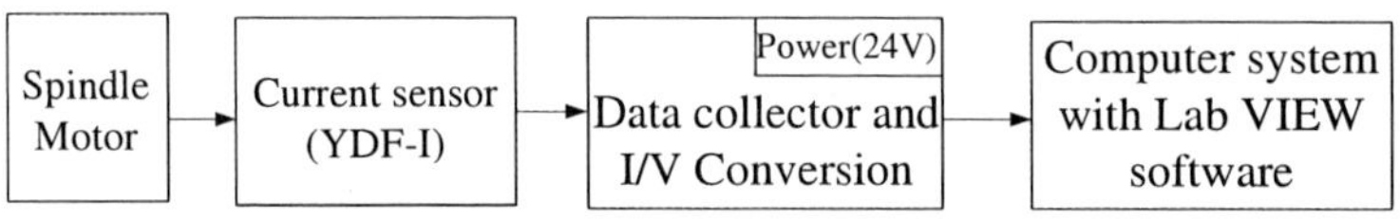

Figure 1. The blade figure of disk milling cutter

Based on the virtual instrument software Lab VIEW developed by NI Company, the wear current detection system of disc forming cutter is developed. The detection system is mainly composed of the equipment selection, channel setting, signal display and preservation, signal analysis, print report forms and alarm components.

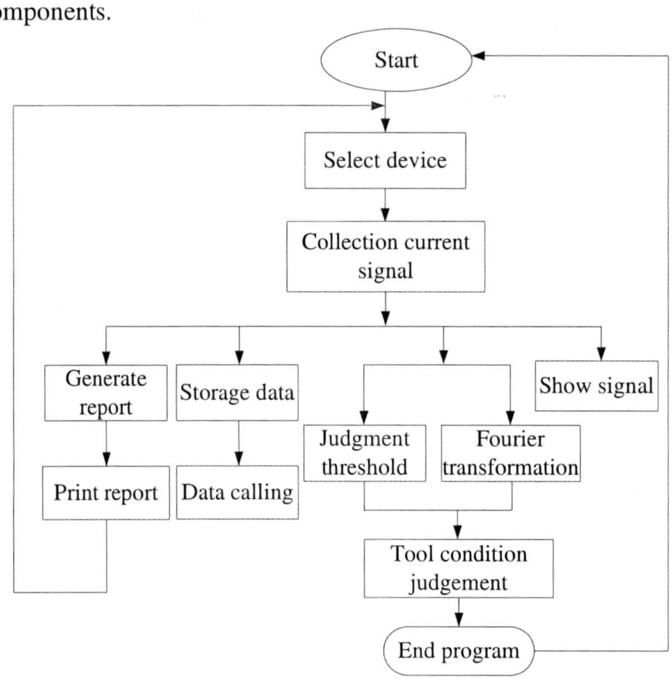

Figure 2. Test system software flow chart

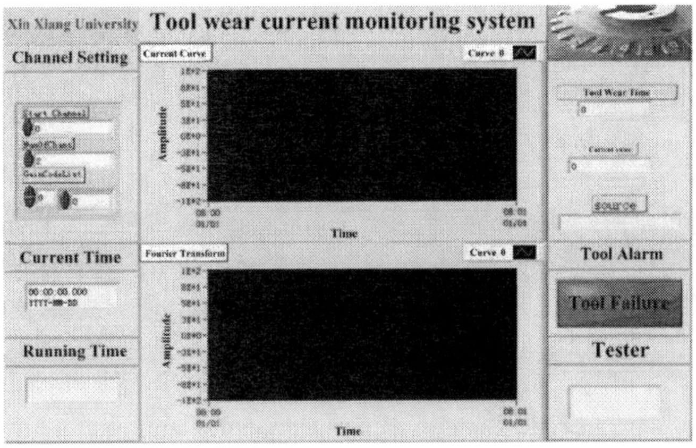

Figure 3. Test system front panel diagram

3. Wear prediction modeling of disk shaped milling cutter

3.1. GA-BP neural network theory

Based on the effect on the chromosome number string, the genetic algorithm (GA) is coded. Then, a group of strings are used to simulate the biological genetic and evolutionary optimization, so as to realize the global search [9]. The Back Propagation algorithm is a feed forward network. Its essence is through the dynamic adjustment of the weights and thresholds to achieve the input and output nonlinear mapping function of height. However, when the initial weight is not suitable, the network will be trapped in local minimum and the convergence rate is too slow [10]. According to the shortcomings of BP neural network, GA algorithm is used to optimize the initial weights and threshold values of input layer, hidden layer and output layer. Furthermore, the wear identification model of GA-BP disc forming cutter is established.

Figure 4. The test site

3.2. Input and target feature vector selection

The Figure 4 is the gear ring cutting test site. In this experiment, the milling machine tool is SKXC-2500W-C. The machining gear parameters are as follows. the work piece material is 50Mn. the surface hardness of gear is HB206-262. The modulus of the gear is 12mm. The number of teeth is 118. the pressure angle is 20°. Tooth height is 100mm. The coefficient of variation is +5mm. Kenna CNC blade model is KC935M. The disc cutter speed is 9r/min, and the feed rate is 140mm/min.

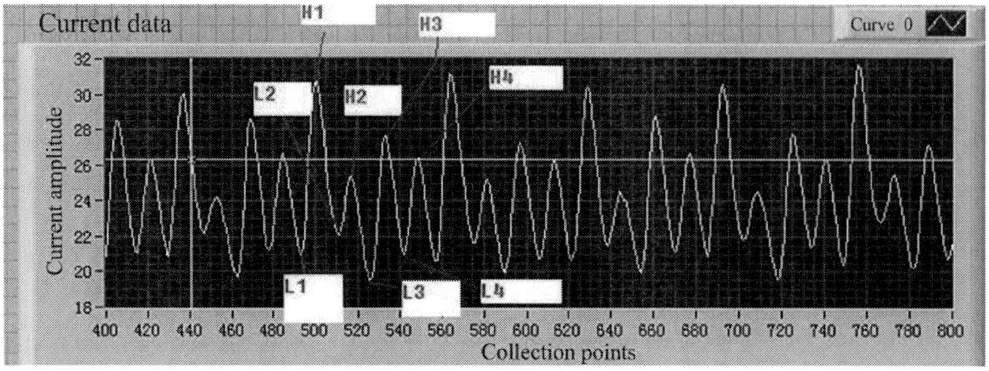

Figure 5. The distribution figure of feature vector

Figure 5 is the distribution of the characteristic vector of the current waveform. In the Figure, the L1, L2, L3, and L4 represent the current value of the lowest point of the current and the next group cutter, the H1, H2, H3 and H4 represent the highest point of the current and the next group cutter. In this paper, L1, H1, L2, H2, L3, H3, L4, H4 are used as input feature vectors of BP neural network. According to the relevant theory of literature 11, this paper defines the formation of the milling cutter as three states which are the primary abrasion (VB=0~0.2mm), normal abrasion (VB=0.2~0.5mm), and sharp abrasion (VB > 0.5mm). In this paper, three kinds of wear states of milling cutters are selected as the feature vectors of BP neural network.

3.3. Sample data acquisition and processing

According to the above experimental conditions, the application of YDF-I-A4-P1-04 current sensor and advantech USB4711 data acquisition card is used to obtain the current waveform of disc cutter in the milling process. Then, the tool wear at different wear time was measured with the ultra depth of field microscope VHX-2000. Because the input variables of BP neural network must belong to [0, 1], it is necessary to normalize the input variables.

$$x' = (x - x_{\min})/(x_{\max} - x_{\min}) \tag{1}$$

In the formula, x' represent the normalized sample data. $x_{\max}$ represents the collected sample data. represents the maximum value of sample data. $x_{\min}$ represents the minimum value of sample data. Table 1 is the normalization of the current amplitude. Table 2 is the normalized test sample data.

Table 1. The normalized sample data of disk milling cutter

NO	L1	H1	L2	H2	L3	H3	L4	H4	Wear state
1	0.2537	0.12799	0.03647	0.048	0.03049	0.04616	0.00326	0.13136	Primary abrasion
2	0.23880	0.072	0.03951	0	0.02711	0.08392	0.07843	0.23306	
3	0.20149	0	0.04599	0.088	0.02372	0.06085	0.00981	0.12711	
4	0.16418	0.00801	0.02127	0.10134	0.07796	0.06504	0.00654	0.19915	
5	0.27612	0.06398	0	0.072	0	0.02518	0.04576	0.33899	
6	0.16791	0.48	0.36170	0.31467	0.18983	0.67768	0.31698	0.42796	Normal abrasion
7	0.1753	0.568	0.35562	0.42667	0.16949	0.61684	0.29085	0.38984	
8	0.14179	0.62401	0.31307	0.41067	0.21356	0.58746	0.26798	0.37712	
9	0.13805	0.51992	0.29787	0.34934	0.21694	0.69237	0.31046	0.43221	
10	0.17164	0.54	0.32523	0.35466	0.17288	0.61893	0.3333	0.37712	
11	0.89551	0.812	0.95137	0.94134	1	0.95043	0.97712	0.73729	Sharp abrasion
12	0.86566	0.80399	0.92401	1	0.96949	0.91476	0.92808	0.86018	
13	0.87686	0.78801	0.89665	0.86399	0.94238	0.99659	0.91503	0.87712	
14	1	1	1	0.90933	0.91865	0.91267	0.91176	0.79237	
15	0.87312	0.89196	0.93920	0.99734	0.98983	0.98819	1	1	

Table 2. Test datas

NO.	L1	H1	L2	H2	L3	H3	L4	H4	Wear state
1	0.2276	0.0200	0.1337	0.0613	0.0102	0	0	0.1992	primary abrasion
2	0.1119	0.6880	0.4316	0.4320	0.2475	0.5812	0.2614	0.4153	Normal abrasion
3	0.8209	0.8880	0.9088	0.9277	0.9831	1	0.9673	0.8729	Sharp abrasion

3.4. The design of neural network model

GA-BP neural network algorithm mainly includes three parts. They are the creation of BP neural network, genetic algorithm to optimize the initial weights and thresholds of BP neural network, and the BP neural network training and prediction [12].

In this paper, a three layer BP neural network with a hidden layer is selected. On the one hand, the three layer BP neural network, which contains a hidden layer, can approach any rational function in theory. Increasing the number of layers can further reduce the error and improve the accuracy, but it can make a complex network, increase the training time. On the other hand, the error precision can be increased by the number of hidden layer nodes to achieve the training effect than increasing the number of layers is easier to observe and adjust. Therefore, the BP neural network with a hidden layer is usually considered.

From the previous section, The BP neural network input layer has 8 neurons, and the output layer has 3 neurons. Therefore, the number of neurons in the hidden layer is 17=2×8+1. The number of weights is 187=8×17+17×3. The threshold number is 20=17+3. The topology of BP neural network is shown in Figure 6.

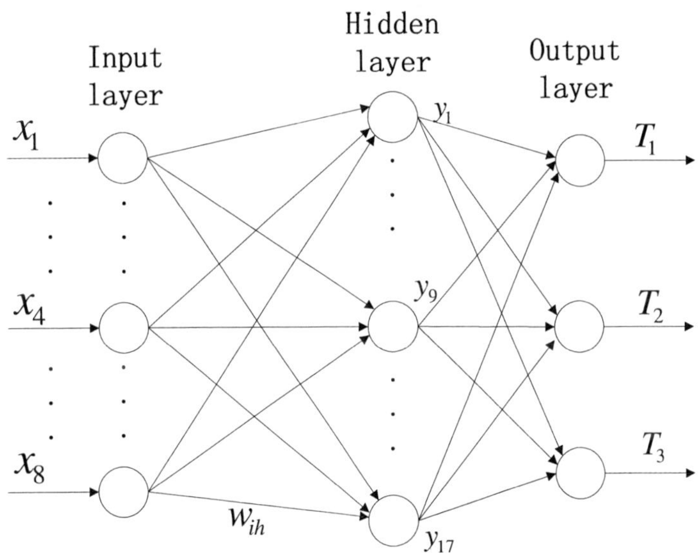

Figure 6. BP neural network topology

The transfer function of hidden layer is the tangent function Tansig (), and the transfer function of output layer is logarithmic function logsig ().The network is trained by using the trainlm() function Levin Berg Marquardt algorithm. The output mode of network is 0 or 1, and the tool wear condition corresponding to the primary abrasion (1, 0, 0), normal abrasion (0, 1, 0), and sharp abrasion (0, 0, 1). BP neural network code is as follows.

```
threshold=[0 1;0 1;z 1;0 1;0 1;0 1;0 1;] % Setting the threshold range
net=newff(threshold,[17,3],{'tansig','logsig'},'trainlm') % Creating The BP neural network
net.trainParam.epochs=1000 % Setting the training times
net.trainParam.show=2 % Setting the display training results frequency
```

net.trainParam.mc=0.9 % Setting the momentum coefficient
net.trainParam.goal=0.01 % Setting performance goals
LP.lr=0.15 % Setting the learning rate
net=train(net,P,T) % Training BP neural network
P and T are the input vector and the target vector. P is the data in table 3.
Genetic algorithm optimization BP neural network mainly includes population initialization, fitness value calculation, individual choice, individual crossover and individual variation. The key part of genetic algorithm code is as follows.
ObjV=sim (net, (bs2rv (Chrom， FieldD))') % Setting objective function
FitnV=ranking (-ObjV) % Setting individual fitness function
SelCh=select (sus, chrom, FitnV, GGAP) % Setting the individual choice rule
SelCh=recombine (xovsp， SwlCh， 0.8) % Setting individual crossover probability
SelCh=mut (SelCh) % Setting probability of individual variation
variable=bs2rv (SelCh， FieldD) % Decimal conversion of offspring
In the Matlab environment, by using the GATBX genetic algorithm toolbox, GA-BP neural network algorithm is established. The algorithm flow is shown in Figure 7.

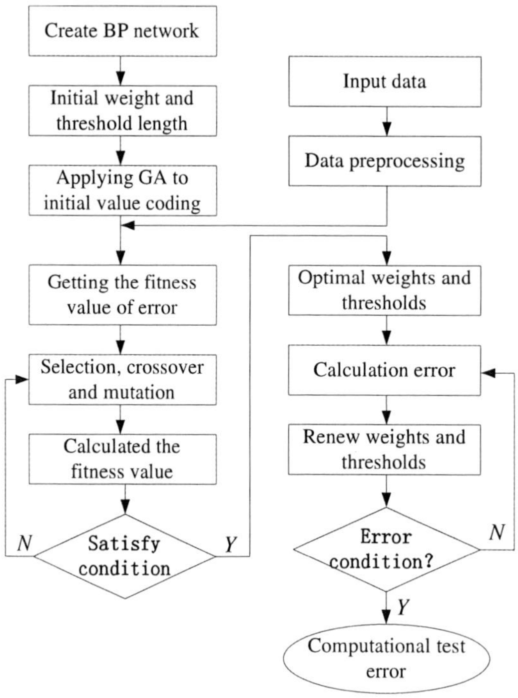

Figure 7. The process for genetic algorithm to optimizing BP neural network algorithm

4. Neural network network training and test results analysis

Based on the 15 sets of normalized sample data in the Table 1, the GA-BP neural network is trained to establish the relationship between the input feature vector and the tool wear state. The weights and thresholds of the training error curve are shown in Figure 8.

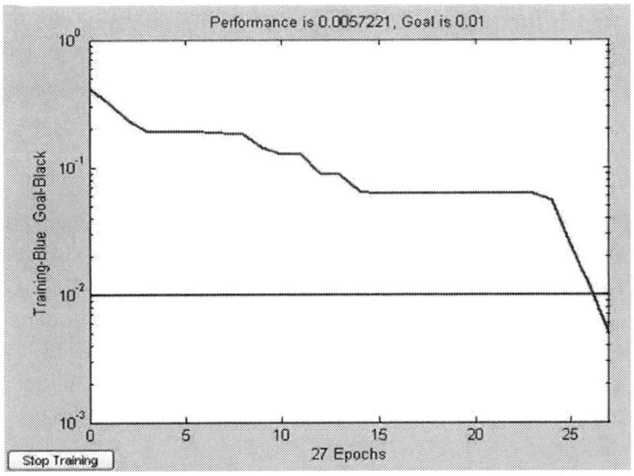

Figure 8. The sketch of training error convergence

In the Figure 8, the fitting accuracy of GA-BP neural network model can reach 0.0057221. The accuracy can meet the needs completely. Then, the 3 groups of new sample data are selected to test the trained GA-BP neural network. The test data are shown in Table 2. Test results and diagnostic rates are shown in Table 3. As can be seen from Table 3, the results show that the tool wear state is consistent with the actual state, and the prediction rate is above 92.78%.

Table 3. Test results

NO.	Ideal output				Actual output				accuracy rate
1	1	0	0	0	0.9945	0.0004	0.0008	0.0077	99.17
2	0	1	0	0	0.0053	0.9634	0.0523	0.0105	95.97
3	0	0	1	0	0.0141	0.0554	0.9640	0.1193	92.78
4	0	0	0	1	0.0045	0.0188	0.0136	0.9982	99.21

5. Conclusion

In the practical application of the disc milling cutter, it is impossible to achieve online wear state prediction. In this paper, based on the current detection method, the Lab VIEW software is used to build the detection system of the disc milling cutter. The neural network normalized training samples of the input features and the target features are obtained. The BP neural network tool wear prediction system is used to be optimized by using the genetic algorithm. The results show that the GA-BP neural network can predict the milling cutter wear state by more than 92.78%, and have a good practical application value.

6. References

[1] Rehom A G, Jiang J, Orban P E. State-of the-art methods andresults in tool condition monitoring: a review [J]. International Journal of Advanced Manufacturing Technology, 2005, 26(7-8): 693-710.

[2] Zhang Chengqiang. The Research of Tool Wear of Disk Milling Cutter [D]. Nanjing: Nanjing University of Technology, 2010.

[3] Qiu Yaner. A preliminary discussion on the monitoring method of tool wear of NC machine tools [J]. Journal of changchun university of science and technology (Higher Education Edition), 2010, 5(4): 180-181.2010.

[4] Zeng Xiangchao, Chen Jie. Study of data processing method for CNC tool wear monitoring [J]. Machinery Design & Manufacture. 2009 (01):18-24.

[5] Gao Hongli, Xu Mingheng, Fu Pan. Tool wear monitoring based on integrated neural networks [J]. Journal of southwest jiao tong university. 2005, 40(5):641- 644.

[6] Yumak O, Ertunc H M. Tool Wear Condition Monitoring in Drilling Processes Using Fuzzy Logic[J]. Lecture Notes in Computer Science, 2006,4234:508-517.

[7] Chen Jie, Zhang Chengqiang. Research on tool wearing diagnosis of disk milling cutter based on BP neural network [J]. Manufacturing technology and machine tools, 2010, 12:34-37.

[8] Qin Guohua, Xie Wenbin, Wang Huamin. Detection and control for tool wear based on neural network and genetic algorithm [J]. Optics and Precision Engineering, 2015, 23(5): 1314-1321.

[9] Tang Jun, Huang Xiaodiao, Fang Chenggang. Optimal design of faceplate structure based on BP neural network and genetic algorithm [J]. Machinery Design & Manufacture, 2011, 11: 27-29.

[10] Li Dasheng, Miao Pengcheng. Application of GA-BP neural network in CNC machine monitoring and control system [J]. Journal of Hunan University of technology, 2013, 27(3):65-69.

[11] Lu Jianzhong, Sun Jianing. Metal cutting theory and cutting tools [M]. Beijing: Machinery Industry Press, 2011:115-139.

[12] Ren Xiaohong, Xu Weidong, Liu Lixin, et al.Thermal error compensation on cnc machine tools based on GA-BP neural network[J]. Manufacturing Automation, 2011, 33 (9):41-43.

[13] Wen Xin, Zhou Lu, Wang Danli, et al. MATLAB neural network application design [M]. Beijing: Science Press, 2000.

Vibration Response of the Launcher under Impact Load

Huazheng Wang[1], Yi Jiang[1], Miao Chen[1], Wei Liu[1]

School of Aerospace Engineering. Beijing Institute of Technology. Beijing, China

E-mail: 805491257@qq.com

Abstract. Vibration of the launcher would cause the initial disturbance of the missile in the launch process. In this paper, the rigid-flexible coupling system model was built on the platform of virtual prototype technology (VPT) based on the improved Craig-Bampton method. Numerical simulation data and experimental results are found to be consistent with each other. Rigid-flexible coupling system was verified to be able to provide valuable results to the research on kinetic, dynamic and vibratory characteristics. Vibration responses of girder and canister under different launching angle are simulated, and results show that increasing launching angle could reduce the vibration of canister and girder.

1. Introduction

The vibration of the launching system would cause the non-zero horizontal angle and angular velocity of the missile, which is an important factor of the initial disturbance of the missile[1]. Elastic deformation has an important effect on the dynamic response of the launcher. Rigid-flexible coupling model would be more accurate in dynamic analysis[2,3].

In the 1960s, based on Hurty and Guyan's idea, Craig and Bampton proposed a component mode synthesis method named the Craig-Bampton (BC) method[4]. However, the BC method is not suitable for the bodies that have large-scale movement and the assumed modes sets can not be in correspondence with the frequency[5,6]. Thus, the improved BC method was proposed by considering the flexible body as substructure.

In this paper, the multi-rigid-body system model and rigid-flexible coupling system model were built on the platform of virtual prototype technology (VPT), based on improved Craig-Bampton method. Simulation data and experimental results are found to be consistent with each other and the rigid-flexible coupling model was more accurate than the multi-rigid-body model. Rigid-flexible coupling system was verified to be able to provide valuable results to the research on kinetic, dynamic and vibratory characteristics. Vibration responses of girder and canister under different launching angle are simulated, and the results show that increasing launching angle could reduce the vibration of canister and girder.

2. Craig-Bampton method and improved Craig-Bampton method

2.1. Original Craig-Bampton method

In the CB method, the linear dynamics equations can be expressed by

$$M_g \ddot{u}_g + K_g u_g = f_g \tag{1}$$

$$M_g = \begin{bmatrix} M_s\, M_c \\ M_c^T\, M_b \end{bmatrix}, \quad K_g = \begin{bmatrix} K_s\, K_c \\ K_c^T\, K_b \end{bmatrix}$$

$$u_g = \begin{bmatrix} u_s \\ u_b \end{bmatrix}, \quad f_g = \begin{bmatrix} f_s \\ f_b \end{bmatrix}$$

where u_g and f_g represent global displacement and force vectors, and M_g and K_g are the global mass and stiffness matrices. The subscript g denotes the global structure. The subscript s , b and c represent substructure, boundary interface and coupling matrices. M_s and K_s are the block diagonal mass and stiffness matrices of substructures.

The global eigenvalue problem is

$$K_g(\phi_g)_i = \lambda_i M_g(\phi_g)_i, \quad i = 1,2,\ldots,N_g \tag{2}$$

where λ_i and $(\phi_g)_i$ are the eigenvalue and eigenvector calculated in the global structure, respectively, and N_g is the number of DOFs of the global structure.

Using the eigenvectors obtained from Equation(2), the global displacement vector u_g is represented by

$$u_g = \Phi_g q_g \tag{3}$$

where Φ_g and q_g are the global eigenvector matrix and its generalized coordinate vector, respectively. The method assume that the body has no large-scale movement, and the rigid displacement of substructures is included in multi-body dynamics. So the BC method is not suitable for the bodies that have large-scale movement. The constraint modes of substructure analysis are the result of static condensation, so the dynamic frequency response of flexible body can not be reflected. The modes of flexible bodies can not be correspondence with the frequency. However, the improved CB method can solve the problem.

2.2. *improved BC method*

Considering each flexible body as a substructure, the equation(3) can be obtained from the original BC method. The structural dynamics equation of substructure can be presented by

$$m_i\ddot{u}_i + c_i\dot{u}_i + k_i u_i = R_i \tag{4}$$

where m_i , c_i , k_i and R_i are the mass, damping, stiffness and external force matrix.

Substituting equation(3) to equation(4) and considering the no damping free vibration, the structural dynamics equation of substructure can be presented by

$$M\ddot{p} + Kp = 0 \tag{5}$$

$$M = \Phi^T m\Phi, \quad K = \Phi^T k\Phi$$

a regularized modes N can be obtained by solving the global eigenvalue problem. The original modes coordinate vector can be presented by

$$q = Np \tag{6}$$

The global displacement vector can be presented by

$$u = \Phi q = \Phi[N_1\, N_2 \cdots N_S] \begin{bmatrix} q_1 \\ \vdots \\ q_s \end{bmatrix} = [\varphi_1\, \varphi_2 \cdots \varphi_s] \begin{bmatrix} q_1 \\ \vdots \\ q_s \end{bmatrix} = \Psi q \tag{7}$$

where φ_i is regularized modes coordinate vector.

The regularized modes of fixed interface is substituted by the eigenvector of the unconstrained body, and the constraint mode is replaced by the interface feature vector. Then the high frequency modes can be selected because the modes sets can be correspondence with the frequency.

3. Experiments and numerical simulation models

IOP Conf. Series: Materials Science and Engineering **398** (2018) 012026 doi:10.1088/1757-899X/398/1/012026

3.1. Multi-rigid-body model

We built the multi-rigid-body model shown in Figure 1. The model was comprised of canister, landing gear, girder, oil cylinder and so on. Constraint relation of parts in the model was shown in Figure 2. The girder was placed on the support springs and leveling oil cylinders, and 2 tons of impact load was acting on the vibration table in -y direction. Three measurement points were built on the model, and the displacement and velocity of the points were monitored through simulation.

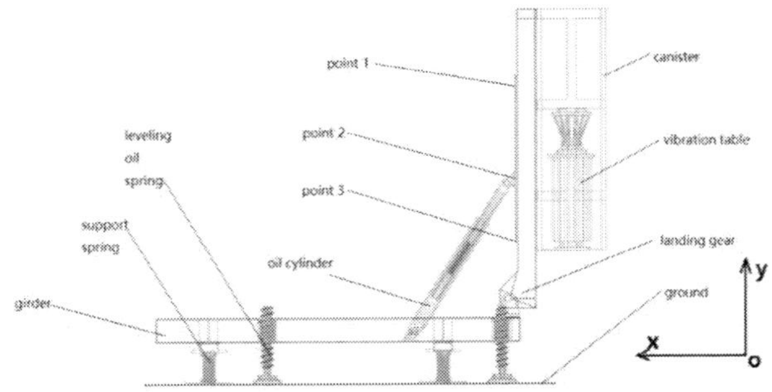

Figure 1. Multi-rigid-body model

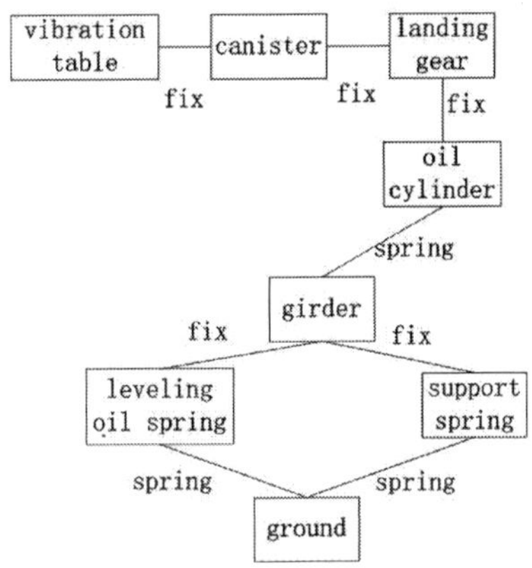

Figure 2. Constraint of parts

3.2. Rigid-flexible coupling model

The rigid-flexible coupling model was comprised of canister, landing gear, girder, oil cylinder and so on as figure 1. For the rigid-flexible coupling model, the canister, landing gear and girder were flexible.

Based on the modal synthesis method, the canister, landing gear and girder were flexible in ANSYS software[7][8][9]. Considering the vibration frequency of the model is small, low order modes are mainly used in simulation. Figure 3 shows the girder model and the mainly used modes of girder.

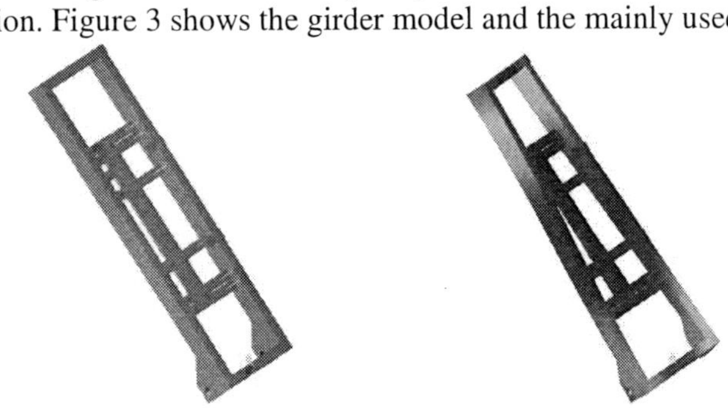

(a) girder model (b) seven order mode

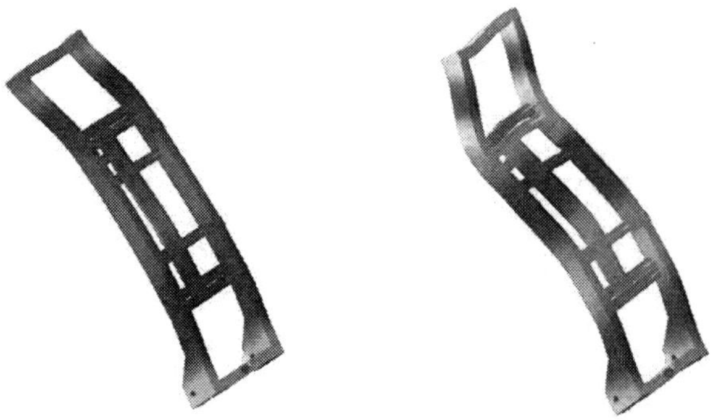

(c) nine order mode (d) eleven order mode

Figure 3. Girder model and mainly used girder modes

3.3. Experiments

The experiments were comprised of canister, landing gear, girder, oil cylinder and so on as shown in figure 1. The launcher was placed on the support springs and leveling oil cylinders, and 2 tons of impact load was acting on the canister with the small vibration table. Three measurement points were built as same as the multi-rigid-body model, and the displacement and velocity of the points were monitored by high speed photographer.

Figure 4 shows the small vibration table of VT series. The small vibration table was used for the impact load.

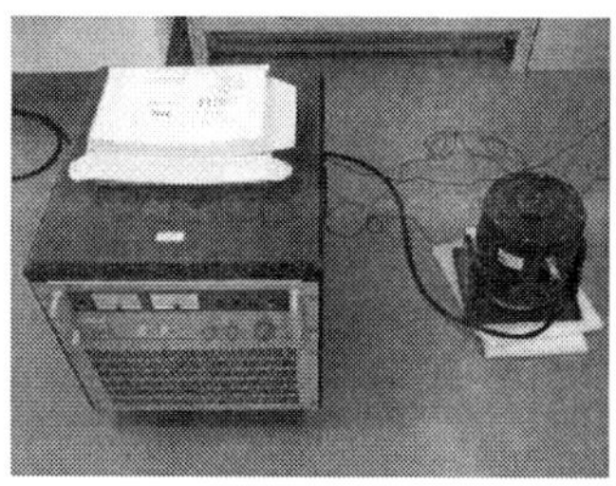

Figure 4. Experiments and experimental equipment

4. Comparison of experiment and simulation data

Launching system models were built based on rigid-flexible coupling method under the launching angle of 60°, 70°, 80° and 90°. Launching system model under the launching angle of 80° is shown in figure 5. Vibration response is shown in figure 6-9.

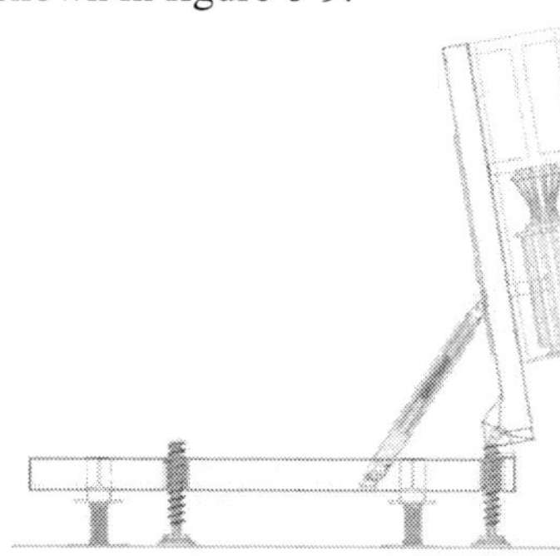

Figure 5. Launching system model under the launching angle of 80°

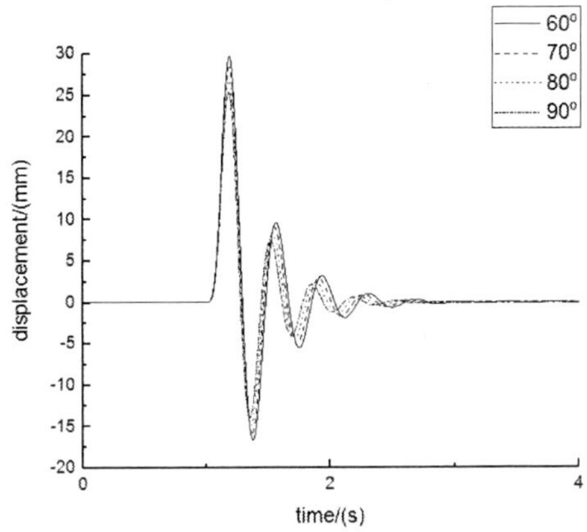

Figure 6. Displacement of canister in x direction

Figure 6 shows the displacement of canister mass center in x and y direction. The amplitude reduces and period of vibration shortens as the launching angle increases. When the launching angle is 60°, the maximum amplitude of canister mass center in x direction is 29mm. When the launching angle is 90°, the maximum amplitude of canister mass center in x direction is 23mm.

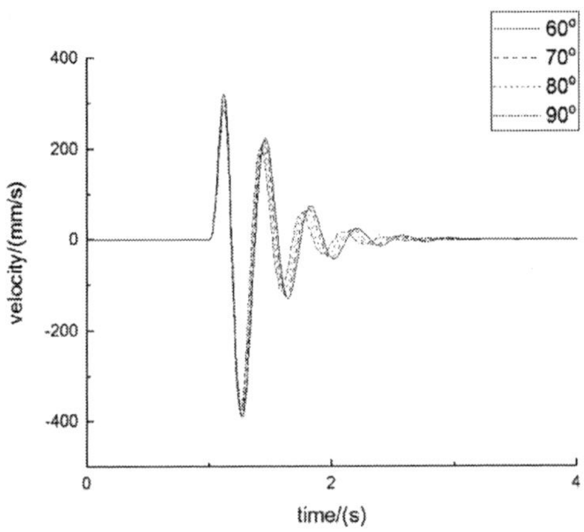

Figure 7. Velocity of canister in x direction

Figure 7 shows the velocity of canister mass center in x and y direction. The amplitude reduces and period of vibration shortens as the launching angle increases. When the launching angle is 60°, the maximum amplitude of canister mass center in x direction is 394mm/s, When the launching angle is 90°, the maximum amplitude of canister mass center in x direction is 376mm/s.

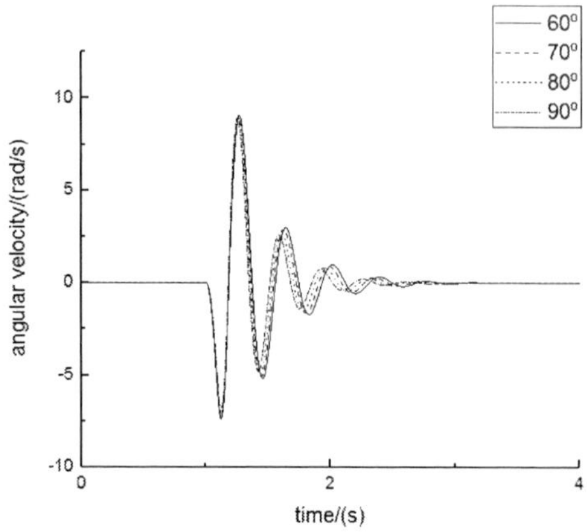

Figure 8. Angular velocity of canister in z direction

Figure 8 shows the pitch angular velocity of canister mass center in z direction. The maximum amplitude under different launching angle is close, and the amplitude value is 9rad/s. But the vibration decays faster as the launching angle gets larger.

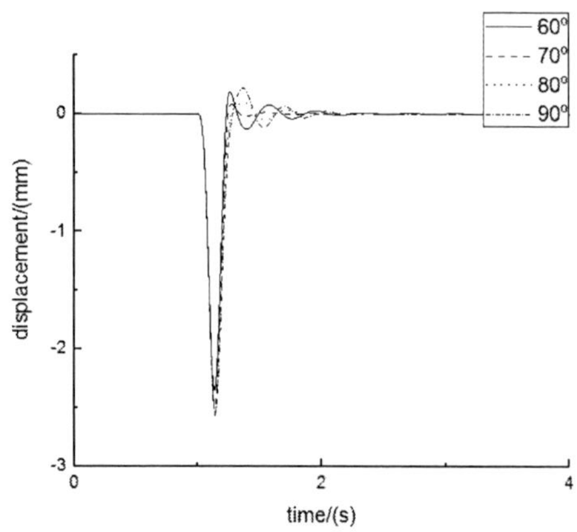

Figure 9. displacement of girder in y direction

Figure 9 shows the displacement of girder mass center in y direction. The maximum amplitude under different launching angle is close, and the amplitude value is 2.6mm. But the vibration decays faster as the launching angle gets larger.

5. Conclusion

The multi-rigid model and rigid-flexible coupling model were built based on improved Craig-Bampton method. By comparison of simulation data and experimental data, it is verified that rigid-flexible model is suitable for analyzing vibration of launcher. Then launching system models were built based on rigid-flexible coupling method under different launching angle. Concerning the canister, the displacement amplitude reduces and period of vibration shortens as the launching angle increases, and the pitch angular velocity amplitude under different launching angle is close, but the vibration decays faster as the launching angle increases. Concerning the girder, the maximum amplitudes under different launching angles are close, but the vibration decays faster as the launching angle gets larger.

6. References

[1] Zhang Zhongli, Dang Yugong, Zhou Mingli, "Simulation Study of Vibration Characteristic for Rocket Launch Container Based on Rigid and Flexible Coupled[J]," Journal of System Simulation. 2012, 24(11), 2249-2254.

[2] Hong Jiazhen, Jiang Lizhong, "Flexible Multibody Dynamics with Coupling Rigid and Deformation Motions[J]," Advances in Mechanics. 2000, 30(1), 15-20.

[3] AI-Bedoor B O, Almusallam A A, "Dynamics of flexible-link and flexible-joint manipulator carrying a payload with rotary inertia [J]," Mechanism and Machine Theory (S0094-114X), 2000, 35(6): 785-790.

[4] Bampton M C C, CRAIG, JR R R, "Coupling of substructures for dynamic analyses[J]," AIAA Journal, 1968, 6(7): 1313-1319.

[5] GAO Xing-dou, BI Shi-hua, CHEN Zhen, "Flexible multi-body dynamics analysis during missile launching based on improved Craig-Bampton method[J]," Journal of Solid Rocket Technology, 2011, 34(5): 559-563.

[6] Feng-yan D, Xing-suo H E, ZHANG J, "Application of a modified Craig-Bampton method in dynamics modeling of multi-body system [J]," Journal of Machine Design, 2004, 3: 013.

[7] SHI Zhen-qiang, XU Pei-min, "Dynamic Modeling of Coupling Rigid-flexible Multibody System Using ADAMS [J]," Journal of Anhui University of Technology, 2007, 24(1), 43-45.

[8] Negrut D, Harris B, "ADAMS theory in a nutshell[M]," US:MDI, 2001: 15-17.

[9] Timothy J. McCoy, "Wind turbine ADAMS model linearization including rotational and aerodynamic effects[J]," AIAA Journal, 2004, 1370: 1-10.

IOP Conf. Series: Materials Science and Engineering **398** (2018) 012028 doi:10.1088/1757-899X/398/1/012028

A Novel Measurement for Monitoring Mechanical Impedance Applied to Breakthrough Detection of Bone-Drill System

P J Ko[1] and M C Tsai[1]
[1]Department of Mechanical Engineering, National Cheng Kung University, No.1, Ta-Hsueh Road, Tainan 701, Taiwan

Abstract. Based on a Chain Scattering Description and the Hilbert Huang Transform, this paper proposes a new measurement for monitoring dynamic changes of mechanical impedance without applying conventionally mechanical sensors. The proposed method is applied to the bone drilling system to detect the moment of bone breakthrough. The single frequency mechanical impedance of bone can be computed from the electrical signal of a DC motor in accordance with the proposed method. The experimental results demonstrate that the moment of bone breakthrough can accurately be detected by monitoring changes of mechanical impedance.

1. Introduction

Bone drilling is frequently performed in medical surgery, such as nail insertions for tooth implantation, orthopedic surgery, neurosurgery, maxillofacial surgery, osteosynthesis, etc. During a bone drilling procedure, a hole through the bone is created by the surgeon with a drilling machine. The surgeon determines the moment of bone breakage from his or her experience. However, bones in the human body are surrounded by blood vessels and nerves. Even the slightest deviation can damage these tissues and threaten the life of patients. Therefore, many studies have been devoted to the techniques of automatically detecting the moment of bone breakthrough since 1996 [1][2][3][4]. Allotta et al. [1] proposed a theoretical model for estimating penetration force during bone-drilling. Then, an upper limit threshold to the first derivative of the penetration force could be set according to the theoretical model, which was applied to detecting the moment of bone breakthrough. Ong and Bouazza-Marouf [2] proposed a different breakthrough detection technique. This technique converted the profiles of drilling force differences between successive samples and/or the drill bit rotational speed into easily recognizable and more consistent profiles by using a modified Kalman filter, which allows a robust and repeatable detection of breakthrough. The established breakthrough detection methodologies in most of the literature were based on the penetration force from the load cell, which suffers from sensor noise. Moreover, given the diversity of human bone density and thickness, a thrust-force signal will have many spurious pulse noises during the drilling process. Therefore, Colla and Allotta [3] presented an alternative breakthrough detection method based on wavelets. They detected bone breakthrough based on wavelet analysis of the thrust force signal. Besides, Lee et al. [4] realized an automatic bone drilling machine that can automatically stop by analyzing the thrust force, drilling torque, and feed rate velocity signals once the system breaks through bone.

Previous studies use the information of load cells or torque sensors mounted on the bone drilling machine. Unfortunately, there is usually an installation issue in the load cell and torque sensor. Moreover, a breakthrough detection technique based on a force/torque signal without using velocity information might easily lead to improper monitoring decisions due to that the bone has an inhomogeneous organic nature of variable impedance. In this paper, through the Chain Scattering Description (CSD) [5] with Hilbert Huang Transform (HHT) technology [6], the equivalent mechanical impedance of bone is detected by measuring the electrical signal in the bone drilling machine, which means the electromechanical actuator is designed to serve as a sensor in order to avoid the installation issues associated with the force sensor. In accordance with monitoring the equivalent mechanical impedance, the evaluation and determination mechanism of breakthrough detection is established, which has the advantage of considering force and velocity information simultaneously to avoid improper monitoring decisions.

2. Experimental bone drilling system

The experimental bone drilling system is shown in figure 1. The servo feed system is a linear motor (Copley Controls Corp. TB2504), which is driven by a velocity-mode driver. The bone drilling system consists of a drill bit and a DC drilling motor. Further, a PID speed controller was designed in a DC drilling motor [7] to effectively strengthen the rigidity and expel the heat to avoid tissue damage caused by high temperatures [8] as shown in figure 2. Given the parameters of the DC servomotor shown in table 1, let the desired bandwidth of the current loop and velocity loop be 1 kHz and 100 Hz, respectively. The controller is obtained as $K_p = 12.747$, $K_i = 6252.5$, and $K_v = 16.721$ with the resulting bandwidths of the current loop and velocity loop are 99.797Hz and 997.63Hz, respectively.

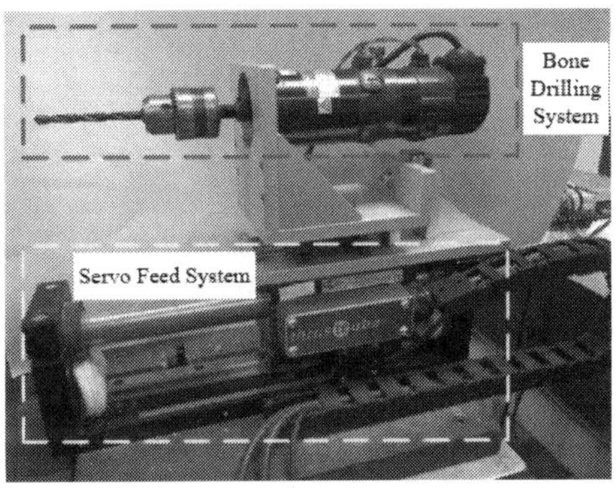

Figure 1. Experimental bone drilling system.

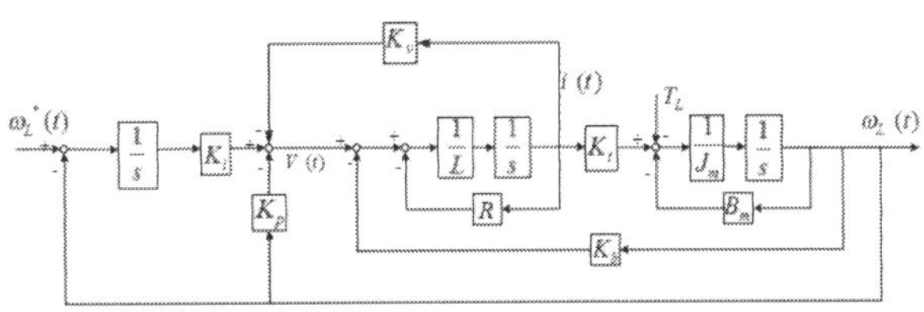

Figure 2. Structure of PID speed controller in DC drilling motor.

Table 1. Parameters of DC drilling motor	
Parameters	Value
Resistance R	7.155 (Ω)
Inductance L	0.0038 (H)
Moment inertia J_m	5.77×10^{-5} ($kg\cdot m^2$)
Damping coefficient B_m	0.00055 ($N\cdot m\cdot s/rad$)
Back EMF constant K_e	0.21 ($V\cdot s/rad$)
Torque constant K_t	0.21 ($N\cdot m/A$)

3. Proposed breakthrough detection method

3.1. Chain scattering description

The CSD of a two-port network originated from the conventional electrical circuit theory [5]. The CSD developed in a network circuit provides a straightforward interconnection in a cascaded way. Thus, many known results that have been developed for a two-port network can then be used in control system analysis and synthesis. Due to its benefits of describing a linear system, the CSD was extended to analyze the characteristics of the proposed bone drilling system.

The bone drilling system applied in this paper is driven by a DC drilling motor. Consider the block diagram of a DC motor shown in figure 3, where Z_m denotes the equivalent mechanical impedance of loading.

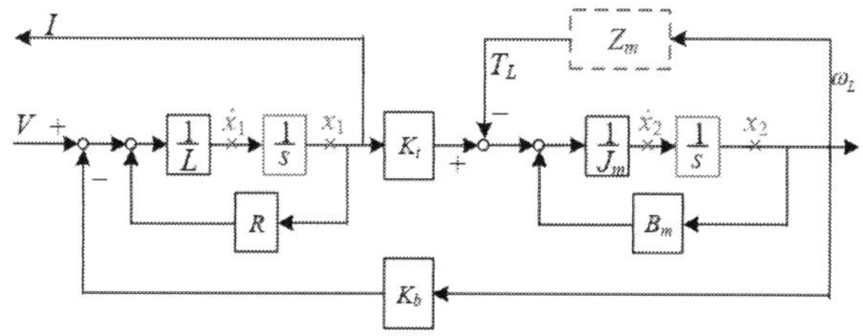

Figure 3. Block diagram of DC motor.

Let V (voltage reference) and T_L (load) be input variables, and I (motor current) and ω_L (motor angular velocity) be the outputs. A (2x2) LFT (linear fractional transformation) representation [5] of figure 3 is depicted in figure 4, as

$$\begin{bmatrix} I \\ \omega_L \end{bmatrix} = P(s)\begin{bmatrix} V \\ T_L \end{bmatrix} = \begin{bmatrix} P_{11}(s) & P_{12}(s) \\ P_{21}(s) & P_{22}(s) \end{bmatrix}\begin{bmatrix} V \\ T_L \end{bmatrix}, \tag{1}$$

where

$$P(s) = \begin{bmatrix} \dfrac{J_m s + B_m}{LJ_m s^2 + (LB_m + J_m R)s + (RB_m + K_b K_t)} & \dfrac{K_b}{LJ_m s^2 + (LB_m + J_m R)s + (RB_m + K_b K_t)} \\ \dfrac{K_t}{LJ_m s^2 + (LB_m + J_m R)s + (RB_m + K_b K_t)} & \dfrac{-(R + Ls)}{LJ_m s^2 + (LB_m + J_m R)s + (RB_m + K_b K_t)} \end{bmatrix}.$$

Through the reconstruction of the relationship between input and output in LFT form, the CSD of figure 5 with the "input" variables T_L, ω and the "outputs" I, V, can be derived from equation (1), as

$$\begin{aligned} \begin{bmatrix} I \\ V \end{bmatrix} &= \begin{bmatrix} P_{12}(s) - P_{11}(s)P_{21}^{-1}(s)P_{22}(s) & P_{11}(s)P_{21}^{-1}(s) \\ -P_{21}^{-1}(s)P_{22}(s) & P_{21}^{-1}(s) \end{bmatrix}\begin{bmatrix} T_L \\ \omega_L \end{bmatrix}, \\ &= \begin{bmatrix} G_{11}(s) & G_{12}(s) \\ G_{21}(s) & G_{22}(s) \end{bmatrix}\begin{bmatrix} T_L \\ \omega_L \end{bmatrix} = G(s)\begin{bmatrix} T_L \\ \omega_L \end{bmatrix} \\ &= \begin{bmatrix} \dfrac{1}{K_t} & \dfrac{J_m s + B_m}{K_t} \\ \dfrac{R + Ls}{K_t} & \dfrac{LJ_m s^2 + (LB_m + J_m R)s + (RB_m + K_b K_t)}{K_t} \end{bmatrix}\begin{bmatrix} T_L \\ \omega_L \end{bmatrix} \end{aligned} \tag{2}$$

where G is a so-called transduction matrix [9].

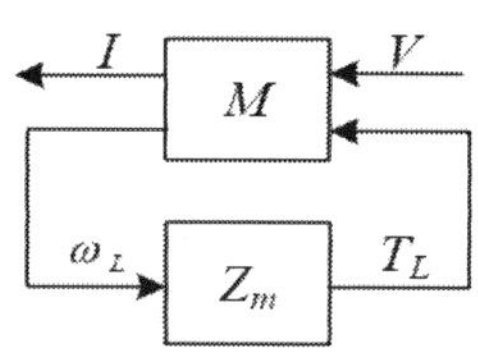

Figure 4. LFT form of DC motor.

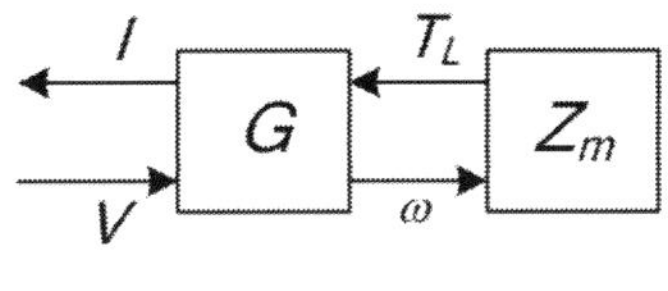

Figure 5. Two-port chain description matrix of a DC motor.

Thus, for any equivalent mechanical impedance of loading $Z_m(s)$, the equivalent electrical impedance of the motor, denoted by $Z_e(s)$, is given by

$$Z_e(s)=\frac{V}{I}=\frac{G_{21}(s)T_L+G_{22}(s)\omega_L}{G_{11}(s)T_L+G_{12}(s)\omega_L}=\frac{G_{21}(s)Z_m(s)+G_{22}(s)}{G_{11}(s)Z_m(s)+G_{12}(s)} \tag{3}$$

Reversely, if the electrical impedance $Z_e(s)$ is measured via V and I at the input-port of figure 5, then the equivalent impedance of mechanical loading at the output-port can be found as

$$Z_m(s)=\frac{-G_{12}(s)Z_e(s)+G_{22}(s)}{G_{11}(s)Z_e(s)-G_{21}(s)} \tag{4}$$

However, equation (4) is impracticable due to an improper transfer function as depicted in equation (2) [10]. In fact, (2) can be realized under a single frequency condition in practical application [9][11]. For instance, under the condition of 10Hz, i.e. s=jω=j10, the transduction matrix of equation (2) becomes

$$G(j10)=\begin{bmatrix}\frac{1}{K_t} & \frac{B_m}{K_t}+j\frac{20\pi J_m}{K_t}\\ \frac{R}{K_t}+j\frac{20\pi L}{K_t} & \frac{(RB_m+K_bK_t)-20\pi LJ_m}{K_t}+j\frac{(LB_m+J_mR)20\pi}{K_t}\end{bmatrix}. \tag{5}$$

Then, the 10Hz equivalent mechanical impedance of the load can be derived by equations (4) and (5) for the evaluation and determination of breakthrough. In the next subsection, the signal processing method, HHT, is adopted to evaluate the required characteristic component of the measured electrical signal for detecting the breakthrough of bone.

3.2. Signal processing

The Fast Fourier Transform (FFT) has been widely utilized in many signal processing applications; however, FFT is not capable of analyzing non-stationary signals, which might lead to inaccurate results, such as side lope and harmonics [12]. Hence, the HHT, an advanced time-frequency signal processing technique that is capable of analyzing non-stationary signals, real-time measurement and preservation of the characteristics of the varying frequency, is applied to analyze the measured electrical signal during the bone drilling process.

Given a tested signal $x(t)$, the EMD (Empirical mode decomposition) [6], a self-adaptive decomposition process for non-stationary and nonlinear signals can be described as

$$x(t)=\sum_{i}^{N}IMF_i(t)+r(t) \tag{6}$$

where N denotes IMF (Intrinsic mode functions) [6] modes and r(t) the residue. The IMFs are simple oscillatory functions with varying amplitudes and frequencies, which are defined to satisfy the following conditions [6]:

1. The number of extrema and the number of zero-crossing must either be equal or differ at most by one.
2. At any point, the mean value of the two envelopes defined by the local maxima and local minima is zero.

In practice, the procedure of EMD is as follows: firstly, identify all the local maxima (minima) and obtain the upper (lower) envelope by connecting all the local maxima (minima) with a cubic spline; secondly, obtain the candidate IMF by subtracting the data from the local mean of the two envelopes;

finally, check if the candidate IMF satisfies the definition of IMF; if not, repeat Steps 1 and 2 until the candidate IMF satisfies the definition of IMF.
A major drawback of the EMD is frequent occurrence of mode mixing. Mode mixing results from signal intermittency, which is defined as a signal of a similar scale residing in different IMF components [13]. Such a drawback makes the IMF lack physical meaning and causes serious aliasing in the time-frequency distribution. To alleviate this problem, EEMD (Ensemble empirical mode decomposition) [13], a major improvement on the EMD method, is applied in this paper.

The EEMD starts by adding white noise $w_m(t)$ into an intermittent tested signal $x(t)$ and converting it into a new continuous tested signal $x_m(t)$. Then, decomposing the tested signal $x_m(t)$ by EMD, the mode mixing can be eliminated due to that the white noise $w_m(t)$ distributes the whole time-frequency space uniformly. However, white noise will remain in each IMF. White noise has a unique property that allows it to cancel itself out; hence, a time-space ensemble mean is applied here. Therefore, the tested signal is decomposed repeatedly by adding in different white noises, and the ensemble means of the corresponding IMFs are acquired as the final IMFs. After the IMFs from EEMD are obtained, the instantaneous amplitude and frequency can be computed by applying HT to each IMF [6] as

$$\hat{F}_i(t) = HT\{F_i(t)\} = \frac{1}{\pi} PV \int_{-\infty}^{\infty} \frac{F_i(\tau)}{t-\tau} d\tau \quad (7)$$

in which the PV indicates the Cauchy principal value [14]. With the Hilbert transform $\hat{F}_i(t)$ of the function $F_i(t)$, the analytic function is

$$z(t) = F_i(t) + j\hat{F}_i(t) = a_i(t)e^{j\theta_i(t)} = a_i(t)e^{j2\pi\int f_i(t)dt} \quad (8)$$

where $a_i(t) = \sqrt{F_i^2(t) + \hat{F}_i^2(t)}$ denotes the instantaneous IMF amplitude, $\theta_i(t) = \tan^{-1}\frac{\hat{F}_i(t)}{F_i(t)}$ the instantaneous phase function, and $f_i(t) = \frac{1}{2\pi}\frac{d\theta_i(t)}{dt}$ the instantaneous IMF frequency.

3.3. Proposed breakthrough detection method
Generally speaking, the medullary cavity in human bone, which is filled with soft yellow marrow, is surrounded by dense strong cortical bone as illustrated in figure 6 [2]. The drill bit is performed on dense strong cortical bone along the soft yellow marrow during bone drilling surgery. Thus, the breakthrough of bone during bone drilling surgery can be detected by monitoring the equivalent mechanical impedance of bone.

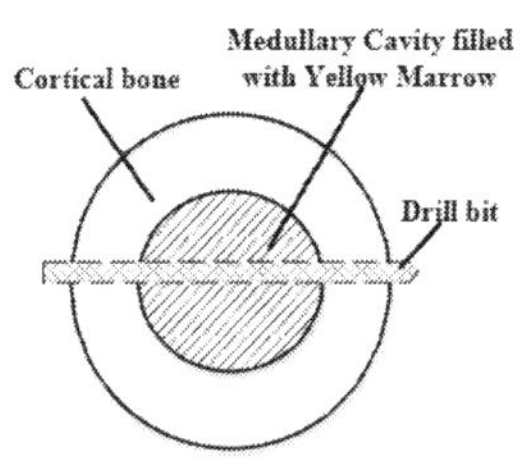

Figure 6. Cross-section of human bone [2].

As mentioned in the previous subsection, the equivalent mechanical impedance of bone can be determined from equation (4) and DC (0Hz) transduction matrix due to the DC drilling motor. However, the bone has inhomogeneous organic nature. It is hard to detect the breakthrough point precisely by monitoring the equivalent mechanical impedance only in a single frequency. Therefore, a small amplitude sinusoidal signal in the specific frequency is injected into the reference velocity command $\omega_L^*(t)$ shown in figure 2 as

$$\omega_{L\_new}^*(t) = \omega_L^*(t) + a\sin(2\pi f t) \tag{9}$$

where a is a small constant value usually set as 1~2% of $|\omega_L^*(t)|$ to maintain the characteristic of the bone drilling system; f denotes the specific frequency. Consequently, the characteristics of specific frequencies become existent in the electrical signal (voltage and current) of the bone drilling system. In accordance with HHT, the measured electrical signals are analyzed for obtaining the DC signal and 10Hz specific frequency sinusoidal signal. Then, the equivalent mechanical impedances can be calculated via a single-frequency transduction matrix and provided for breakthrough detection.

4. Experiment results

The porcine bone demonstrates a good likeness with human bone [15]; hence, the proposed breakthrough detection method is validated by the experiment results of drilling porcine bone in this paper. The prototype of the experimental bone drilling system is shown in figure 1.

According to equation (9), a 10Hz sinusoidal signal is injected into the reference velocity command $\omega_L^*(t)$ after 1 sec as depicted in figure 7, where the amplitude a is 2% of $|\omega_L^*(t)|$. Therefore, the characteristic of a specific frequency becomes existent in the corresponding voltage $V(t)$ and current $i(t)$ as shown in figures 8 and 9. Moreover, the corresponding voltage $V(t)$ and current $i(t)$ are extracted and computed by applying the HHT as shown in figures 10 and 11; note that only IMF 5 and the residue are depicted here for brevity. Then, IMF 5 and the residue of the voltage $V(t)$ and current $i(t)$, respectively, are implemented for equation (2) to compute the equivalent mechanical impedances at the condition of 10Hz and DC.

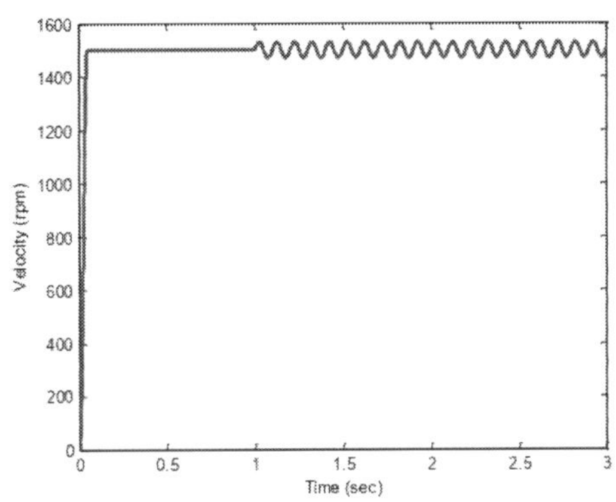

Figure 7. Reference velocity command injected with 10Hz sinusoidal signal.

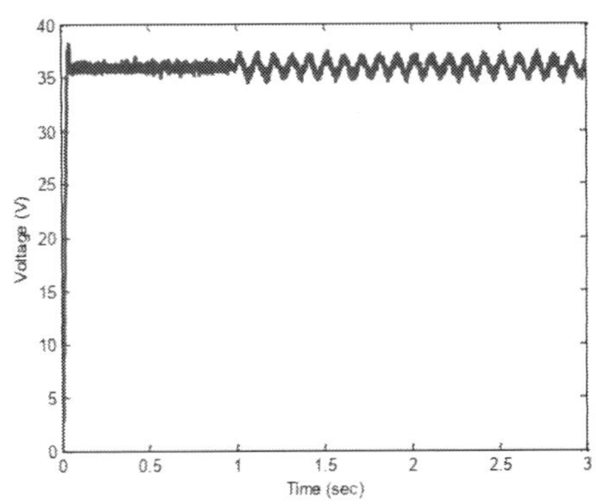

Figure 8. Driven voltage $V(t)$ of bone drilling system.

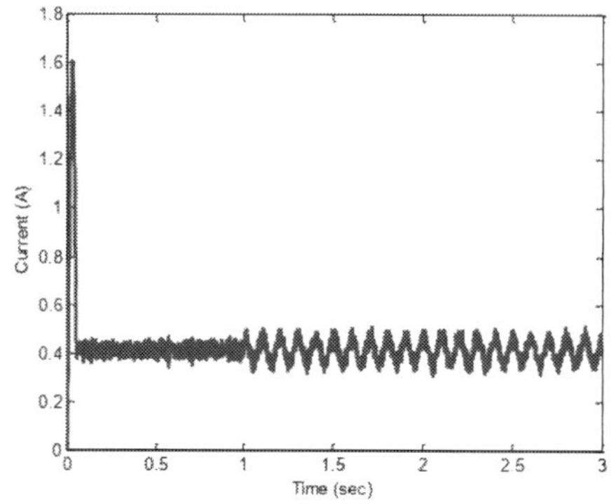

Figure 9. Measured current *i(t)* from bone drill system.

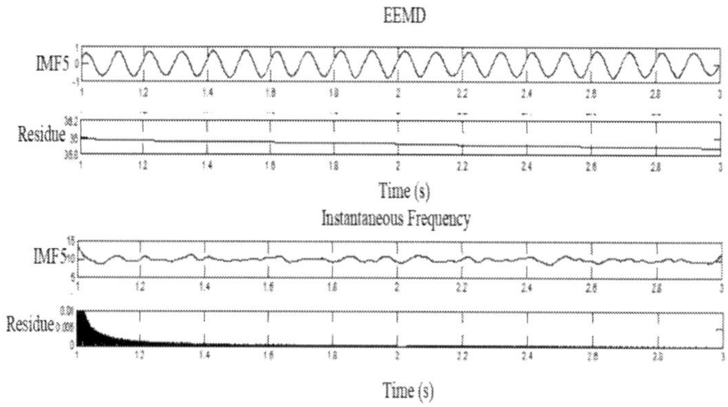

Figure 10. IMFs of driven voltage *V(t)* and corresponding instantaneous frequency.

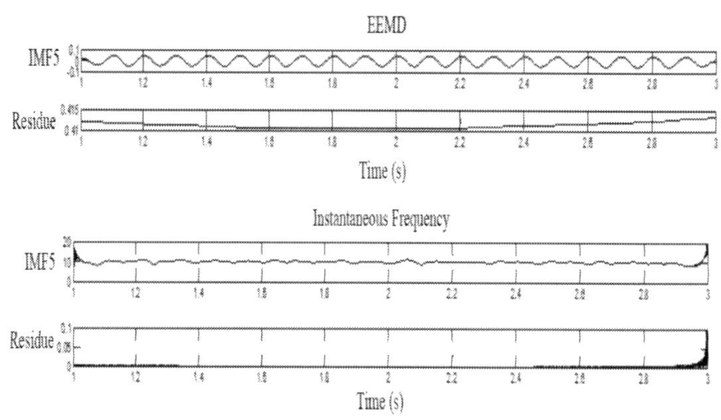

Figure 11. IMFs of measured current *i(t)* and corresponding instantaneous frequency.

Through monitoring the 10Hz and DC equivalent mechanical impedance, respectively, during the bone drilling process, the experiment results are illustrated in figure 12. From figure 12 (a), the DC equivalent mechanical impedance rises rapidly after 10 seconds, which indicates that the drill bit is in contact with bone. Afterward, the DC equivalent mechanical impedance dramatically decreases at 24 seconds and 65 seconds, which means that the bone drilling system penetrates the layers of the bone, respectively. The result of the specified frequency signal (10Hz) as shown in figure 12 (b) is similar to figure 12 (a). Therefore, the moment of bone breakthrough is detected when second peak occurs in both DC and 10Hz equivalent mechanical impedance.

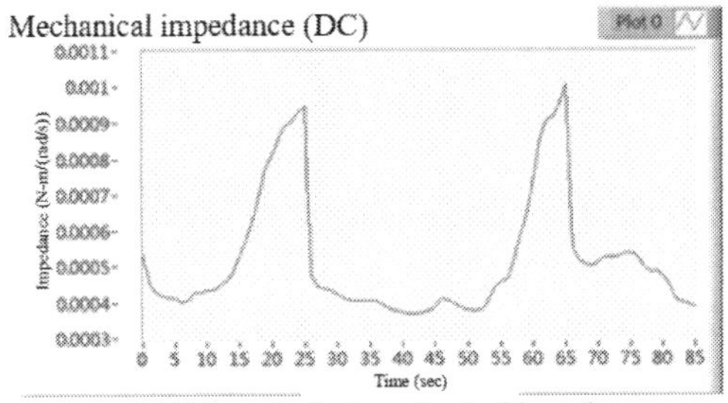

(a) DC equivalent mechanical impedance

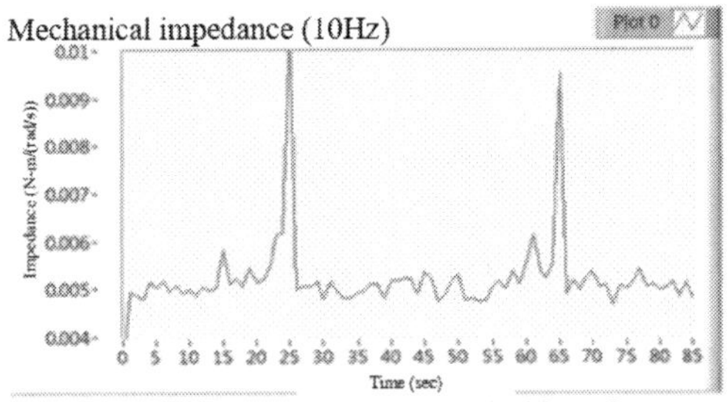

(b)10Hz equivalent mechanical impedance

Figure 12. Experiment results of monitoring equivalent mechanical impedance of bone at DC and 10Hz.

5. Conclusions

This paper has proposed a new approach for evaluating the equivalent mechanical impedance of the mechanical load of a DC motor without applying conventionally mechanical sensors. The proposed method employs the electrical signal of a DC motor to compute equivalent mechanical impedance based on Chain Scattering Description and Hilbert Huang Transform method. The proposed method was applied to the bone drilling system to detect the moment of bone breakthrough. The idea is to monitor the mechanical impedance of bone, which can establish the mechanism of breakthrough detection. The experimental results demonstrate that the proposed method can accurately evaluate the moment of bone breakthrough.

6. Acknowledgments

This work was supported by Ministry of Science and Technology of Taiwan, under project MOST 105-2218-E-006-008. The authors are grateful to Mr. Ming-Wei Shen, who contributed to pre-research of this study.

7. References

[1] Allotta B, Belmonte F, Bosio L and Dario P 1996 Study on a Mechatronic Tool for Drilling in the Osteosynthesis of Long Bones: Tool/Bone Interaction, Modeling and Experiments *Mechatronics* **6** 447-459.

[2] Ong F R and Bouazza-Marouf K 1998 Drilling of Bone: A Robust Automatic Method for the Detection of Drill Bit Break=Through *Proceedings of the Institution of Mechanical Engineerings, Part H: Journal of Engineering in Medicine* **212** 209-221.

[3] Colla V and Allotta B 1998 Wavelet-Based Control of Penetration in a Mechatronic Drill for Orthopaedic Surgery *IEEE International Conference on Robotics and Automation* **1** 711-716.

[4] Lee W Y, Shih C L and Lee S T Force Control and Breakthrough Detection of a Bone-Drilling System *IEEE/ASME Transactions on Mechatronics* **9** 20-29.

[5] Tsai M C and Gu D W 2014 *Robust and Optimal Control: A Two-Port Framework Approach* London: Springer-Verlag.

[6] Huang N E and Shen S S P 2014 Hilbert-Huang Transform and Its Applications *Interdisciplinary Mathematical Sciences* **16.**

[7] Ellis G 1991 *Control System Design Guide: Using your Computer to Develop and Diagnose Feedback Controllers* San Diego: Academic Press.

[8] Eriksson R A and Albrektsson T 1984 The Effect of Heat on Bone Regeneration: An Experimental Study in the Rabbit Using the Bone Growth Chamber *Journal of Oral & Maxillofacial Surgery* **42** 70.-711.

[9] Ling S F and Xie Y 2001 Detecting Mechanical Impedance of Structure Using the Sensing Capability of a Piezoceramic Inertial Actuator *Sensors and Actuators A* **93** 243-249.

[10] Sung S W, Lee J and Lee I B 2009 *Process Identification and PID Control* Wiley-IEEE Press.

[11] Liang C, Sun F and Rogers C 1996 A Electro-Mechanical Impedance Modeling of Active Material Systems *Smart Mater. Struct.* **5** 171-186.

[12] Rai V K and Mohanty A R 2005 Condition monitoring techniques for rolling element bearings: an overview *in Proceedings of National Conference on Role of NDE in Modern main Tenance Management* 8-18.

[13] Wu Z and Huang N E 2009 Ensemble empirical mode decomposition: a noise assisted data analysis method *Advances in Adaptive Data Analysis* **01** 1-41.

[14] Paget D F and Elliott D 1972 An algorithm for the numerical evaluation of certain Cauchy principal value integrals *Numerische Mathematik* **19** 373-385.

[15] Pearce A I, Richards R G, Milz S, Schneider E and Pearce S G 2007 Animal models for implant biomaterial research in bone: a review *European Cell and Materials* **13** 1-10.

IOP Conf. Series: Materials Science and Engineering **398** (2018) 012029 doi:10.1088/1757-899X/398/1/012029

Condition Monitoring and Fault Diagnosis for Wind Turbine Gearbox Based on Waterfall

Hongwei Xin[1], Xinjian Feng[2], Ye Xin[1], Ye Tian[3],Yutong Lin[1], Yingjie Wu[1], Jianguo Wang[1]

School of Automation Engineering, Northeast Electric Power University, Jilin,Jilin132012,China
China Datang Corporation Science and Technology Research Institute Northwest Branch,Xi'an,Shanxi710021,China
Jilin CPI New Energy Co., Ltd., Changchun, Jilin130012,China

13147769933@163.com

Abstract. Threshold analysis is a common method in condition monitoring of wind turbine gearboxes. It means set alarm thresholds to monitor working state of gearboxes, such as RMS, kurtosis and peak-peak and so on. However, these methods influenced by noise cause false alarms easily. Waterfall has the advance in reflecting the relationship that frequency spectrum changes over time. Before getting frequency spectrum information of each part from the fast Fourier transform, the vibration data of the gearbox is split into several equal parts according to the time. Then frequency spectrum of each part is chronologically stacked into a waterfall. Through the analysis of practical gearbox vibration data, it can obviously differ the waterfall of different state, such as normal state, stop, noise and failure. Waterfall can directly and accurately monitor the working status of the wind turbine gearbox.

1. Introduction

Wind energy, as a clean and renewable energy source, has huge reserves and attracts many countries' attention[1]. Due to geographical constraints and conditions of wind energy resources, the wind turbine transmission system malfunctions easily[2]. Therefore, the condition monitoring of gearbox and the evaluation of the running state can achieve the purpose of reducing maintenance costs and improving economic benefits.

Scholars at home and abroad have carried out fruitful researches. [3] uses K proximity anomaly detection technology to realize fault early warning of wind turbine. In [4], the wind speed, speed and vibration time frequency parameters are fused by support vector machines (SVM) to realize the condition monitoring of wind turbines. In [5], the parameters such as peak value, kurtosis, margin, skewness and power spectrum center of gravity are used as the characteristic parameters of fault vibration. However, the above researches achieve the alarm basically through the threshold setting, and are unable to monitor the evolution of the fault process because of interference from noise.

Waterfall is a visual method to monitor the change of spectrum to achieve the purpose of condition monitoring. The waterfall effectively expresses the change of signal frequency in a certain period of time, contains more abundant information than the simple time-domain and frequency-domain analysis and is often used in study of non-stationary signal[6]. In fault diagnosis, waterfall is useful to understand the occurrence and development of fault and achieves the purpose of condition monitoring and fault diagnosis [7].

2. Gearbox vibration signal model

(1) Parallel wheel fault signal model

The impact caused by meshing of gear fault point will produce amplitude modulation and frequency modulation to the meshing vibration, and the modulation frequency is the characteristic frequency of fault gear, and the carrier frequency is the meshing frequency and its frequency doubling [8]. The mathematical expression of vibration signal is as follows:

$$x(t)=\sum_{k=0}^{\infty}a_k(t)\cos[2\pi kf_m t+b_k(t)+\theta_k] \tag{1}$$

$$a_k(t)=c\sum_{n=0}^{\infty}A_{kn}\cos[2\pi nf_g t+\phi_{kn}] \tag{2}$$

$$b_k(t)=c\sum_{l=0}^{\infty}B_{kl}\sin[2\pi lf_g t+\varphi_{kl}] \tag{3}$$

Among, respectively, $a_k(t)$ is the AM function and $b_k(t)$ is the FM function, A_{kn} is the AM intensity, B_{kl} is the FM intensity, c is the non-dimensional constant, f_m is the gear meshing frequency, f_g is the fault characteristic frequency, θ_k, ϕ_{kn}, φ_{kl} are the initial phase.

(2) Planetary train fault signal model

The vibration model of the planetary train fault point is the same as that of the parallel wheel. In the process of signal transmission, the amplitude is modulated again by the influence of the planetary wheel and the transmission path [9]. The mathematical expression of the vibration signal is as follows:

Sun gear fault:

$$x(t)=[1-\cos(2\pi f_s^{(r)}t)][1+A\cos(2\pi f_s t+\phi)]\cos[2\pi f_m t+B\sin(2\pi f_s t+\varphi)+\theta] \tag{4}$$

Planetary wheel fault:

$$x(t)=[1-\cos(2\pi f_c t)][1+A\cos(2\pi f_p t+\phi)]\cos[2\pi f_m t+B\sin(2\pi f_p t+\varphi)+\theta] \tag{5}$$

Ring gear fault:

$$x(t)=[1+A\cos(2\pi f_r t+\phi)]\cos[2\pi f_m t+B\sin(2\pi f_r t+\varphi)+\theta] \tag{6}$$

Among, respectively, $f_s^{(r)}$ is the sun wheel absolute rotation frequency, f_c is the planetary frame rotation frequency. Respectively, f_s, f_p, f_r is the characteristic frequencies of sun gear, planet gear and gear ring. f_m is the gear engagement frequency. A, B is respectively the amplitude modulation and the frequency modulation intensity. ϕ, θ are the initial phase.

According to the formula (1) - (6), due to the modulated signal of gear fault, the sideband which is $f_m \pm mf_g$ appears in spectrum. Among, the m is integer, which represents for frequency doubling of fault characteristic frequency. Therefore, monitoring the change of sideband can realize the condition monitoring of gearbox.

3. Construction of waterfall

Let $x(t)$ be equal to gearbox vibration signal, f_s is a sampling frequency, T is a sampling time.

(1) The vibration signal $x(t)$ is divided into n parts according to time (The size of the n is set according to the requirement), denoted by $x_1(t), x_2(t)\ldots\ldots x_n\ (t)$,Respectively represent $\frac{1}{n}T$, $\frac{2}{n}T\ldots\ldots T$ vibration signals of time.

(2) According to fast Fourier transform, the spectrum of $x_1(t)$, $x_2(t)\ldots\ldots x_n\ (t)$ are calculated, denoted by $X_1, X_2\ldots\ldots X_n$ 。

(3) $F = \left(0, \frac{1}{2T}, \frac{2}{2T} \cdots\cdots \frac{Tf_s - 1}{2T}\right)$ (7)

$T_1 = \left(\frac{1}{n}T, \frac{1}{n}T \cdots\cdots \frac{1}{n}T\right)$ (8)

$T_2 = \left(\frac{2}{n}T, \frac{2}{n}T \cdots\cdots \frac{2}{n}T\right)$ (9)

……

$T_n = (T, T \cdots\cdots T)$ (10)

Drawing the spectrum of T_1 time with (F, X_1, T_1) as a point; Drawing the spectrum of T_2 time with (F, X_2, T_2) as a point; And so on, drawing the spectrum of T_n time with (F, X_n, T_n) as a point, they form the waterfall.

4. Condition monitoring and fault diagnosis of gearbox

4.1 Gearbox parameters

Taking the gearbox of 1.5MW wind turbine in a wind farm in Inner Mongolia as an example. The gearbox adopts two stage planetary parallel structure and the structural parameters are shown in Table 1, and the sampling frequency of vibration data is 2560Hz, and the sampling points of each time period are 16384.

Table 1. Gearbox parameters

Gear Name	First solar wheel	Primary planetary gear	First stage inner gear ring	Two stage solar wheel	Two stage planetary gear	Two stage inner ring gear	Low speed shaft gear	High speed shaft gear
Number	1	3	1	1	3	1	1	1
Tooth Number	24	39	102	24	39	102	102	27

4.2 Condition monitoring

4.2.1 Normal condition. In theory, when gearbox is normal, the low frequency section is mainly the rotation frequency and its frequency doubling; In intermediate frequency section, due to gear machining and installation error, the sideband appears at the meshing frequency, the spacing of the sideband is rotation frequency; In high frequency section, it's low energy and small amplitude.

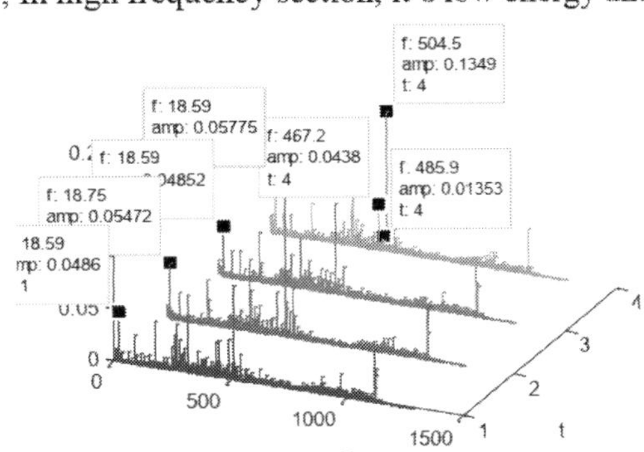

Figure. 1 Normal gearbox

When high speed shaft rotational frequency is 18.59Hz, the high speed shaft gear meshing frequency is 501.93Hz. In Figure.1, there is mainly 18.59Hz and its harmonic in low frequency section of waterfall, there is mainly meshing frequency and its sideband in intermediate frequency section, in high frequency section, amplitude is lower, there is mainly meshing frequency doubling. In Table 2, it can see that the energy is low and amplitude is small at the meshing frequency under different speed, the spacing of sideband is about high speed frequency.

Table 2. Frequency and amplitude parameters

Time	Rotating frequency/ Hz	Meshing frequency/Hz	Meshing frequency amplitude/(ms^{-2})	Left sideband/Hz	Left sideband amplitude/(ms^{-2})
1	18.59	505.2	0.09088	486.4	0.03854
2	18.75	508.1	0.147	489.4	0.01272
3	18.59	502.3	0.08847	483.9	0.02031
4	18.59	504.5	0.1349	485.9	0.01353

4.2.2 Fault condition. When gearbox faults, due to periodic impact, the spectrum peak increases. The vibration signal is the modulation signal, and the fault characteristic frequency is modulated to the meshing frequency and the natural frequency. The modulation result is that large energy and high amplitude sidebands appear at meshing frequency and natural frequency, and the spacing of sideband is the fault characteristic frequency.

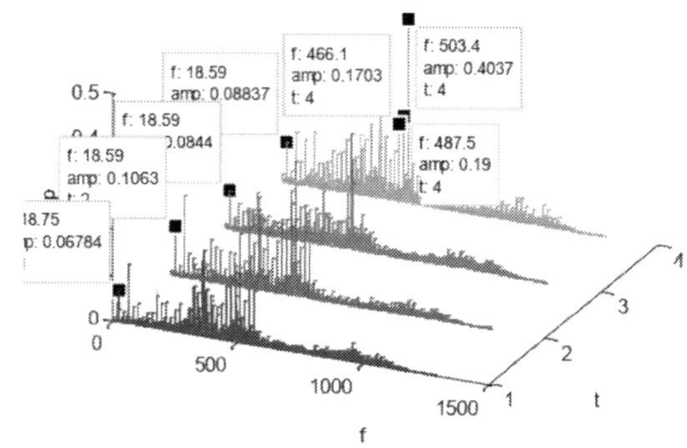

Figure. 2 Waterfall of the state of high-speed shaft gear failure

When gearbox high speed shaft speed is 18.59r/s, its meshing frequency is 501.93Hz. In Figure. 2, the peak increases. When its frequency is 503Hz, high amplitude sideband appears, it is about 10 times its normal value. In Table 3, the spacing of high amplitude sideband is about 18.59 r/s, which is consistent with the high speed shaft gear fault.

Table 3. Meshing frequency and sideband

Time	Rotating frequency/ Hz	Meshing frequency/ Hz	Meshing frequency amplitude/(ms^{-2})	Left sideband/Hz	Left sideband amplitude/(ms^{-2})
1	18.75	503.8	0.2413	487.8	0.1545
2	18.59	503.4	0.2963	487.7	0.229
3	18.59	502.7	0.2553	486.7	0.1277
4	18.59	503.4	0.4037	487.5	0.19

4.3 Fault development monitoring and diagnosis

Taking the gearbox vibration data of wind turbine in a wind farm in Inner Mongolia as an example. The sampling frequency of vibration data is 2560Hz, and the sampling points of each time period are 16384. Selecting gearbox failure early, mid and late data draw waterfall. In Fig. 3, the amplitude and spacing of sideband change at 1st moment, corresponding to the initial stage of fault evolution. At 2nd moment, the sideband becomes gradually clear and the amplitude increases. At 3th moment, There are obvious sideband clusters, the amplitude and spacing are clearly visible; At 4th moments, the sideband forms, the fault evolution completes. the speed is 18.75r / s, the peak is obvious at 505.9Hz, the sidebands stand in great numbers, the spacing is 18.8Hz. According to Table 4, it can judge that the high speed shaft gear faults because the meshing frequency is 505.9Hz and the fault characteristic frequency is 18.8Hz.

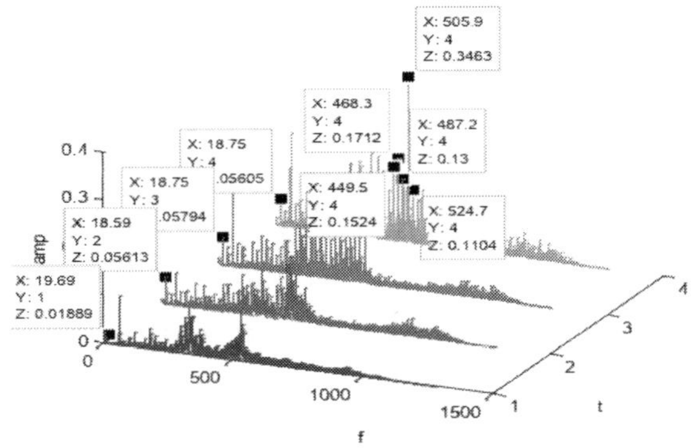

Figure. 3 Waterfall of fault development

Table 4. Fault characteristic frequency of gear box

Name	High speed shaft rotating frequency	Meshing frequency of primary planetary gear trains	Meshing frequency of two stage planetary gear trains	Meshing frequency of three parallel shafts
f / Hz	18.75	18.37	96.43	506.25

5. Summary

(1) There are obvious differences in normal and fault state of gear box with waterfall, and this method can be used for gearbox condition monitoring and fault location.

(2) The spectrum waterfall map can easily observe how frequency and amplitude vary with time. It is very prominent to observe the change of the signal in a period of time in the waterfall.

6. Acknowledgement

The research work was supported by Science & Technology Development Plan of Jilin province -Major Science & Technology Bidding(No. 20150203001SF).

7. References

[1] P. Tchakoua, R. Wamkeue, M. Ouhrouche; Wind turbine condition monitoring: state-of-the-art review, new trends and future challenges[J]. Energies, 2014, 7(4): 2595-2630.

[2] Q.H. Huang, X.S Du., H.C. Liu, et al. Remote on-line vibration test for multi-megawatt wind turbine[J]. Journal of Chongqing University, 2015, 38(1): 68-73.

[3] Y.J. Guo, L. Song, T.J. Xu, et al. Research on wind turbine gearbox fault warning method under variable working conditions[J]. China mechanical engineering, 2014, 25(10): 1346-1351.

[4] X.L. An, M.H. Zhao, D.X. Jiang, et al. Direct-drive wind turbine fault diagnosis based on support vector machine and multi-source information[J]. Power system technology. 2011, 35(4): 117-122.

[5] S.M. Zhang, D. Mao, B.Y. Wang. Application of big data processing technology in fault diagnosis and early warning of wind turbine gearbox[J]. Automation of electric power systems, 2016, 40(14): 129-134.

[6] Y. Zhou, L.P. Pan. Waterfall analysis technology research of hydropower unit on-line monitoring system[J]. Mechanical & electrical technique of hydropower, 2008, 31(6): 25-28.

[7] G.A. Yang. Fundamentals of signal processing[M]. Beijing: China petraochemical press company limited, 2012.

[8] Z.P. Feng, F.L. Chu. Vibration spectral characteristics of distributed gear fault of planetary gearboxes[J]. Proceedings of the CSEE, 2013, 33(2): 118-125.

[9] Z.P. Feng, L.L. Zhao, F.L. Chu. Vibration spectral characteristics of localized gear fault of planetary gearboxes[J]. Proceedings of the CSEE, 2013, 33(5): 119-127.

A Dynamic Relation Model Analysis of High-speed Impact and Sabot Strength

H Y Wu, L ZH Wang Y Q Li and M Zhao

School of Mechanical Engineering, Xi'an Jiaotong University,Xi'an,710049,China

E-mail: wuhy@xjtu.edu.cn

Abstract. To understand how a projectile penetrates a sabot during a high-speed impact testing, which led to the failure of the impacting experiment, a dynamic model was built based on the interactions between the projectile and the sabot; and the bearing strengths of the sabot with varying impact speeds were analyzed. The correctness of the model was verified by an experiment. It shows that, with all the other conditions kept the same, the maximum stress between the sabot and the projectile increases with the decrease of the sabot mass, the increase of the projectile and impact speed.

1. Introduction

In earlier days, high-speed impact problems mainly appear in the military field, and now they emerges in civilian areas [1]. Air gun is a useful experiment tool for high-speed launching and high pressure loading. It is commonly used for the study of mechanics of materials, penetration, and high-speed impact[2]. The launching tube of the gun normally has a large diameter to meet the broader size experiment requirements. In this condition, a small-size projectile should be supported by a sabot, which traps the propellant gas and keeps the projectile centered in the launching tube while steadily moving forward. The sabot is not an indispensable component, the mass of the sabot is required to be designed as low as possible[3]. When firing, the high pressure gas forces the sabot to carry the projectile down the lauching tube, the contact surface of the sabot suffers huge stress. The material of the sabot should have high strength and good toughness to ensure its structure integrity[4].

Earlier sabots were made of metallic materials, such as red copper, copper alloy, pure iron, aluminum, aluminium alloy, etc. A large number of new materials who are light weight yet have high mechanical strength and good toughness have been appeared, such as various resin materials, fiber glass, carbon fiber, etc. [5]. The majority of the research on the sabot has been focused on the sabot separation process[6,7,8], the structural analysis of the composite sabot[9,10], the motion analysis during sabot opening process[11], sabot opening lift force[12], etc. While this paper focuses on how to keep the projectile and sabot always be a whole to ensure the high impact experiment doing successfully. Polyurethane foam is a good candidate for normal impact experiment which can fulfill the loading requirements with low cost. The strength of polyurethane foam depends on the foaming compactness. If the strength is insufficient, the projectile may penetrate it and causes the experiment failure. The relationships among the required strength of the sabot, the projectile mass and the experimental velocity should be analyzed and built to ensure a successful impact test.

2. The phenomenon of sabot penetrates projectile

The first-class air gun is commonly used during an impact testing. The diameter of the sabot is approximately equal to the diameter of the launching tube, so that it can carry various projectiles with different sizes (Fig. 1).

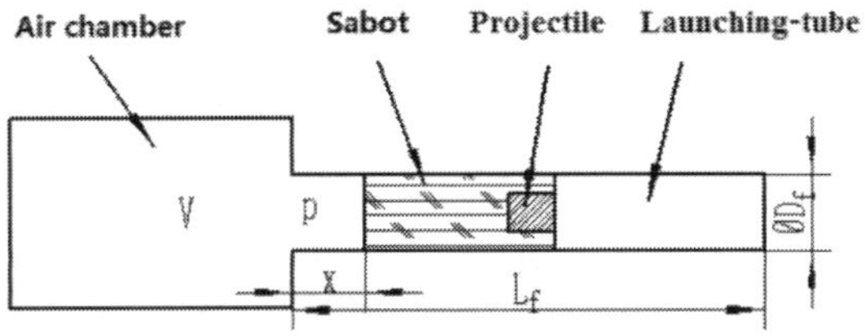

Figure 1. Sketch of projectile and sabot in air gun

When firing, the sabot and the projectile are accelerated instantly by the compressed air, a significant pressure is applied to their contact area. Fig. 2 shows a home-made sabot with density of 47kg/m^3. When the velocity was higher than 145m/s, the projectile penetrated its support sabot and caused the failure of the high speed impact experiment(Fig. 3).

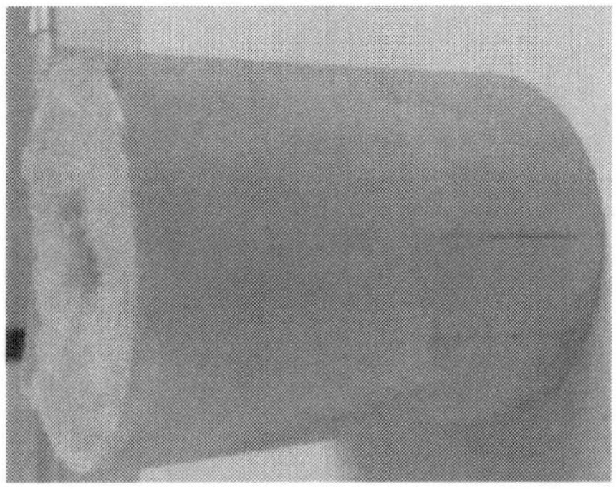

Figure 2. Self-made foam sabot

Fig. 3 showed the moment of the projectile penetrated through the sabot, which was captured by a high speed camera. After recording the sabot at t_1, the projectile rolled into the shot range at t_2. In this condition, there was no way for projectile to hit the target plate with scheduled trajectory and speed. It hit the sabot separator and was rebounded. This led to a complete failure of the experiment. The free flying projectile would easily hit on the sabot separator, and both of them were damaged (Fig. 4), the projectile may not pass through the sabot separator successfully. And the irregular bouncing of projectile may cause severe safety issues. This penetration problem must be solved to ensure the high speed impact examination can be carried out normally.

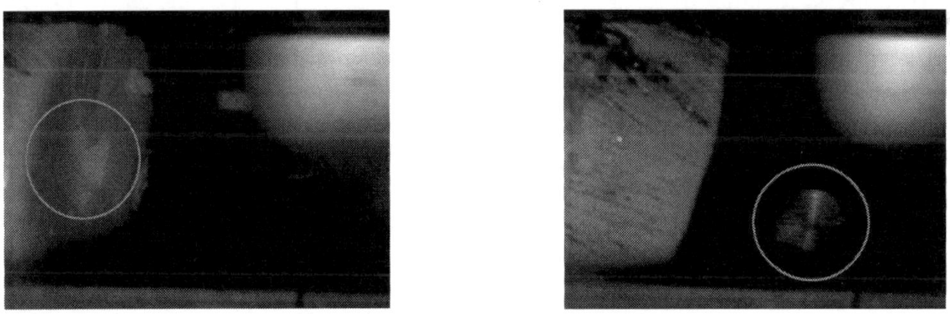

(a) Sabot at t_1 moment (b) Rolling flying projectile at t_2 moment

Figure 1. High speed camera recording projectile penetrating sabot

Figure 2. Impact damage on the sabot separator and the projectile

It is necessary to find out the relationship between the shear strength of the sabot, the projectile velocity and their mass to ensure a successful high speed impact experiment.

3. Dynamic analysis of interaction force between sabot and projectile

Define m_1 and m_2 as the mass of the sabot and the projectile respectively, ρ_1 and ρ_2 as the density of the them, D_f, S_f and L_f as the diameter, sectional area and effective length of the launching tube, p as the gas pressure at some moment, x as the distance between the projectile and its initial position.
Sabot velocity v is an important parameter in air gun test. It can be estimated by the following formulas. When the gas pushes the projectile, the gas expansion process can be regarded as the ideal gas adiabatic expansion process[13].

$$pV = \frac{M}{\mu_g} RT$$
$$p_{cs} V_{cq}^{\gamma} = p(V_{cq} + S_f x)^{\gamma} \tag{1}$$

In the formula: p—gas pressure; V—space volume behind projectile; μ_g-- the mocular number of gas; M—injected gas mass; R—gas constant; T—gas temperature; γ--gas adiabatic index; S_f -- cross section area of launching tube, $\pi D_f^2/4$; p_{cs}--gas injection pressure of gas chamber; V_{cq}-- initial chamber volume; p_{21}—pressure exerted on sabot by projectile; S_{dw}-- the cross sectional area of the projectile.
The stress analysis of the sabot is carried out when the high pressure air is suddenly released(Fig. 5). The sabot is subjected to the force exerted by compressed air(F_{01})and projectile(F_{21}).

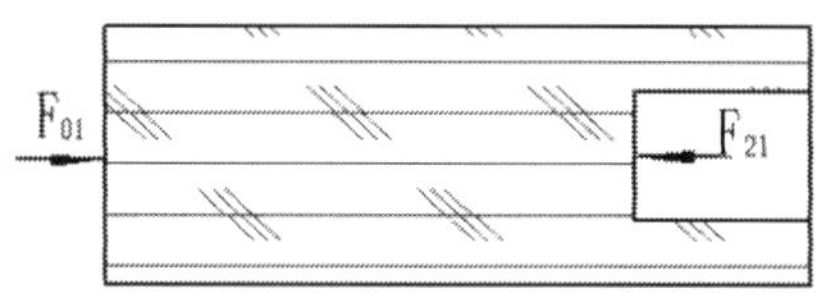

Figure 5. Force analysis of sabot

Without considering the specific forms of various energy loss, the mass of the sabot is increased from m_1 to φm_2 in order to replace various loss factors. So the motion equation of the sabot is as follows:

$$\varphi m_1 \frac{dv}{dt} = F_{01} - F_{21} = pS_f - p_{21} S_{dw} \tag{2}$$

where φ is called virtual mass coefficient, which is calculated as follows in the interior ballistics theory:

$$\varphi = K + \frac{1}{3} \frac{M}{m} \tag{3}$$

The factor K is determined by experiment, which is about 1~1.10. The letter m represents the sum mass of sabot and projectile and M is the mass of the gas.

$$M = \frac{p_{cs}V_{cq}}{RT}\mu_g \tag{4}$$

Eq.(5) is got by calculating Eq.(1) and Eq.(2).

$$\frac{dv}{dt} = \frac{1}{m_1\varphi}\left(\frac{S_f p_{cs} V_{cq}^{\gamma}}{(V_{cq}+S_f x)^{\gamma}} - p_{21}S_{dw}\right) \tag{5}$$

After formula transformation, Eq.(6) can be obtained.

$$vdv = \frac{1}{m_1\varphi}\left(\frac{S_f p_{cs} V_{cq}^{\gamma}}{(V_{cq}+S_f x)^{\gamma}} - p_{21}S_{dw}\right)dx \tag{6}$$

According to the motion consistency, the sabot and projectile can be regarded as one part, the change law of stresses exerted on them are similar to the chamber pressure in launching tube[14], which increases sharply at first then decreases gently (Fig.6). In order to simplify the calculation, it is assumed that the contact stress at the bottom of the projectile increases linearly during the launching process, and k is the linear slope. It is assumed that v_m is the maximum speed and L_m is the moving distance of sabot when the stress between sabot and projectile reaches the peak value. The relation law between p_{21} and x can be expressed as Eq.(7).

$$p_{21}(x) = kx, \quad x \le L_m \tag{7}$$

Eq.(7) is put into Eq.(6), and Eq.(8) is obtained through integral.

$$\int_0^{v_m} vdv = \frac{S_f p_{cs} V_{cq}^{\gamma}}{m_1\varphi}\int_0^{L_m}\frac{dx}{(V_{cq}+S_f x)^{\gamma}}dx - \frac{1}{m_1\varphi}\int_0^{L_m} kxS_{dw}dx, \quad x \le L_m \quad v_m^2 = \frac{2p_{cs}V_{cq}}{m_1\varphi(\gamma-1)}[1-\frac{V_{cq}^{\gamma-1}}{(V_{cq}+S_f L_m)^{\gamma-1}}] - \frac{kS_{dw}L_m^2}{2m_1\varphi} \tag{8}$$

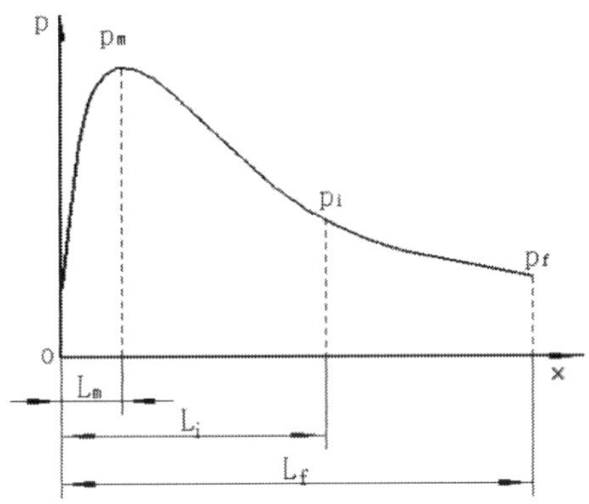

Figure 3. Bottom pressure and travel curve of projectile

The projectile is analyzed without considring other forces (Fig. 7). The force from the sabot acting on the projectile F_{12} is equal to its reaction force F_{21} in numerical value and opposite in direction. The projectile is embedded in the sabot then move together with the same speed after firing. The mass loss of projectile is set as φm_2 to represent various loss factors. Then Eq.(9) is obtained as follows:

$$\varphi m_2\frac{dv}{dt} = F_{12}$$
$$\varphi m_2\frac{dv}{dt}\frac{dx}{dx} = \varphi m_2 v\frac{dv}{dx} = p_{12}(x) = p_{21}(x) = kS_{dw}x, \quad x \le L_m \tag{9}$$

When the projectile and sabot move L_m, Eq.(10) can be get as follows:

$$k = \frac{\varphi m_2}{S_{dw}}\left(\frac{v_m}{L_m}\right)^2 \tag{10}$$

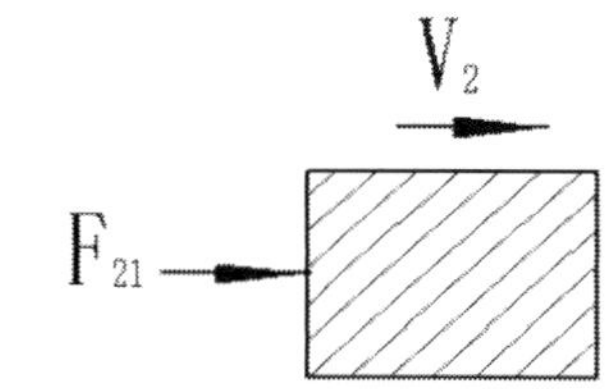

Figure 7. Force analysis of projectile

Taking Eq.(10) into Eq.8, Eq.(11) can be acquired as follows:

$$v_m^2 = \frac{2p_{cs}V_{cq}}{m_1\varphi(\gamma-1)}[1-\frac{V_{cq}^{\gamma-1}}{(V_{cq}+S_fL_m)^{\gamma-1}}]\cdot\frac{1}{1+\frac{m_2}{2m_1}} \tag{11}$$

According to Eq.(11), the relation curves between v_m and L_m, v_m and m_1 can be drawn as Fig. 8(a)~(b). Some conclusions can be deduced from the relation curves:

- There are direct proportion relation between v_m and L_m. If L_m increases, the maximum velocity v_m will increase.
- If the total mass of projectile and sabot is constant, the bigger m_1, the bigger v_m.

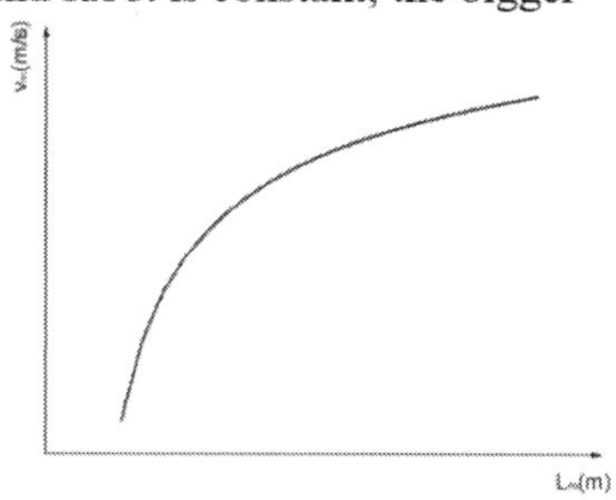

(a) The relation curve between v_m and L_m

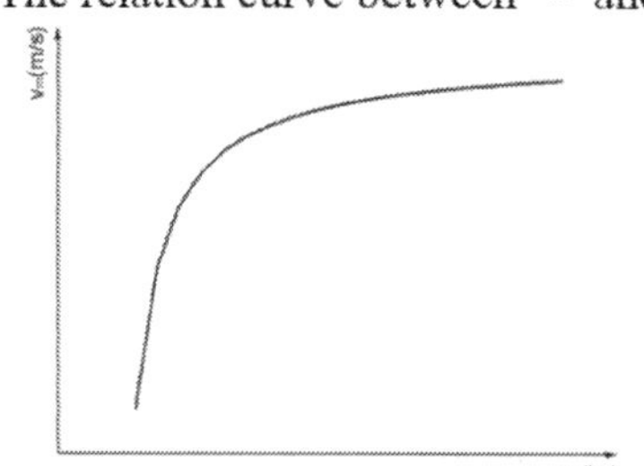

(b) The relation curve between v_m and m_1

Figure 8. The relation curves deduced from Eq.(11)

4. Experimental verification

From Fig.6, when projectile travels L_m and the velocity of it reaches v_m, the pressure exerted on it becomes the maximum. With the distance increasing, the pressure in launching tube decreases, but the projectile is still accelerated during the whole launching process. Its velocity v_e at the end position of gun barrel must be higher than v_m. The pressure calculated by v_e will be higher than the actual maximum pressure. Replacing v_m as v_e to analyze the strength of sabot will be more reliable.

The experimental device is shown in Fig. 9. The diameter of the metal projectile is 25mm, the air gun's tube diameter is 65mm, the sabot's length l is 100mm, the design impact speed v_e is 150m/s, the p_{cs} value is 0.132MPa after calculating by above equations. The calculation parameters of this high speed impact experiment are shown in Table 1.

Figure 9. Impact test device

Table 1. The parameters of high-speed impact experiments

Name	Value
μ_g ($kg\,gmol^{-1}$)	2.8e-2
γ	1.41
V_{cq} (m^3)	0.08482
T (K)	300
R ($J/(mol\,gK)$)	8.31
κ	1.1
p_{cs} (Pa)	0.29E6
S_f (m^2)	3.3e-3
S_{dw} (m^2)	4.9e-4
m_1 (kg)	0.013
m_2 (kg)	0.128
v_m (m/s)	150

According to Table 1, Eq.(3) and Eq.(11), the distance L_m can be calculated by the velocity at the end of launching tube. Then the pressure between projectile and sabot can be achieved based on Eq.(9) and Eq.(10). This pressure will be higher than the calculating result with v_m. The results are obtained as follows: $p_{12} = p_{21} = 2.68MPa, L_m = 3.05m$. In other words, when the sabot and the projectile as a whole run to the middle position of the launch tube, the pressure is expected to reach the maximum of 2.68MPa. Because the damage between projectile and sabot is obviously shear failure, the shear stress is calculated as follows:

$$\tau = \frac{p_{12} S_{dw}}{\pi d_{dw} (l - l_k)} \tag{12}$$

Where d_{dw} is the diameter of projectile, l is the length of sabot, l_k is the depth of hole in projectile.

If we put the related parameters of projectile and sabot ($d_{dw} = 0.025m, l = 0.1m, l_k = 0.03m$) into Eq. (12), we can get τ ($\tau = 0.24MPa$). The shear test is carried out, the compressive strength and shear strength of the sabot are 0.15MPa and 0.02MPa. They are less than 0.24MPa, which means that the damage will occur. This calculation result is consistent with the examination phenomenon(Fig. 3), the projectile penetrated through the sabot and formed a hole.

The material of sabot was changed to increase its density and shear strength. Three new sabots were made, Fig.10(a) and (b) were polyurethane foams,whose density were $200^{kg/m^3}$ and $300^{kg/m^3}$. Fig.10(c) was made by ethylene vinyl acetate (short for EVA) foam. Their shear strength were 0.12^{MPa},0.18^{MPa} and 0.4MPa respectively and higher than the calculation value. The three material were all used and the experiments were all successful. The density of EVA foam was only 40 kg/m^3, and it was the best choice for sabot among the three material.

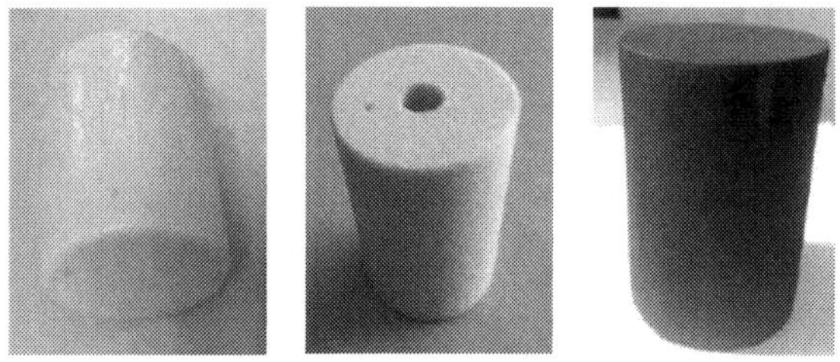

(a) Polyurethane foam (b) Polyurethane foam (c)EVA foam

Figure 10. Sabot with big density

5. Conclusions

In view of the separation of projectile and sabot in high-speed impact experiment, the dynamics of high speed impact and the bearing strength of the sabot were analyzed and deduced. The model of the interaction between the sabot and the projectile is established, and the correctness of the model is verified. According to the relationship model, the following conclusions can be acquired:

- Under the same conditions, the lighter the sabot, the larger the maximum stress between sabot and projectile.
- Under the same conditions, the heavier the projectile, the larger the maximum stress between sabot and projectile.
- The higher the impact speed, the larger the maximum stress between sabot and projectile.

This model is acquired by theoretical derivation, but it still needs more data to verify its validity, and the difference between theoretical data and real data should be researched deeper.

6. Acknowledgments

This work is supported by the Major Project of High-end CNC Machine Tools and Basic Manufacturing Equipment of China (2014ZX04011021), Project of Shaanxi Province(2016KTCQ01) .

7. References

[1] Rob R, Sivaiah A Et Al, Sivaiah A, Nageshwar R V and Syed A H, Simulation studies on the effect of projectile nose shape impacting on aluminum plates, *Int J Mech Eng & Robto Res*,3(1),2014:119-122.
[2] Yan L, Experimental research on high impact test,*Taiyuan: North University of China*, 2008:p.7-8
[3] Xu H X, Study on the structure of elastic material of armor piercing projectile, *Nanjing: Nanjing University of Science and Technology*, 2007:p.3.
[4] Yang SH L, Plastic and composite material sabot, *Missile Tech*, 1998(2):p.42-45.
[5] Tang Z X, Lin G B, Sun S H and Lin W. Application of new material technology in sabot, *Ship Elec Eng*, 35(3), 2015: p.158-160.
[6] Celmins, I. Subscale simulation of sabot separation. Proceedings - *28th Int Symposium on Ballistics, BALLISTICS 2014, Atlanta, GA, United states*, 2014: p.860-869.
[7] Takeuchi, F and Matsuo A. Numerical investigation of sabot separation process in a ballistic range using moving overlapped grid method[C]. *Proceedings -51st AIAA Aerospace Sciences Meeting including the New Horizons Forum and Aerospace Exposition 2013, Grapevine, TX, United states*,2013:p.025003.
[8] Huang Z G, Wessam M E and Chen ZH. Numerical investigation of the three-dimensional dynamic process of sabot discard. *J Mech Sci & Tech,* 28 (7) , 2014: p.2637-2649.

[9] Minnicino, M A. Structural dynamics of despun kinetic energy projectiles launched with composite sabots. *Proceedings-27th International Symposium on Ballistics, BALLISTICS 2013. Freiburg, Germany,*2013:p.428-439.
[10] JH Choi. A Study on the Fabrication of the Composite Sabot for a Kinetic Energy Projectile, Brain Research,9(3) ,2006:p.15
[11] Acharya, R S and Naik S D, Motion analysis during sabot opening process. *Defence Sci J.* 57(2) , 2007:pp.229-241.
[12] Mikhail, A G and Heavey K R. Sabot opening lift force: Analysis, CFD, and test.*38th Aerospace Sciences Meeting and Exhibit.Reno, NV, United states*, 2000:p.1-9.
[13] Wang J G, Principle and technology of gas gun,*Beijing: National Defence Industry Press,* 2001: p.43.
[14] Jin Zh M, Yuan Y X and Song M. Modern interior ballistics,*Beijing: Beijing Institute of Technology press*, 1992

IOP Conf. Series: Materials Science and Engineering **398** (2018) 012031 doi:10.1088/1757-899X/398/1/012031

Penetration of Hemispherical Projectile Into the Three Layer Targets with Different Materials and Layouts

Mehran Moradi and Ali Tavana

Department of Mechanical Engineering
Isfahan University of Technology
Isfahan, 84156-83111, Iran

e-mail: moradi@cc.iut.ac.ir

Abstract. In this paper, the penetration of a hemispherical projectile into the targets with three adjacent layers (without air gap) with different arrangement and materials has been studied. To study the set parameters, some ballistic tests as well as FEM simulations had done and the numerical results compared with experimental ones. While defining ballistic limit for the projectile-target system, the effect of layers arrangement sequence and material in their ballistic operation was perused. The result of the experiments showed that if the aluminum plates (harder material) are placed in the two first layers and CU (softer material) is placed in the third layer, the resistance against penetration will have the maximum value. Experimental results show that by decreasing the initial velocity of projectiles, velocity downfall is increased while the amounts of absorbed energy do not change significantly. The projectile velocity has a significant effect on the fracture mode. Increasing the velocity make the fracture accomplish in the Plugging mode while decreasing the velocity get the fracture to take petaline mode.

1. Introduction

The penetration of projectile into the target is a complicated mechanical process. This subject has been under consideration by researchers due to its military and non-military applications. The most important applications of this subject include: military applications like the design of structures that are resistant against the penetration of projectile, anti-bullet armors, design of the light weight armor systems (personal armor) to protect police officers, governmental and military individuals, design of missiles and rockets, and non-military applications like manufacturing of machines for transporting dangerous materials, strengthening the body of helicopters and aircrafts, design of ships body, protecting spacecrafts, securing important centers and design of protective coverings for nuclear reactors. In the above mentioned applications the impact loads arising from the stroke of external objects such as missiles and rockets or internal reasons like increasing the applied pressure are of the major importance.

In 1973 Zaid et al. [1] studied multi-layer targets and concluded that single plate of equal thickness have more strength in comparison with laminated (multi-layer) shields when the thicknesses in both states are equal while by increasing the thickness of targets, the resistance of multi-layer targets would be larger than solid targets.

In 1978 Backman and Goldsmith [2] during a series of ballistic experiments showed that the resistance of multi-layer objects is less than monolithic targets. In 1979 Marom and Bodner [3] studied analytically and experimentally the resistance of adjacent multi-layer and monolithic targets with identical thickness against penetration for Al6061-T6 materials. They deduced that in comparison with

IOP Conf. Series: Materials Science and Engineering **398** (2018) 012031 doi:10.1088/1757-899X/398/1/012031

solid targets, multi-layer targets have more resistance. In 1983 Corran et al. [4] stated that in comparison with monolithic targets, adjacent multi-layer targets have more resistance if the arrangement of layers causes to change the response of targets from shear state to bending or membrane extension state. In 1990 Nurick and Walters [5] perused the resistance against penetration for multi-layer targets and they concluded that in comparison with solid targets, ballistic limit for multi-layer targets has greater value by 4% to 8% when the thickness in both states is identical.

In 1991 Woodward et al. [6] studied the penetration through adjacent multi-layer sheets. They demonstrated that near the ballistic limit, the amount of energy absorbed by thin steel targets is independent of three components of being laminated, bending and tension while the material properties have the main effect in this case. They also stated that in penetration process, layers with specific gap have more efficiency in comparison with laminated layers with no gap. In 1988 Radin and Goldsmith [7] during a series of ballistic experiments that was performed with ductile aluminum demonstrated that the resistance of multi-layer targets with gap is less than the resistance of multi-layer targets with no gap. In 1999 Ben-Dor et al. [8] studied the effect of arrangement of adjacent multi-layer targets with different materials in the penetration of a projectile with semi-spherical head. They introduced distortion pressure and stated that for increasing the resistance of a target, the layers should be arranged so that this ratio of distortion pressure is increased. In 2000 Chao et al. [9] perused two-layer targets during a numerical study. In 2002 Borvik et al. [10] studied penetration of projectiles with flat, semi-spherical and conical heads through steel sheets using ballistic experiments. In 2005 Elek et al. [11] perused multi-layer targets using a thorough analytical study. During this study they perused the effect of sequence of layers arrangement and number of layers. Using their analytical model they specified that the maximum resistance of a target is obtained when the thickness of first layer approaches to zero and this is equivalent to using a solid target. This study declared that increasing the number of layers results in the reduction of projectile penetration into the target. In 2007 Dey et al. [12] studied the resistance of multi-layer targets with different arrangement sequences against projectile penetration and they concluded that despite of the projectile geometry, the resistance against the penetration for two-layer targets in comparison with solid targets is significantly greater. In 2008 Gupta et al. [13] studied the behavior of laminated aluminum sheets with various thicknesses and solid sheets with equivalent thicknesses. They stated that by increasing the number of layers, the downfall in ballistic limit velocity is less in comparison with solid target. Referring to results printed by other researchers one can say that for determining whether multi-layer targets are more resistant or solid targets, different parameters should be assessed including impact velocity, target material and thickness, projectile head geometry, projectile stiffness, the number and arrangement of layers, the gap between layers and etc. Accordingly for achieving a precise result, all of the possible states should be studied. In the present study, the resistance against penetration for three layers targets with different materials under the impact of a cylindrical projectile with semi-spherical head is perused experimentally. The main object of this research is to compare the ballistic operation of multi-layer targets with different materials using eight different arrangements. The results of experiments and numerical simulations are compared and then the amount of error related to numerical simulation is calculated.

2. Experimental Work

The equipments used for experiment are: gas gun with caliber of 8.7 mm and length of 2 m, pressurized Helium gas used as the firing instrument in the tests and a chronograph that is placed after target to increase accuracy in measuring residual velocity of projectile. There is a fixture around all of the targets as depicted in Fig. 1.

To prevent deviation of projectile, targets are placed in front of the gun at a distance of 30 mm. To obtain the impact velocity, gun volleys three to six times before targets are placed within the fixtures. When the projectile crosses between two optical sensors, which are placed in front of the target, chronograph saves the time of the projectile stroke. According to the velocity formula, the velocity of impact is calculated.

Figure 1. Targets support.

2.1. Preparation of projectile and samples

The targets are circular sheets with diameter of 240 mm, thickness 1 mm and the materials chosen are unalloyed copper (cu) and aluminum 1100 (al). Due to using three-layer targets, eight possible modes of layers arrangement are considered.

The material of projectiles used in the experiments is silver steel which are hardened to 54 Rockwell to prevent them from sustaining plastic deformation while impacting with targets. The dimension and geometry of all projectiles is depicted in Fig. 2. It should be noted that all projectile have a mass of 11.30 g.

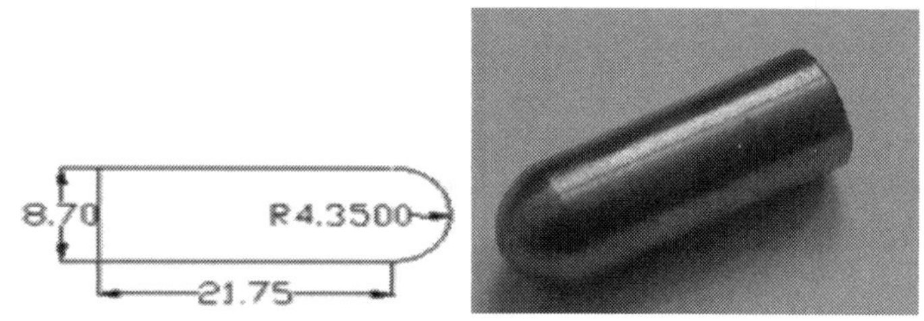

Figure 2. Projectile dimension and geometry

2.2. Tensile test and material analysis

To obtain the properties of projectiles and targets, tensile tests and material analysis are implemented. According to the ASTM-E8 standard [14] that is used for non-ferrous sheets, three samples are formed for each target using a milling machine. After tensile test, the yield stresses for aluminum and copper are specified as 108.54 Mpa and 81.92 Mpa respectively. On the other hand for specifying the material analysis for aluminum and copper targets, a sample is prepared and analyzed with material analysis machine and the percentage of each constitutive material is obtained.

3. Experimental Results

Ballistic limit velocity is the minimum velocity need to create complete penetration in the target so that the projectile can cross the target and also the residual velocity after crossing reaches zero. Thus to compute the ballistic limit velocity, projectile is shot with a velocity greater than ballistic limit and then shooting velocity is decreased so that the projectile could not penetrate the target thoroughly. Then the velocities between these two limits are tested and by averaging the minimum velocity of complete penetration and maximum velocity that projectile could not cross the target, the ballistic limit velocity of the projectile-target system is obtained.

The results show that by decreasing the initial velocity of projectiles, velocity downfall is increased while the amounts of absorbed energy do not change significantly. The fracture mechanism varies between Plugging and petaline. Usually the fracture mechanism occurs between these two modes. While a projectile impacts the target, the back face of the target bulges near the impact zone. If the fracture mode occurs in form of Plugging, the cut zone that is called plug can be removed from the target wholly with no crack (Figs. 3.6 and 3.7). If the fracture mode occurs in form of petaline, the bulged zone is divided into the zones like petals and there are cracks around the hole (Figs. 3.1, 3.3, 3.5 and 3.8).

These experiments declared that in high velocity, the dominant fracture mechanism is plugging and by decreasing the velocity, the fracture mechanism tends to petaline. It should be noted that in the petaline mode there are small plug removed from the target but their size is smaller relative to plugging mode (Fig. 4).

In fact by decreasing the impact velocity, developed cracks are enlarged, the number of petals is decreased and their sizes are enlarged while the size of the removed plugs is shrunk. When the fracture mechanism is between petaline and plugging, the cracks are fine and the number of petals is generous while the size of petals is too small and their shape is like a rectangle. For example in sixth target (Fig. 3.6), the last layer has sustained petaline mode with 4 petals near the ballistic limit and by increasing the velocity, the number of petals is increased and their size is shrunk so that in velocities far from the ballistic limit, the fracture mode tends to Plugging mode and the diameter of plugs is enlarged and tends to diameter of cylindrical part of the projectile. The diagrams of the residual velocities versus initial velocities of projectile are plotted in Fig. 5 to Fig. 12.

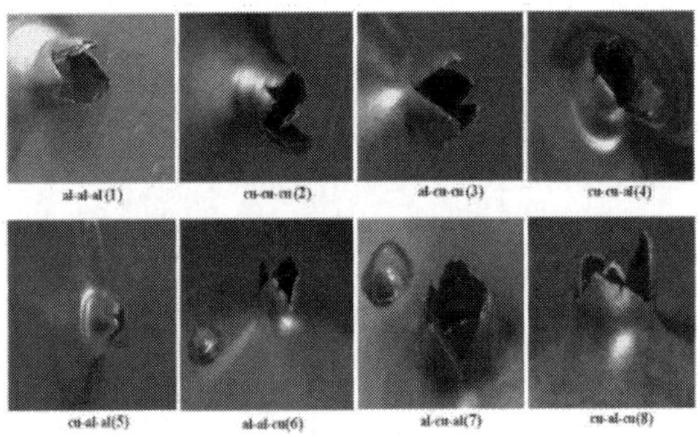

Figure 3. Fracture mechanism of targets; (3.1): Vi =169.17 m/s, (3.2): Vi = 165.46 m/s, (3.3): Vi = 180.91 m/s, (3.4): Vi =159.94 m/s, (3.5): Vi = 174.48 m/s, (3.6): Vi =169.17 m/s, (3.7): Vi =180.91 m/s, (3.8): Vi =165.46 m/s

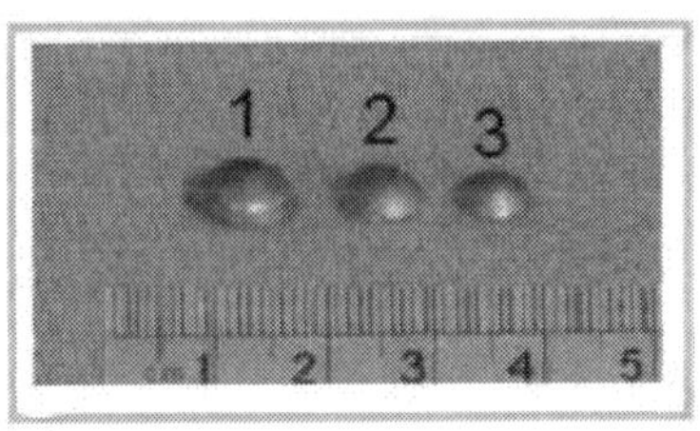

Figure 4. Plugs created due to different firing velocity.

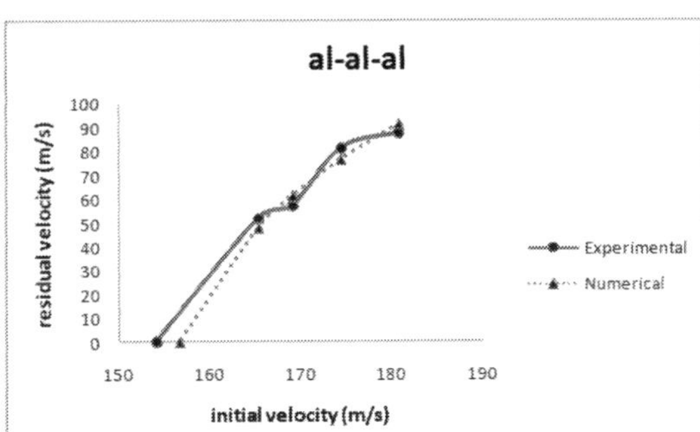

Figure 5. Diagram of residual velocity - initial velocity for target No. 1

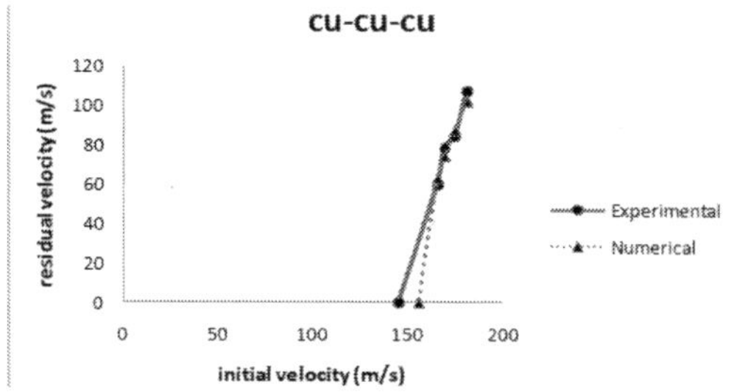

Figure 6. Diagram of residual velocity - initial velocity for target No. 2

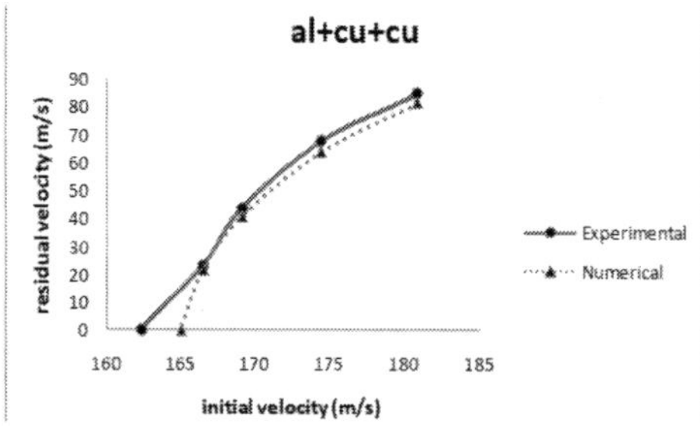

Figure 7. Diagram of residual velocity - initial velocity for target No. 3

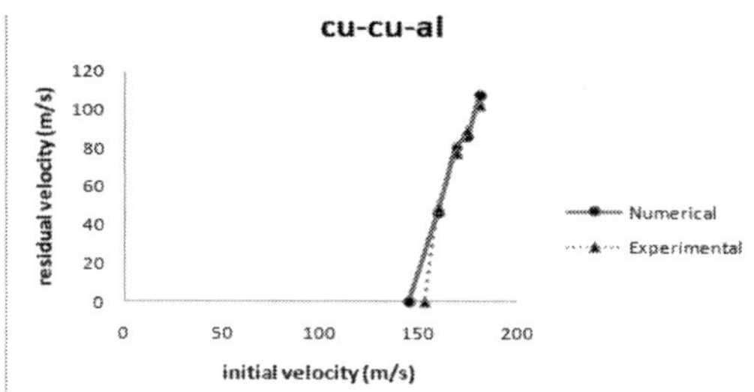

Figure 8. Diagram of residual velocity - initial velocity for target No. 4

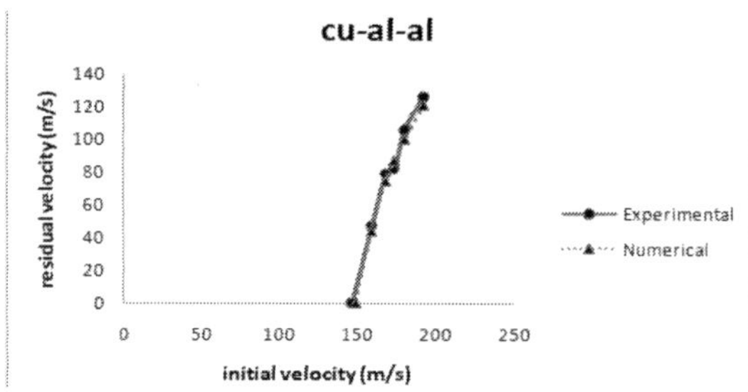

Figure 9. Diagram of residual velocity - initial velocity for target No. 5

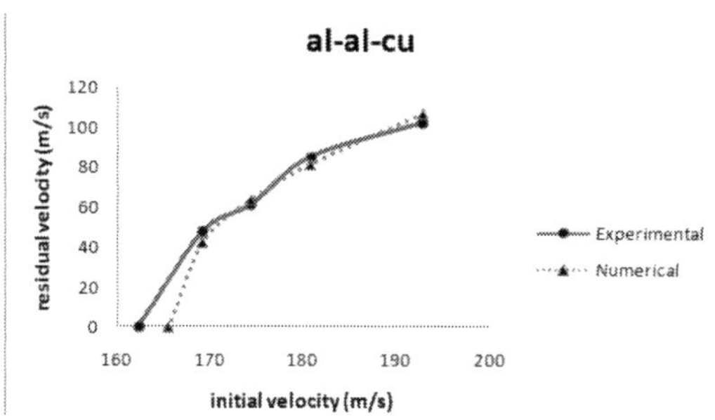

Figure 10. Diagram of residual velocity - initial velocity for target No. 6

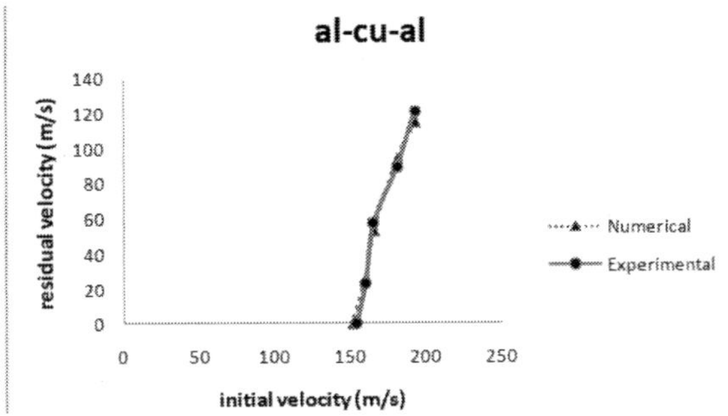

Figure 11. Diagram of residual velocity - initial velocity for target No. 7

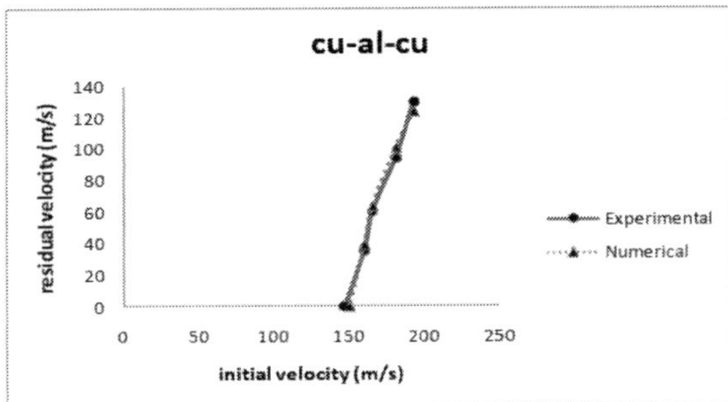

Figure 12. Diagram of residual velocity - initial velocity for target No. 8

4. Perusing the effect of different parameters

To study sequence of layers arrangement, 8 possible arrangements of layers are considered. The difference between maximum and minimum ballistic velocity is about 16.67 m/s. The maximum ballistic velocity belongs to targets number 3 and 6 and is about 162.3 m/s and the minimum ballistic limit velocity belongs to targets number 2 and 4 and is about 145.63 m/s.

4.1. Comparing target number 1 (al-al-al) with target number 2 (cu-cu-cu)

The results show that if all of the layers are chosen from the hard material (in this case; aluminum), the resistance against penetration is more in comparison with the state that all of the layers are chosen from a more ductile material (copper). Consequently target number 1 with ballistic limit velocity of 154.11 m/s is more resistant than target number 2 with ballistic limit velocity of 145.63 m/s.

4.2. Comparing target number 3 (al-cu-cu) with target number 4 (cu-cu-al)

The results in this state show that if the harder layer is placed as the first layer, its resistance is more in comparison with the state that it is placed as the last layer (from left to right). Consequently target number 3 with ballistic limit velocity of 162.3 m/s is more resistance than target number 4 with ballistic limit velocity of 145.63 m/s.

4.3. Comparing target number 5 (cu-al-al) with target number 6 (al-al-cu)

The results in this state show that if the layer that is more ductile is placed after two harder layers, its resistance against penetration is more in comparison with the state that it is placed as the first layer (from left to right). Consequently target number 6 with ballistic limit velocity of 162.3 m/s is more resistance than target number 5 with ballistic limit velocity of 147.13 m/s.

4.4. Comparing target number 7 (al-cu-al) with target number 8 (cu-al-cu)

The results in this state show that if the layer that is more ductile is placed between two harder layers, its resistance against penetration is more in comparison with the state that harder layer is placed between two layers that are more ductile (from left to right). Consequently target number 7 with ballistic limit velocity of 154.11 m/s is more resistance than target number 8 with ballistic limit velocity of 147.13 m/s.

5. Conclusion

In the present study the penetration of projectile with semi-sphere head into the targets with one layer and three layers is perused experimentally as well as numerically and the numerical results compared with experimental ones. In most cases, the error was arising from numerical simulation was observed to be less than 10%. The result of this study can be briefed as below:

The sequence of layers arrangement according to the used material effects the ballistic operation and can change it.

Projectile with semi-sphere head, pierces targets with fracture mode between petaline and Plugging and the projectile velocity has a significant effect on the fracture mode. Increasing the velocity has the fracture take the Plugging mode and decreasing the velocity make the fracture happen in petaline mode.

Decreasing initial velocity of projectile increases the radius of plastic deformation zone near the impact zone.

It is declared in the present study that targets uses layers with different materials are more resistant in comparison with targets that are homogeneous.

6. References

[1] A.I.O. Zaid, A. El-kalai, and F.W. Travis, An examination of the perforation of mild steel plate by a flat-ended cylindrical projectile, Int. J. Mech. Sci., 15(2) (1973) 129-143.

[2] M. E. Backman, and W. Goldsmith, The mechanics of penetration of projectiles into targets, International Journal of Engineering Science, Vol. 16, (1978)1-99.

[3] I. Marom, and S. R. Bodner, Projectile perforation of multi-layered beams, International Journal of Mechanical Science, Vol. 21(1979) 489-504.

[4] R. S. J. Corran, P. J. Shadbolt, and C. Ruiz, Impact loading of plates an experimental investigation, International Journal of Impact Engineering, Vol. 1 (1983) 3-22.

[5] GN. Nurick and CE. Walters, The ballistic penetration of multiple thin plates separated by an air gap, Proceedings of SEM conference on experimental mechanics, The Society of Experimental Mechanics, Inc., (1990) 631–7.

[6] R. L. Woodward, S. R. Tracey, and I.G. Crouch, The response of homogeneous and laminated metallic sheet material to ballistic impact, J. de Physique IV, 01(C3) (1991) C3-277-C3-282

[7] J. Radin, and W. Goldsmith, Normal projectile penetration and perforation of layered targets. International Journal of Impact Engineering, 7(2) (1988) 229-259.

[8] G. Ben-Dor, A. Dubinsky, T. Elperin, On the order of plates providing the maximum ballistic limit velocity of a layered armor, International Journal of Impact Engineering, Vol. 22 (1999) 741-755.

[9] C.K. Chao, and R.C. Chang, Damage evaluation of two-layer targets impacted by a projectile, International Journal of Impact Engineering, Vol. 24 (2000) 299-311.

[10] T. Borvik, M. Langseth, O.S. Hopperstad, and K.A. Malo, Perforation of 12mm thick steel plates by 20mm diameter projectiles with flat, hemispherical and conical noses Part I: Experimental study, International Journal of Impact Engineering, Vol. 27 (2002) 19-35.

[11] P. Elek, S. Jaramaz, and D. Mickovic, Modeling of perforation of plates and multi-layered metallic targets, International Journal of Solids and Structures, Vol. 42 (2005) 1209-1224.

[12] S. Dey, T. Borvik, X. Teng, T. Wierzbicki, and O.S. Hopperstad, On the ballistic resistance of double-layered steel plates: An experimental and numerical investigation, International Journal of Solids and Structures, Vol. 44 (2007) 6701-6723.

[13] N.K. Gupta, M.A. Iqbal, G.S. Sekhon, Effect of projectile nose shape, impact velocity and target thickness on the deformation behavior of layered plates, International Journal of Impact Engineering, Vol. 35 (2008) 37-60.

[14] ASTM, MetalsMechanical Testing, Elevated and Low Temperature Tests Metallography, Volume 03.01 (1998) 152-158.

IOP Conf. Series: Materials Science and Engineering **398** (2018) 012032 doi:10.1088/1757-899X/398/1/012032

A Fuzzy Satisfactory Optimization Method Based on Stress Analysis for a Hybrid Composite Flywheel

Liping Yang[1], Curran Crawford[2], Zhengyi Ren[1]

[1] Research Institute of Energy Storage Technology and Application, Harbin Engineering University, Harbin, Heilongjiang, 150001, China
[2] Department of Mechanical Engineering, University of Victoria, Victoria, BC, V8W 2Y2, Canada

E-mail: yanglipingylp02@ 126.com

Abstract. Considering the cost evaluation and the stress analysis, a fuzzy satisfactory optimization (FSO) method has been developed for a hybrid composite flywheel. To evaluate the cost reasonable, the cost coefficients of the flywheel components are obtained through calculating the weighted sum of the scores of the material manufacturability, the structure character and the material price. To express the satisfactory degree of the energy, the cost and the mass, the satisfactory functions are proposed by using the decline function and introducing a satisfactory coefficient. To imply the different significance of the objectives, the object weight coefficients are defined. Based on the stress analysis of composite material, the circumferential and radial stresses are considered into the optimization formulation. The simulations of the FSO method with different weight coefficients and storage energy density optimization (SEDO) method of a flywheel are contrasted. The analysis results show the FSO method can satisfied different requirement of the designer, the FSO method with suitable weight coefficients can replace the SEDO method.

1. Introduction

As the core components of a Flywheel Energy Storage System (FESS), the flywheel structure is very important not only for storage capacity, but also for safety and manufacturing cost of the FESS. Since the accurate data of the cost of a flywheel is not available in cost optimization design, Krack M proposed that the cost could be normalized by the price of composite material [1]. Lin C calculated the manufacturing cost of a motor rotor (similar to a flywheel) considering the effect of the dimension and the punching cost [2]. Yan X designed per unit mass price of glass fiber as 1[3]. However, it can be found that the cost of a flywheel is also affected by the structure and manufacturing process. Hence, how to evaluate the cost of a flywheel should be deeply researched. As the structure of the flywheel decides how much kinetic energy can be stored by the whole system, many researchers used the kinetic energy as the optimization object to design a flywheel [4], [5]. However, too much mass will bring inconvenience to the assembly and transportation of the flywheel especially for the application in mobile machine. Hence, Storage energy density (SED) be used as another optimization object [6]-[8]. Along with the development of the composite material and the magnetic suspended bearing technologies, the cost of the flywheel becomes an important factor affecting the application of the FESS. Hence, how to obtain big SED and low cost simultaneously is becoming more and more important.

On the other hand, the safety of a high velocity flywheel is very important. Curtiss D calculated an equation which expresses the relationship between the stresses and flywheel parameters [9]. Krack M used the equation as a constraint for the optimization design to calculate the maximum angular velocity [1]. Ha S also considered the stress into the design optimization to calculate the optimum radius under the condition of the maximum angular velocity being fixed [10]. But all the researchers did not use the stress equation for calculating the maximum angular velocity and the maximum outer radius simultaneously.

In this study, a new flywheel cost evaluation method is developed. Then, to express the satisfactory degrees for different objectives reasonably, the FSO method based on fuzzy satisfactory function and the object weight coefficients is proposed. Moreover, to guarantee the safety of the high velocity flywheel, based on the analysis of the relationship between the maximum angular velocity, maximum outer radius and stresses, FSO method based on the calculation of stresses is proposed. At last, the simulation results contrast of different optimization design methods shows the effective of the proposed method.

2. Cost Evaluation for Flywheels

The traditional cost is calculated only according to the price of materials, it is clear that this evaluation method has great limitations. We know that although being made by same material, the flywheels with different structure have different costs. It is also obvious that some material is easy to be machined, but some is not. Hence, we proposed that the cost of a flywheel is influenced not only by the material price, but also by the manufacturing process and the material machinability. To evaluate the cost more reasonably, the flywheel is divided into m components according to the material character and the flywheel structure. For every component, there are some main indicators being used to evaluate the flywheel cost, such as material manufacturability, structure character and material price. Material manufacturability indicates the material is easy or not to be formed and machined. Structure character means the structure is easy or not to be manufactured. Therefore, the evaluation indicator set of the ith (i=1,2,…,m) component of a flywheel is described as $\psi_i=\{f_1^i, f_2^i, f_3^i\}$={material manufacturability, structure character, material price}.

To evaluate the performance of each element f_j^i, the score of $x_{ij}\in[0,100]$ (i=1,2,…,m; j=1,2,3) is given to f_j^i. The higher the score for f_1^i is, the more difficulty the material manufacturing process is. On the other hand, the higher the score for f_2^i is, the more complicated the structure is. Similarly, the higher the score for f_3^i is, the more expensive the material price is.

According to the expert evaluation, the weight vector is: $W=(w_1,w_2,w_3)$, $\sum_{j=1}^{3}w_j=1$, $0<w_j<1$.

Therefore, the cost coefficient of the ith (i=1,2,…,m) component is

$$C_{di}=\sum_{j=1}^{3}\frac{w_j x_{ij}}{100} \tag{1}$$

According to the cost coefficient and mass distribution, the flywheel equivalent cost can be calculated as

$$D_r=\sum_{i=1}^{m}C_{di}\rho_i V_i(d_i,h_i) \tag{2}$$

where ρ_i is the density of the ith component, d_i and h_i are the outer diameter and the height of the ith component, $V_i(d_i,h_i)$ is the volume of the ith component, which is the function about d_i and h_i.

3. Multi-objective Optimization Based on FSO Method

High storage energy and low cost usually are the purpose of the flywheel optimization problem. Moreover, less mass will makes the transportation more convenient and will decrease the rotational inertia which will makes the flywheel rotate more stable. Therefore, the maximum storage energy, the cost and the mass are the optimization objects of the flywheel. In order to express the optimization objects, $M_r(X)$, $D_r(X)$ and $E_k(X)$ are defined as the maximum kinetic energy, the equivalent cost and the mass of the flywheel. $X=(\omega,d_i,h_i)$ (i=1,2,…,m) is the decision vector, where ω is the maximum angular velocity of the flywheel, d_i and h_i are outer diameter and height of the ith component. According to the flywheel working principle, $M_r(X)$, $D_r(X)$ and $E_k(X)$ can be represented as

$$M_r(X)=\sum_{i=1}^{m}\rho_i V_i(d_i,h_i) \tag{3}$$

$$D_r(X)=\sum_{i=1}^{m}C_{di}\rho_i V_i(d_i,h_i) \tag{4}$$

$$E_k(X)=\frac{1}{2}J(d_i,h_i)\omega^2 \tag{5}$$

where $J(d_i,h_i)$ is the rotational inertia which is the function of d_i and h_i.

For different designers, their satisfactory degrees for the objective values are different. For this reason, the fuzzy satisfactory degree functions are introduced. For multi-objective optimization problem in fuzzy environment, there are various kinds of membership functions. To simplify the calculation and represent the desirability of designers, the triangle-like membership functions are used as the fuzzy satisfactory degree functions for the flywheel optimization objectives [11]. In order to adjust the triangle-like membership functions easily, a satisfactory coefficient of $\beta(0.5<\beta\le 1)$ is used to describe membership functions. Hence, the satisfactory degree functions of the optimization objects are represented as

$$\mu_M(X)=\begin{cases}1 & M_{r\min}\le M_r(X)\le(1-\beta)\ M_{r\max}+\beta M_{r\min}\\ \dfrac{M_r(X)-(1-\beta)\ M_{r\max}-\beta M_{r\min}}{\beta(M_{r\max}-M_{r\min})} & (1-\beta)\ M_{r\max}+\beta M_{r\min}<M_r(X)\le M_{r\max}\end{cases} \tag{6}$$

$$\mu_D(X)=\begin{cases}1 & D_{r\min}\le D_r(X)\le(1-\beta)\ D_{r\max}+\beta D_{r\min}\\ \dfrac{D_r(X)-(1-\beta)\ D_{r\max}-\beta D_{r\min}}{\beta(D_{r\max}-D_{r\min})} & (1-\beta)\ D_{r\max}+\beta D_{r\min}<D_r(X)\le D_{r\max}\end{cases} \tag{7}$$

$$\mu_E(X)=\begin{cases}\dfrac{E_k(X)\text{-}E_{k\min}}{\beta(E_{k\max}\text{-}E_{k\min})} & E_{k\min}\le E_k(X)\le\beta E_{k\max}+(1-\beta)\ E_{k\min}\\ 1 & \beta E_{k\max}+(1-\beta)\ E_{k\min}<E_k(X)\le E_{k\max}\end{cases} \tag{8}$$

where $M_{r\max}$ and $M_{r\min}$, $D_{r\max}$ and $D_{r\min}$, $E_{k\max}$ and $E_{k\min}$ are upper and lower bounds of $M_r(X)$, $D_r(X)$ and $E_k(X)$ respectively. As an example, the fuzzy satisfactory degree function of flywheel mass is shown in Fig. 1.

To realize the multi-objective optimization along with the different significance of the objectives, the weight coefficients of ε ($0<\varepsilon\le 1$) and γ ($0<\varepsilon+\varepsilon\gamma\le 1$) are defined. The ε is the coefficient of the energy significance, and the γ is the coefficient of the cost significance relative to the energy. Accordingly, the multi-objective optimization formulation based on the satisfactory function is represented as

$$\begin{aligned}&\text{Maximize:}\quad \varepsilon[\mu_E(X)+\gamma\mu_D(X)]+[1-(1+\gamma)\varepsilon]\mu_M(X)\\ &\qquad\qquad\quad X_{\min}\le X\le X_{\max}\\ &\text{Subject To:}\quad 0<\varepsilon<1\\ &\qquad\qquad\quad 0<\varepsilon(1+\gamma)\le 1\end{aligned} \tag{9}$$

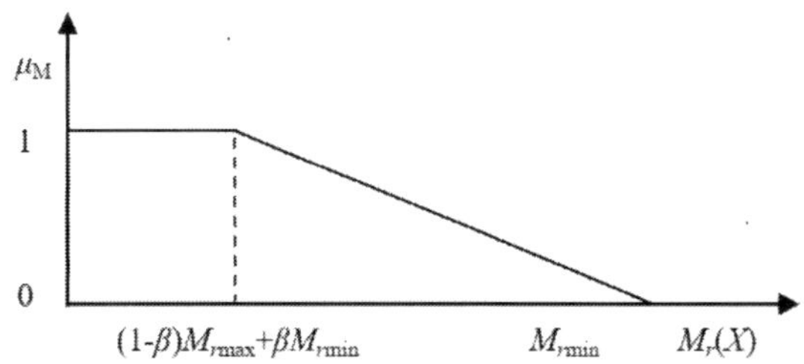

Figure 1. Fuzzy satisfactory membership function of mass.

4. Stress Analysis for Flywheel Optimization Design

As the safety of the flywheel is very important, the influence of the composite material stress should be fully considered. Since the stresses on the outer ring have the greatest influence on the flywheel safety, the relationship between the stress and the material parameters and the size of this component is analysed. If defining $\alpha = R_i/R_o$ and $x = r/R_o$, where R_i and R_o are the disk inner and outer radius, r ($R_i \le r \le R_o$) is the radius of select point, according to the research of [9], the radial stress σ_r and circumferential stress σ_θ can be derived as

$$\sigma_\theta = \rho v_e^2 (c_1 \lambda x^{\lambda-1} - c_2 \lambda x^{-\lambda-1} - c_3 x^2) \quad (10)$$

$$\sigma_r = \rho v_e^2 (c_1 x^{\lambda-1} + c_2 x^{-\lambda-1} - c_4 x^2) \quad (11)$$

where ρ is material density, v_{θ_r} is Poisson ratio of contraction in radial direction due to circumferential θ tension, v_e is the maximum disk rim velocity $\lambda = \sqrt{E_0/E_r}$, E_0 and E_r are radial elastic modulus and circumferential elastic modulus, $c_1 = \frac{3+v_{\theta_r}}{9-\lambda^2}\frac{\alpha^{3+\lambda}-1}{\alpha^{2\lambda}-1}$, $c_2 = \frac{3+v_{\theta_r}}{9-\lambda^2}\frac{\alpha^{2\lambda}-\alpha^{3+\lambda}}{\alpha^{2\lambda}-1}$, $c_3 = \frac{\lambda^2+3v_{\theta_r}}{9-\lambda^2}$, $c_4 = \frac{3+v_{\theta_r}}{9-\lambda^2}$.

As $v_e = \omega R_o$ and the stresses should be less than the allowable stresses, from (10) and (11), the following formulas are derived as

$$\omega R_o \le \sqrt{[\sigma_\theta]/\rho\sigma_{\theta in}} \quad (12)$$

$$\omega R_o \le \sqrt{[\sigma_r]/\rho\sigma_{rin}} \quad (13)$$

where $\sigma_{\theta in} = c_1 \lambda x^{\lambda-1} - c_2 \lambda x^{-\lambda-1} - c_3 x^2$ and $\sigma_{rin} = c_1 x^{\lambda-1} + c_2 x^{-\lambda-1} - c_4 x^2$ are the circumferential stress coefficient and radial stress coefficient, $[\sigma_\theta]$ and $[\sigma_r]$ are the circumferential allowable stress and radial allowable stress.

The maximum of $\sigma_{\theta in}$ and σ_{rin} can be calculated for a kind of composite material. Therefore, for a fixed value of α, the multi-objective optimization formulation considering stresses constraints is represented as

$$\begin{aligned} \text{Maximize:}\quad & \varepsilon[\mu_E(X)+\gamma\mu_D(X)]+[1-(1+\gamma)\varepsilon]\mu_M(X) \\ & X_{\min} \le X \le X_{\max} \\ & 0<\varepsilon<1 \\ \text{Subject To:}\quad & 0<\varepsilon(1+\gamma)\le 1 \\ & \omega R_o \le \min\{\sqrt{[\sigma_\theta]/\rho\sigma_{\theta in\max}}, \sqrt{[\sigma_r]/\rho\sigma_{rin\max}}\} \\ & R_i/R_o=\alpha \end{aligned} \quad (14)$$

where X_{min} and X_{max} are the minimum value and maximum value of X, $\sigma_{\theta in\max}$ and $\sigma_{rin\max}$ are the maximum of σ_{rin} and $\sigma_{\theta in}$ respectively.

5. Comparison of Simulation Results

To analyse the presented multi-objective fuzzy satisfactory optimization, a composite material flywheel structure is designed as shown in Fig. 2.

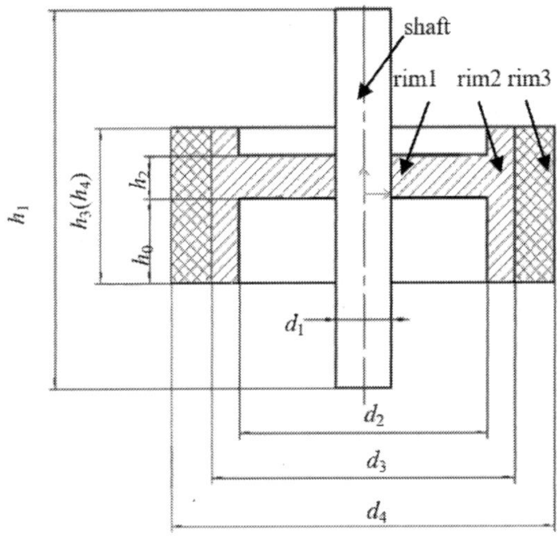

Figure 2. A flywheel structure.

According to the material and the structure, the flywheel is divided into 4 components: shaft, rim1, rim2 and rim3. The material of shaft is steel. The material of rim1 and rim2 is aluminium alloy. The material of rim3 is carbon fiber. Because outer ring of this flywheel is rim3, there are $R_i = d_3/2$, $R_o = d_4/2$. The variables include diameters of the 4 components d_1, d_2, d_3, d_4 (m) and the maximum angular velocity ω (rad/s). Setting $\alpha = 0.5$, there are $X = (\omega, d_1, d_2, d_4)$. According to the design requirement, there are $X_{min} = (200, 0.06, 0.15, 0.6)$ and $X_{max} = (2000, 0.08, 0.25, 0.8)$. Other structure parameters of the flywheel are $h_1 = 0.5$ m, $h_2 = 0.1$ m, $h_3 = h_4 = 0.3$ m. The densities of steel, aluminium alloy and carbon fiber are 7800Kg/m^3, 2700Kg/ m^3 and 1780Kg/ m^3 respectively.

For material manufacturability, the manufacturing processes of the steel and aluminium alloy are obviously different from the process of the carbon fiber. Because the aluminum alloy material belongs to the non-ferrous metal, the hardness is inferior to the steel, the processing difficulty must be smaller than the steel. Due to the high requirement of the winding technology of carbon fiber flywheel, the technology maturity is not as good as the metal manufacturing process. Therefore, the material manufacturability scores of the 4 components are: $x_{11} = 30$, $x_{21} = 20$, $x_{31} = 20$, $x_{41} = 80$.

On the other hand, the structure character of the shaft is relatively complex. It not only requires strict dimensional accuracy, rigorous surface roughness and surface hardness, but also requires high shock resistance. On contrast, the structure character of rim1 and rim2 mainly reflects in the complex shape. Additionally, the demands on dimensional accuracy and surface roughness of the inner hole surface of rim1 are more rigorous than the outer circle surface of rim2. The rim3 is winded by carbon fiber which has regularly shape and lower dimensional accuracy. Hence, the score of structure character of rim3 is lowest in the scores of the 4 components. Therefore, the score of the structure character of the 4 components are: $x_{12} = 90$, $x_{22} = 50$, $x_{32} = 40$, $x_{42} = 10$.

Based on the market prices of steel, aluminium alloy and carbon fiber, the scores of the material price are: $x_{13} = 5$, $x_{23} = 12$, $x_{33} = 12$, $x_{43} = 100$.

According to the importance degree, the weight set of the evaluation indicators is: *W*= (0.3, 0.2, 0.5). Therefore, from (1), the cost coefficient of the *i*th (*i*=1,2,...,4) component is calculated respectively as: $C_{d1}=0.295, C_{d2}=0.22,\ C_{d3}=0.20, C_{d4}=0.76$.

According to (2) the equivalent cost of the flywheel is: $D_r=(2.3V_1+0.59V_2+0.54V_3+1.35V_4)\times10^3$.

According to (3), (4), (5) , the $M_r(X)$, $D_r(X)$ and $E_k(X)$ can be represented as

$$E_k(X)=\frac{\pi\times10^3}{4}\omega^2(3.63d_1^4-0.54d_2^4+0.28d_3^4+0.53d_4^4)$$

$$D_r(X)=\frac{\pi\times10^3}{4}(1.091d_1^2-0.1d_2^2-0.24d_3^2+0.4d_4^2)$$

$$M_r(X)=\frac{\pi\times10^3}{4}(3.63d_1^2-0.54d_2^2+0.28d_3^2+0.54d_4^2)$$

The parameters of the carbon fiber are v_{θ_r}=0.3, $E_0=203.0$ GPa, $E_r=11.2$ GPa, $\rho=1780$ Kg/m^3 [12]. Hence, λ can be derived as: $\lambda=\sqrt{E_0/E_r}=4.257$. According to (12) and (13), the values of c_1, c_2, c_3 and c_4 can be calculated as: c_1 = -0.36 , c_2 = -0.0014 , c_3 = -2.085 , $c_4=-0.362$ .Hence, $\sigma_{\theta inamx}$ and $\sigma_{rin\max}$ are calculated as $\sigma_{\theta inamx}$ = 0.617 , $\sigma_{rin\max}=0.056$.

Since the longitudinal and transverse tensile strength of carbon fiber are $[\sigma_\theta]$=3500MPa and $[\sigma_r]$ =56MPa [10], we can obtain: $\sqrt{[\sigma_\theta]/(\rho\sigma_{\theta in\max})}=1786$, $\sqrt{[\sigma_r]/(\rho\sigma_{rin\max})}=749.5$.

5.1. Comparison of FSO with Different β, ε and γ

To verify the FSO method, the optimum results of different β, ε and γ are listed in Table 1.

Table 1. The Optimum Results of FSO with Different Coefficients.

Indicators	ε=0.8, γ=0.2		ε=0.3, γ=1		ε=0.3, γ=2	
	β=0.8	β=1	β=0.8	β=1	β=0.8	β=1
ω (rad/s)	1949	1899	2000	2000	2000	2000
d_1(mm)	60	60	60	60	60	60
d_2(mm)	250	250	194.4	250	250	250
D_4(mm)	769	789	648.1	600	649.9	600
E_k(MJ)	839	886	446.6	323	447	323
D_r	158	167	113.5	95.2	112.8	95.2
M_r(Kg)	359	381	261.4	213	252	213
μ_E	1.0	0.80	0.54	0.31	0.54	0.31
μ_D	0.153	0.23	0.995	1	1	1
μ_M	0.21	0.27	0.946	1	0.94	1

In order to analyse the optimization effect, the values of the object which has the maximum weight among the 3 objects are compared. In the case of ε =0.8 and γ=0.2, the optimum value of energy of β=1 is 5.6% higher than the value of β=0.8. In the case of ε =0.3 and γ=1, the optimum value of mass of β=1 is 18.5% lower than the value of β=0.8. Moreover, in the case of ε =0.3 and γ=2, the optimum value of cost of β=1 is 15.6% lower than the value of β=0.8. All these shows that when β=1, the design requirements of the satisfaction degree is more stringent. On the other hand, in Table I, when ε=0.3 and β=1, the optimum results with different γ are same. But when ε =0.3 and β=0.8, the optimum cost of γ=1 is 0.62% bigger than the cost of γ=2, and μ_D of γ=1 is 0.5% smaller than μ_D of γ=2. It is obvious that the importance of cost increases with the increase of β.

5.2. Comparison of FSO with and without the Stress Constraints

When β=1, the optimum results of FSO with or without the stress constraints are listed in Table 2. Considering the stress constraints, the values of $\omega d_4/2$ under different ε and γ is smaller than the allowable value of 749.5 rad·m/s. On contrast, the optimum results of $\omega d_4/2$ without stress constraints are all bigger than 749.5 rad·m/s. It is obvious that the optimization with the stress constraints can guarantee the safety of the flywheel.

Table 2. The Optimum Results with or without Stress Constraints.

Indicators	ε =0.8, γ=0.2		ε =0.6, γ=0.5	
	With	*without*	*with*	*without*
ω(rad/s)	1899	2000	2000	2000
d_1(mm)	60	60	60	60
d_2(mm)	250	208	250	250
d_4(mm)	789	800	749.4	800
E_k(MJ)	886	1039	796	1036
D_r	167	172.5	149	171
M_r(Kg)	381	399	341	391
$\omega d_4/2$(rad·m/s)	749.2	800	749.4	800

5.3. Simulation Contrast of FSO and SEDO

In order to provide further insight on the FSO, the optimum results of FSO of β=1 and SEDO with same constraints are listed in Table 3.

Table 3. The Optimum Results of FSO and SEDO.

Indicators	FSO		SEDO
	ε =0.8, γ=0.2	ε =0.5, γ=0.5	
ω(rad/s)	1899	2000	2000
d_1(mm)	60	60	60
d_2(mm)	250	250	250
d_4(mm)	789	748.3	748.3
E_k(MJ)	886	791	791
D_r	167	149	149
M_r(Kg)	381	340	340
SED(MJ/Kg)	2.325	2.327	2.327

For the optimum E_k and M_r of FSO, the SED of ε =0.8 and γ=0.2 is 0.08% smaller than that of SEDO. But on the contrary, the SED of ε =0.5 and γ=0.5 is 0.34% bigger than that of SEDO. It can be seen that the optimum ESD can be acquired by the FSO method when the values of the weight coefficients are suitable.

6. Conclusion

An equivalent cost evaluation method was proposed. To evaluate the cost more reasonably, material manufacturability, structure character and material price were considered as the main indicators for the flywheel cost evaluation. Having introduced the fuzzy satisfactory degree functions, the energy coefficient and the cost coefficient, the FSO method was proposed. The triangle-like membership functions are used as the fuzzy satisfactory degree functions for the flywheel optimization objectives. A satisfactory coefficient was used to adjust the triangle-like membership functions. The relationship among the outer diameter, maximum velocity and the flywheel stresses was deduced. Using the maximum of the circumferential stress and radial stress as the constraints, the FSO method considering the stresses constraints was proposed. To verify the effect of FSO, the FSO methods with different satisfactory coefficient and weight coefficients were simulated. The analysis result demonstrated the FSO could realize the different design target. The simulation contrast showed the FSO with stress constraint can ensure the radial stress less than the allowance radial stress.

Additionally, the contrast of the SEDO and the FSO indicated that the FSO could reach a higher SED when the values of parameters are suitable.

7. Acknowledgments

The authors wish to thank the Postdoctoral Scientific Research Developmental Fund of Heilongjiang Province (LBH-Q16049) and National High Technology Research and Development Program 863 of China (2013AA050802) for providing an opportunity to carry out this research work.

8. References

[1] Krack M, Secanell M and Mertiny P 2010 Cost Optimization of Hybrid Composite Flywheel Rotors for Energy Storage *Struct Multidisc Optim* **41** 779–95.
[2] Lin C and Hwang C 2017 Multi-objective Optimization Design Using Amended Particle Swarm Optimization and Taguchi Method for a Six-phase Copper Rotor Induction Motor *Engineering Optimization* **49** 693-708.
[3] YAN X 2012 Cost Optimization Design of Hybrid Composite Flywheel Rotor *Journal of mechanical engineering* **4**(12). 118-26.
[4] Huang J and Fadel G 2000 Heterogeneous Flywheel Modeling and Optimization *Materials and Design* **21**(2) 111-25.
[5] Krzyszt D and Jan T 2009 Two Methods for Optimization of Flywheel *Engineering Optimization* **41**(4) 351-63.
[6] Ha S, Jeong H and Cho Y 1998 Optimum Design of Thick-walled Composite Rings for an Energy Storage System *Journal of Composite Materials* **32**(9) 851-73.
[7] Ha S, Kim J and Han Y 2008 Design of a hybrid composite flywheel multi-rim rotor system using geometric scaling factors *Compos Mater* **42**(8) 771-85.
[8] Kwon J, Kim S, Nasir S and Ha S 2011 Design and Fabrication of Hybrid Composite Flywheel Rotor *World Academy of Science* **60**1869-73.
[9] Curtiss D, Mongeau P and Puterbaugh R 1995 Advanced Composite Flywheel Structural Design for a Pulsed Disk Alternator *IEEE Transactions On Magnetics* **31**(1) 26-31.
[10] Ha S, Kim S, Nasir S and Han S 2012 Design optimization and fabrication of a hybrid composite flywheel rotor *composite structure* **94** 3290-9.
[11] Tiwari R, Dharmar S and Rao J 1987 Fuzzy Goal Programming-An Additive Model *Fuzzy Sets and Systems* **24** 27-34.
[12] Ha S, Yang H and Kim D 1999 Optimal Design of a Hybrid Composite Flywheel with a Permanent Magnet Rotor *Journal of Composite Materials* **33**(16) 1544-75.

IOP Conf. Series: Materials Science and Engineering **398** (2018) 012033 doi:10.1088/1757-899X/398/1/012033

Synthesis and Characterizations of Nickel (II) Oxide Sub-Micro Rods via co-precipitation Methods

Hui Hui Lim[1], Bahman Amini Horri[1,2] and Babak Salamatinia[1,a]

[1]Discipline of Chemical Engineering, School of Engineering, Monash University Malaysia, Jalan Lagoon Selatan, Malaysia
[2]Department of Chemical Engineering, Faculty of Engineering, University of Surrey, Guildford GU2 7XH, United Kingdom

babak.salamatinia@monash.edu

Abstract. Co-precipitation method has been used for the synthesis of single crystalline Nickel (II) Oxide sub-micro rods. A two-step method was used to synthesise NiO sub-micro rods, in which the first step was the co-precipitation process Nickel (II) Nitrate with Ammonium Oxalate to form Nickel (II) Oxalate (NiC_2O_4), which is the precursor for the precipitates. The precipitate precursor then underwent the calcination process to form NiO sub-micro rods. Synthesized materials undergo Field Emission Scanning Electron Microscope (FESEM), particle size measurement via Zetasizer, elemental analysis via Energy-Dispersive X-ray Spectroscopy (EDX), Fourier Transform Infrared Spectroscopy (FT-IR) and Thermogravimetric Analysis (TGA). The morphology of NiO is found to have uniform rod-like crystals with mean z-diameter between 1600 nm to 2600 nm using Zetasizer.

1. Introduction

Nickel (II) oxide (NiO) is an important p-type semiconductor that exists in cubic lattice structures [1]. NiO particles are suitable materials in many applications, such as lithium battery, fuel cell, supercapacitor, photoelectrodes, gas sensor, chemical catalyst, and electrochromic film [2]. This is mainly because NiO exhibits excellent electrochemical performance, optical electronic performance, photoelectric catalytic, magnetic, thermal, mechanical, and chemical properties [3]. Various quasi-one-dimensional (wires, rods, ribbons) of NiO have attracted significant interest in the past years due to their ability to combine several essential diagnostics, imaging, delivery, and dosage properties, and can still be functionalised. These materials offer considerably higher surface areas than the bulk materials as well as able to function as both structural and functional components in devices [4]. Due to the many advantages of the NiO morphology, many unique synthesis methods have been developed over the years.

NiO nanowires, with a diameter range of 16 to 50 nm, have been synthesised on alumina substrate using the vapor-liquid-solid (VLS) technique with gold particles as the catalyst [5]. In addition, similar NiO nanowires can also be synthesised via hydroxylation and dehydration of nickel foil [6]. In this method, NiO particles are synthesised by soaking nickel foils in a lithium hydroxide solution for a day at room temperature before annealing the foils at high temperature. Mesoporous NiO nanowires were synthesised via hydrothermal treatment of nickel chloride with ethylene glycol at 400 °C [7].These mesoporous NiO nanowires have high specific surface areas and high specific capacitance, which become a high lithium storage capacity in lithium-ion cells.

Other than NiO wires, other methods to produce NiO rods have also been developed. NiO nanorods were successfully synthesised using the microwave–assisted method, which was similar to this method. Microwave heating was used with nickel (II) acetate solution to form nickel (II) hydroxide that acted as the precipitate precursor to form NiO [8]. In another study, the microwave-assisted method was used to synthesise polycrystalline NiO nanorods from CH3COO–Ni–OC2H4OH as precipitated precursors [9]. Meanwhile, polycrystalline and single crystalline rod-shaped NiO have synthesised via thermal dehydration of nickel (II) hydroxide that acted as the precipitated precursor [10].
In comparison with other synthesis method, co-precipitation method is a reliable method as the synthesised oxide powders were well-defined and less-agglomerated [11]. Co-precipitation method is believed to be a simple and well-establish method for large-scale synthesis. Various metal oxide particles have been successfully synthesised using the co-precipitation method, such as iron oxide [12], nickel oxide [13], nickel ferrite [14], zinc oxide [15] and tin oxide [16]. Hence, the aim of this research was to synthesise a NiO sub-micro rods via co-precipitation method and to characterise the physiochemical properties of the particles. Particles morphology (FE SEM), particles size measurement (Zetasizer), elemental analysis (EDX), thermal analysis (TGA) and structural analysis (FT-IR) of NiO sub-micro rods were applied to evaluate the performance of the synthesised particles.

2. Materials and methods

All chemicals used in this study were of analytical grade and were used without any modification, unless explained in the methodology. All chemicals are clearly described in Table 1.

Table 1. List of chemicals

Materials	Chemical Formula	Assay
Nickel (II) Nitrate	$Ni(NO_3)_2.6H_2O$	99.999 % trace metal basis
Ammonium Oxalate Monohydrate	$(NH_4)2C_2O_4.H_2O$	$\geq$ 99 %
Ethylene Glycol	$C_2H_6O_2$	$\leq$ 100 %

A two-step method was used to synthesise NiO sub-micro rods, in which the first step was the co-precipitation process of nickel (II) salt with basic oxalate salt to form NiC2O4, which is the precursor for the precipitates. The precipitate precursor then underwent the second steps, which was the calcination process, to form NiO sub-micro rods.
In the co-precipitation step, 12 g/L concentration of nickel (II) salt and 8 g/L concentration of oxalate salt were prepared. Meanwhile, ethylene glycol, as the co-solvent, was used at approximately 40 % of the total amount of reactants in this co-precipitation process. The oxalate salt was stirred and heated up in a beaker (reactor) until the set temperature was reached. Once the desired temperature was reached, the nickel (II) salt was poured into the oxalate salt solution and the timer was started. When the reaction time ended, all the contents of the reactor were immediately poured into centrifuge tubes to centrifuge out the precipitates to prevent further reactions. Once the precipitates have been separated, they were washed with distilled water for three times (pouring distilled water and centrifuging) to ensure that the precipitates were free of co-solvents and unreacted reactants. Each centrifuging step took 10 min or more at 4,500 rpm. The washed precipitates were then dried in the oven at 50 °C for 24 h.
Then, the dried precipitates were ready for calcination, which was the second step for the synthesis of NiO. The calcination process was fixed at 400 °C for 8 h, with a ramping rate, R of 2 °C/min. The calcination process was conducted in a tubular furnace, which was consistently pumped with air to ensure that excess oxygen was available for the calcination reaction. Characterisation tests were performed on the dried precipitates, before and after calcination, for comparison. The characterisation

tests performed in this study were FESEM, TGA, FT-IR, XRD, Zetasizer, ImageJ, and Raman Spectroscopy.

3. Results and discussions

Table 2 lists the co-precipitated NiO sub-micro rods that were obtained at various co-precipitation temperatures and dwelling times. Table 2 also lists the mean size of the sub-micro rods in terms of the mean major axis (length of rod), minor axis (rod diameter), and aspect ratio. The mean size of these sub-micro rods were obtained using ImageJ software from a group of randomly selected samples, with a sample size of 25.

Table 2. Dimensions of NiO sub-micro rods obtained from various co-precipitation temperatures and dwelling times

Sample	Mean Major Axis (μm)	Mean Minor Axis (μm)	Mean Aspect Ratio, AR
NiO50 °C(60 min)	0.883	0.245	3.883
NiO75 °C(30 min)	1.677	0.465	3.738
NiO75 °C(60 min)	2.413	1.077	2.277
NiO100 °C(6 min)	1.346	0.254	5.296
NiO100 °C(7 min)	1.723	0.392	4.724
NiO100 °C(8 min)	1.615	0.432	4.014
NiO100 °C(9 min)	1.882	0.444	4.433
NiO100 °C(10 min)	1.514	0.424	3.788
NiO100 °C(30 min)	3.199	1.243	2.605
NiO100 °C(60 min)	3.946	1.852	2.187

3.1 Morphology (FE SEM)

3.1.1 Effect of co-precipitation temperature. Figure 1. shows FESEM images of NiO synthesised at different temperatures: (a) 50 °C, (b) 75 °C, and (c) 100 °C at a constant dwelling time at 60 min. It was found that the resulting NiO particles were in the form of smooth, cylindrical rod-like crystals. Under higher magnification, a rough structure was observed on the surface of the rod, which could be the outcome of the degradation of organic groups and phase transformation [9].

These NiO rods underwent two stages, which were the nucleation stage of atom formed by metal salt reduction and the crystal growth stage from nuclei through atomic growth [17]. When results from various temperatures at constant reaction time were compared, it was observed that the size of the NiO sub-micro rods had increased with increasing temperature. This can be observed from the increasing mean major axis (0.883 μm, 2.413 μm and 3.946 μm), and mean minor axis (0.245 μm, 1.077 μm and 1.852 μm), as tabulated in Table 2. However, the mean aspect ratio, AR had decreased to 3.883, 2.277 and 2.187 when the co-precipitation temperature was increased. The decreasing trend of AR was because crystals tend to elongate along the c-axis in the crystal lattice when the particles size increases [18].

Figure 1. FESEM images of samples: (a) NiO 50°C(60 min), (b) NiO 75 °C(60 min), and (c) NiO 100 °C(60 min) at constant co-precipitation dwelling time of 60 min

3.1.2 *Effect of co-precipitation dwelling time.* Figure 2 shows the comparison between FESEM images of NiO sub-micro rods synthesised at a constant 100 °C, with different co-precipitation dwelling times. Figure 2 (a) shows the FESEM images of NiO sub-micro rods synthesised at the shortest time, which was at 6 min. No precipitate had formed for dwelling time of shorter than 6 min as no nuclei was formed at the beginning of the co-precipitation reaction, even when the saturation of metal atom concentration was achieved. Crystals growth stage would only start when metal atom concentration had decreased lower than the nucleation concentration [17].

At 6 min, the NiO sub-micro rods have the highest aspect ratio of 5.296 with the smallest mean minor axis at 0.254 μm. Figure 2 (b), Figure 2 (c), Figure 2 (d), and Figure 2 (e) show FESEM images of NiO sub-micro rods synthesised at 7 min, 8 min, 9 min, and 10 min of dwelling time, respectively. These NiO sub-micro rods had shown increments in mean minor axis values (0.392 μm, 0.432 μm, 0.444 μm, and 0.424 μm, respectively), and a decreasing trend of mean aspect ratio values (4.724, 4.014, 4.433, and 3.788, respectively). Additionally, these NiO sub-micro rods had more uniform particles compared to the rods formed at only 6 min of dwelling time.
Figure 2 (f) and Figure 2 (g) show FESEM images of NiO sub-micro rods after longer dwelling times of 30 min and 60 min. The mean minor axis of the NiO sub-micro rods had significantly increased to 1.243 μm and 1.852 μm, respectively with these times. When the dwelling time was increased, more nuclei that were formed during nucleation could reach the crystal growth stage. During this stage, atoms or molecules in the solution will be integrated into the surface of the crystals, contributing to the increase in crystal size [19]. Hence, the size of the NiO sub-micro could increase as the dwelling time increases.

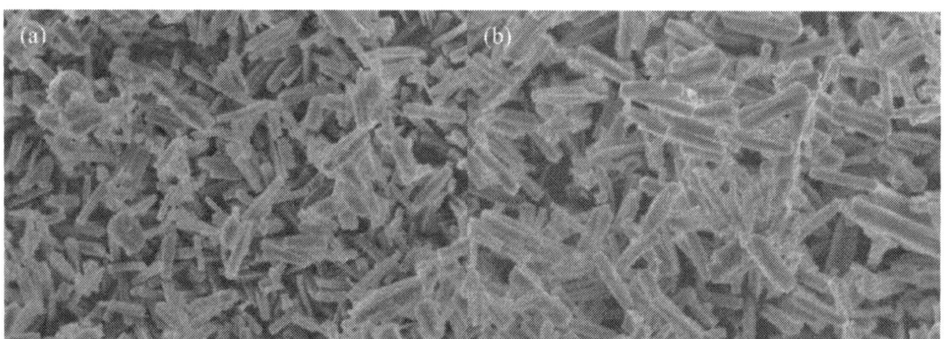

Figure 2. FESEM images of: (a) NiO100(6), (b) NiO100(7), (c) NiO100(8), (d) NiO100(9), (e) NiO100(10), (f) NiO100(30), and (g) NiO100(60) at constant co-precipitation temperature of 100 °C

3.2 Particle size measurement

Figure 3 shows the particles size distribution for various co-precipitation durations at 100 °C, as obtained from the Zetasizer. The particles size distributions for all reaction times were symmetrical bell-shaped curves, which were similar to a normal distribution curve [20]. The particles size distribution has justified that by increasing the reaction time, the size of the NiO sub-micro rods would concurrently increase. Hence, the mean z-diameter was found to be at 1623 nm, 2083 nm, and 2639 nm, with respective to the increased time. From the particle size distribution, it was found that a longer reaction time would give a narrower size range, causing the distribution curve to have a sharper peak. NiO sub-micro rods, which resulted after 60 min of reaction time, had the smallest size range from 1990 – 3580 nm, followed by the rods after 30 min (1484 – 3091 nm), and 10 min (825 – 3580 nm). These results showed that a longer reaction time would give more uniformly sized NiO sub-micro rods.

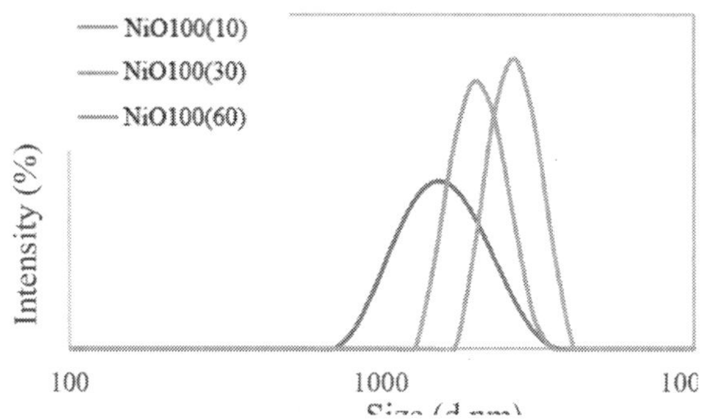

Figure 3. Intensity of particle size distribution of NiO sub-micro rods at different reaction dwelling times

3.3 Elemental analysis (XRD)

Elemental analysis was conducted by scanning FESEM images using an EDX to obtain the elemental composition of the synthesised material. NiC_2O_4 and NiO that were synthesised at a reaction temperature of 100 °C for 10 min were used to conduct the elemental analysis. EDX reported that NiC_2O_4 was made up of mainly oxygen with 58.93 atomic %, 21.64 atomic % of carbon, and 19.43 atomic % of nickel. Using the chemical formula for NiC_2O_4, the theoretical atomic % was determined to be 57.14 %, 28.57 %, and 14.29 % for oxygen, carbon, and nickel, respectively, which showed similar trends with the EDX results.

Since NiO was made up of 1:1 atomic ratio of nickel to oxygen, the theoretical atomic % for nickel and oxygen was 50 %. EDX reported similar results for the synthesised NiO, which consisted of 47.78 atomic % of nickel and 52.22 atomic % of oxygen. The EDX results had also shown that the calcination reaction had caused the NiC_2O_4 to decompose into high purity NiO, with no side reaction that could have produced by-products like nickel or nickel carbonate [21]. This can be further justified by conducting the FT-IR and thermal analyses.

3.4 FT-IR

Figure 4 shows the FT-IR spectra of NiC2O4 and NiO synthesised at 100 °C for 10 min. The FT-IR spectra of NiC2O4 exhibited a weak, broad band at 3390 cm-1 that corresponded with the O-H stretching vibration, and a strong broad band at 1600 cm-1 that correlated to the H-O-H bending vibration of water [22]. Another two sharp peaks were observed at wavenumber 1360 cm-1 and 1310 cm-1, which indicated the C-O vibration in the CO32- group and the C-O stretching of oxalate ion, respectively [21].

Meanwhile, the FT-IR spectra of NiO showed one peak at 550 cm-1 that can be attributed to the Ni-O bond stretching vibration [23]. No NiC2O4 characteristic bonds had appeared in the NiO spectra, indicating that all NiC2O4 had decomposed into NiO during calcination. Hence, the synthesised NiO was of high purity, with no traces of impurities.

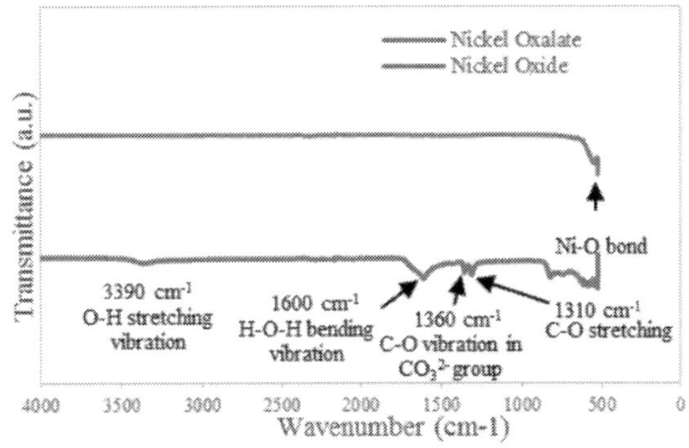

Figure 4. FT-IR Spectra of NiC2O4 and NiO

3.5 *Thermal analysis (TGA)*

TGA curve of NiC_2O_4 and NiO in air at a heating rate of 10 °C/min are shown in Figure 5. TGA curves of NiC_2O_4 clearly showed two well-defined steps, indicating the dehydration and decomposition of NiC_2O_4 upon heating. The first significant weight loss of 17.4 % occurred between 180 °C and 235 °C, which could attributed to the dehydration process of NiC_2O_4 dehydrate to anhydrous oxalate [24]. This step showed an endothermic characteristic with a DTA peak at 225 °C.
On the other hand, the second step weight loss occurred between 315 to 340 °C, which showed an exothermic DTA peak at 330 °C. The weight loss of 34.9 % during the second step could be attributed to the decomposition of anhydrous oxalate to NiO [24]. Hence, a minimum calcination temperature of 340 °C was applied to form NiO from NiC_2O_4 dihydrate. TGA curve of NiO showed no/minimal changes in weight loss, which indicated that the synthesised NiO was thermally stable and did not experience change in state upon extreme heating.

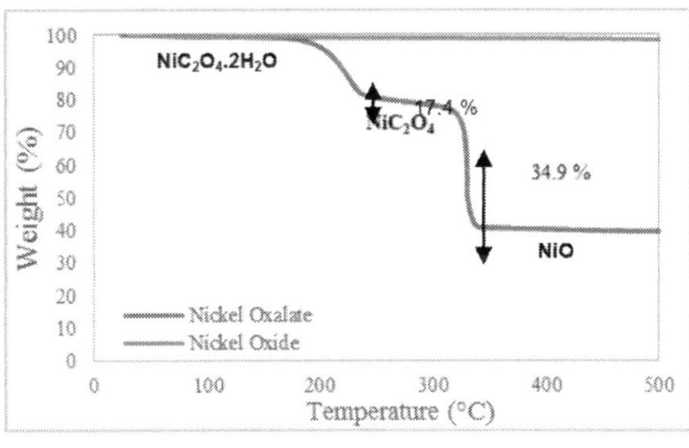

Figure 5. TGA curve of NiC2O4 and NiO in air Conclusion

High purity NiO sub-micro rods have been successfully synthesised using the co-precipitation method. The surface morphology of NiO showed it to be in the form of rod-shaped single crystals, with crystal size that depended on the co-precipitation time and temperature. FESEM images showed that the crystal size had increased with increasing co-precipitation temperature and dwelling time. When the co-precipitation temperature was increased from 50 to 75 oC and 100 oC at a constant dwelling time of 60 min. On the other hand, increasing the dwelling time at 100 oC had led to increasing rod diameter. These results were further proven using zetasising which showed increases in the mean z-diameter of the particles with increasing co-precipition dwelling time. EDX found that the atomic percentages of both oxalate and oxide had tallies with the theoretical atomic percentages. Furthermore, FT-IR was conducted to identify the chemical bonding of the elements. The NiC2O4 sample had peaks at 3390 cm-1 (O-H stretching vibration), 1600 cm-1 (H-O-H bending vibration), 1360 cm-1 (C-O vibration), and 1310 cm-1 (C-O stretching). However, these peaks were not found in the NiO sample and only one peak was found at 550 cm-1 showing no traces of impurities. Lastly, the TGA result of NiC2O4 showed two well-defined steps indicating the dehydration and decomposition of NiC2O4. In the meantime, the TGA result of NiO showed minimal changes in weight loss, thus indicating a thermally stable NiO.

4. References

[1] M. Tadic, D. Nikolic, M. Panjan, and G. R. Blake, Magnetic properties of NiO (nickel oxide) nanoparticles: Blocking temperature and Neel temperature, Journal of Alloys and Compounds, 647, pp. 1061-1068 (2015).

[2] M. El-Kemary, N. Nagy, and I. El-Mehasseb, Nickel oxide nanoparticles: Synthesis and spectral studies of interactions with glucose, Materials Science in Semiconductor Processing, 16, 6, pp. 1747-1752 (2013).

[3] M. M. Ba-Abbad, P. V. Chai, M. S. Takriff, A. Benamor, and A. W. Mohammad, Optimization of nickel oxide nanoparticle synthesis through the sol–gel method using Box–Behnken design, Materials & Design, 86, pp. 948-956 (2015).

[4] T. Ahmad, A. Ganguly, J. Ahmed, A. K. Ganguli, and O. A. A. Alhartomy, Nanorods of transition metal oxalates: A versatile route to the oxide nanoparticles, Arabian Journal of Chemistry, 4, 2, pp. 125-134 (2011).

[5] N. Kaur, E. Comini, N. Poli, D. Zappa, and G. Sberveglieri, Nickel Oxide Nanowires Growth by VLS techniqu for gas sensing application, Procedia Engineering, 120, pp. 760-763 (2015).

[6] K. Sekiya, K. Nagato, T. Hamaguchi, and M. Nakao, Morphology control of nickel oxide nanowires, Microelectronic Engineering, 98, pp. 532-535 (2012).

[7] D. Su, H.-S. Kim, W.-S. Kim, and G. Wang, Mesoporous Nickel Oxide Nanowires: Hydrothermal Synthesis, Characterisation and Applications for Lithium-Ion Batteries and Supercapacitors with Superior Performance, Chemistry – A European Journal, 18, 26, pp. 8224-8229 (2012).

[8] M. A. Bhosale and B. M. Bhanage, Rapid synthesis of nickel oxide nanorods and its applications in catalysis, Advanced Powder Technology, 26, 2, pp. 422-427 (2015).

[9] X. Song and L. Gao, Facile Synthesis of Polycrystalline NiO Nanorods Assisted by Microwave Heating, Journal of the American Ceramic Society, 91, 10, pp. 3465-3468 (2008).

[10] N. Srivastava and P. C. Srivastava, Synthesis and characterization of (single- and poly-) crystalline NiO nanorods by a simple chemical route, Physica E: Low-dimensional Systems and Nanostructures, 42, 9, pp. 2225-2230 (2010).

[11] Z. Shao, W. Zhou, and Z. Zhu, Advanced synthesis of materials for intermediate-temperature solid oxide fuel cells, Progress in Materials Science, 57, 4, pp. 804-874 (2012).

[12] N. D. Kandpal, N. Sah, R. Loshali, R. Joshi, and J. Prasad, Co-precipitation method of synthesis and charactrization of iron oxide nanoparticles, Journal of Scientific & Industrial Research, 73, pp. 87-90 (2014).

[13] R. Kumar, A. Sharma, N. Kishore, and N. Budhiraja, Preparation and Characterization of NiO Nanoparticles Co-Precipitation Method, International Journal of Engineering, Applied and Management Sciences Paradigms, 6, 1 (2013).

[14] K. Velmurugan, V. S. K. Venkatachalapathy, and S. Sendhilnathan, Synthesis of Nickel Zinc Iron Nanoparticles by Coprecipitation Technique, Materials Research pp. 299-303 (2010)

[15] A. J. Ahamed and P. V. Kumar, "Synthesis and characterization of ZnO nanoparticles by co-precipitation method at room temperature, Journal of Chemical and Pharmaceutical Research, 8, 5, pp. 624-628 (2016).

[16] L. I. Nadaf and K. S. Venkatesh, Synthesis and Characterization of Tin Oxide Nanoparticles by Co-precipitation Method, IOSR Journal of Applied Chemistry, 9, 2, pp. 1-4 (2016).

[17] J. Lai, W. Niu, R. Luque, and G. Xu, Solvothermal synthesis of metal nanocrystals and their applications, Nano Today, 10, 2, pp. 240-267 (2015).

[18] R. Ho, D. A. Wilson, and J. Y. Y. Heng, Crystal Habits and the Variation in Surface Energy Heterogeneity, Crystal Growth & Design, 9, 11, pp. 4907-4911 (2009).

[19] P. Cubillas and M. W. Anderson, Synthesis Mechanism: Crystal Growth and Nucleation, in Zeolites and Catalysis, Synthesis, Reactios and Applications, 1, J. Cejka, A. Corma, and S. Zones, Eds. Weinheim: Wiley-VCH Verlag GmbH & Co. KGaA (2010)

[20] E.Weisstein. (2016). Normal Distribution. Available: http://mathworld.wolfram.com/NormalDistribution.html

[21] B. Małecka, A. Małecki, E. Drożdż-Cieśla, L. Tortet, P. Llewellyn, and F. Rouquerol, Some aspects of thermal decomposition of NiC2O4·2H2O, Thermochimica Acta, 466, 1, pp. 57-62 (2007).

[22] S. V. Ganachari, R. Bhat, R. Deshpande, and V. A, Synthesis and characteization of nickel oxide nanoparticles by self-propagatinglow temperature combustion method, Recent Research in Science and Technology, 4, 4, pp. 50-53 (2012).

[23] A. E.-A. A. Said, M. M. A. El-Wahab, S. A. Soliman, and M. N. Goda, Synthesis and Characterization of Nano CuO-NiO Mixed Oxides, Nanoscience and Nanoengineering, 2, 1, pp. 17-28 (2014).

[24] D. Zhan, C. Cong, K. Diakite, Y. Tao, and K. Zhang, Kinetics of thermal decomposition of nickel oxalate dihydrate in air, Thermochimica Acta, 430, 1, pp. 101-105 (2005).

Acknowledgments

This research was funded by Ministry of Science, Technology and Innovation (MOSTI) Science Fund Project (03-02-10-SF0259). We thank Monash University Malaysia for supported this research by providing an equipped research environment. We would also like to show our gratitude to the laboratories technician/staff in Monash who provide expertise that greatly assisted this research.

IOP Conf. Series: Materials Science and Engineering **398** (2018) 012035 doi:10.1088/1757-899X/398/1/012035

Evaluation of Softening Zone in X80 Pipeline Steels under Tension-bending Load

Liu Ligang[1], Zhang Chaofan[1], Liu Yu[2], Sun Shumao[1], and Yu Hui[1]

[1] College of Mechanical Engineering, Yanshan University, Qinhuangdao 066004, China
[2] China Petroleum Pipeline Research Institute, Langfang, 065000, China

E-mail: lgliu@ysu.edu.cn

Abstract. Softening of heat-affected zone of X80 pipeline steel is common in weld joints. The finite element analysis model was established based on ABAQUS software in this paper, then the numerical simulation of the strain capacity of X80 steel pipe was carried out under tension-bending load. Researches show that the softening zone width has little influence on service performance of the pipelines under tension-bending load. When the softening rates are 5%, 10% and 15%, the peak values of load are similar. However, structural instability of the pipeline with softening rate 15% is worse and more serious. The pipeline with design coefficient of 0 can withstand greater deformation under tension-bending load, and it can keep good performance.

1. Introduction

In recent years, high-strength pipelines with anti large deformation have been widely applicated in long-distance oil and gas transportation, and pipeline welding technology has been rapidly developed [1-3]. The heat-affected zone (HAZ) grains are coarse due to the heat welding input , after the phase transformation and recrystallization appear in some places which lead to the softening of HAZ [4-5]. The welding heat-affected zone is an uneven region where cracks or failure commence, of which the deformation behavior is extremely complex under external load [6-7].

The utilize of numerical simulation method can realize data rich, diverse and accurate. A.BASTOLA et al. [8] presented details of an experiment research method for small-scale and full-scale experiments on X80 line pipe specimen containing girth welds. The tensile tests indicated that the yield stress overmatch of approximately 20% for the weld metal, and that the HAZ region is about 7.5% softer compared with base metal at 1% applied strain, leading to the conclusion that HAZ softening may limit the acceptable critical size in an ECA (Engineering Critical Assessment) for X80 pipes. By using the finite element software, Fengping Yang et al. [9] simulated the crack propagation with fluid-solid coupling method. They found that when the plastic uniform elongation was equal to 0.054 and the damage strain energy per volume was equal to 0.64 J/mm3, the X90 Φ1219 × 16.3 mm gas pipeline with 12 MPa internal pressure produced by lean gas could arrest cracking. Satoshi Igi et al. [10] examined the influence of internal pressure on the tensile strain capacity of pipelines and discovered that the critical tensile strain drastically decreased under a high internal pressure condition. The results demonstrated that the tensile strain capacity of pressurized pipes could be predicted by using the R-curve from SENT tests and the crack driving force curves from FE analyses. However, little researches have been focused on the service performance of pipeline considering more complicated cases such as tension-bending load and the resistance to deformation is not completely understood.

IOP Conf. Series: Materials Science and Engineering **398** (2018) 012035 doi:10.1088/1757-899X/398/1/012035

The finite element model was established based on ABAQUS software in this paper, then the numerical simulation of the strain capacity of X80 steel pipe was carried out under tension-bending load considering the influence of internal pressure. By using the above model, the effect of softening zone width on steel pipe's deformation resistance was investigated. Then the softening behaviors with different softening ratios and different design coefficients were evaluated.

2. Materials and methods

2.1. Test materials

X80 pipeline steel was used as research object in the work, and the specimens were taken from the welded joint. The chemical compositions(wt%) of the experimental steel are 0.063 C; 0.28 Si; 1.83 Mn; 0.011 P; 0.0006 S; 0.03 Cr; 0.03 Ni; 0.016 Ti; 0.061 Nb; 0.059 V; 0.22 Mo and remainder Fe.

2.2. Test methods

2.2.1. Tensile test. The stress-strain curve of the experimental steel was obtained by tensile test machine as shown in Figure 1. The mechanical properties of X80 steel pipe shown in Table 1.

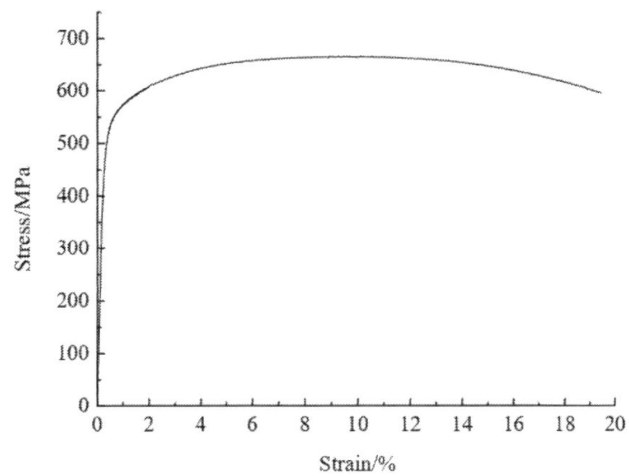

Figure 1. The stress-strain curve

Table 1. Mechanical properties of X80

Material	Yield strength ($R_{t0.5}$/MPa)	Tensile strength (R_m/MPa)	Elastic modulus (E/MPa)	Poisson ratio
X80	530	670	206000	0.3

2.2.2. Metallographic test. The metallographic sample removed from welded joint was mechanically grinded polished and then examined in optical microscopes. The microstructures were revealed by using 4% Nital etching for 10s. As shown in Figure 2, the geometry of welded joint can be seen clearly.

IOP Conf. Series: Materials Science and Engineering **398** (2018) 012035 doi:10.1088/1757-899X/398/1/012035

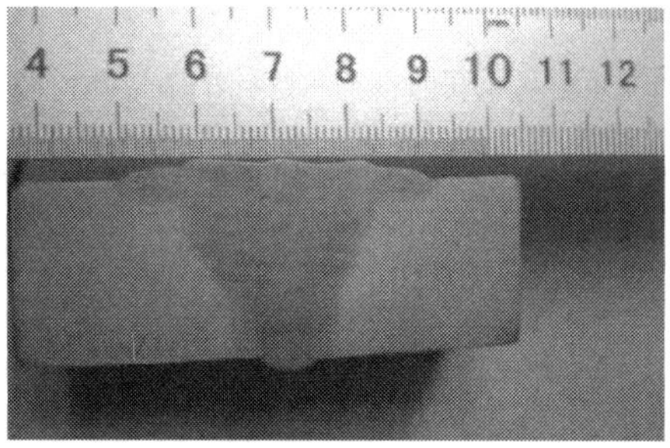

Figure 2. The geometry of welded joint

2.2.3. Vickers hardness test. According to the geometric contour of welded joint, we took a point at about 1.2mm from the top surface of the cover, and draw another point symmetrically to the center line of the weld, then draw a straight line along these two points. In the specimen thickness direction, another straight line parallelled with the first line was drew 3mm from the first line, so do another 7 parallel lines. Vickers hardness tests are conducted on the 9 lines with spacing of 0.03mm. Hardness values of each point on the center line are in Figure 3. Researches show that the hardness of the weld zone is large, and the hardness suddenly decreases in HAZ either of the weld joint which is the softening zone, while the hardness of the base metal is also large. Form the hardness values of different positions, we can obtain the geometry of weld joint shown in Figure 4.

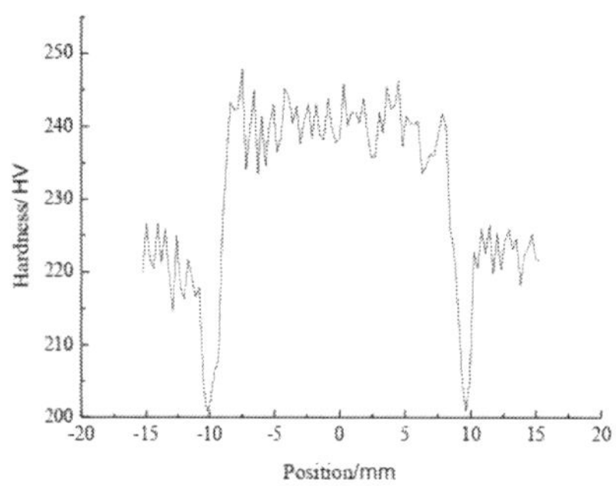

Figure 3. Hardness values of points on the center line

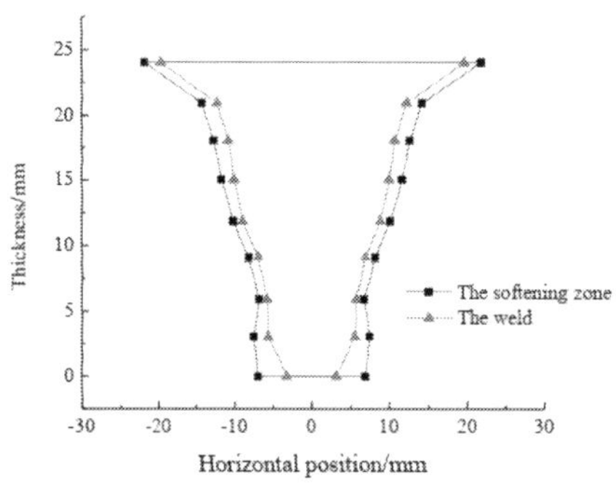

Figure 4. The geometry of welded joint

2.2.4. Bending Test. The bending test conditions are as follows: the demensions of the welded pipe are Φ1219mm × 26.4mm, design coefficient is 0; the length is 10m; the overmatching strength of for weld metal is 10%. The bending test was carried out at the Renqiu branch of Petro China. The degree of ellipticity near the weld was measured with the actual bending angle of 5.5 degrees.

3. Numerical simulation

3.1. Geometric model
The 1/2 two-dimensional model of weld joint is shown in Figure 5, which is built according to the geometry of the etched weld joint.

3.2. Meshing
Actually, it is unnecessary to mesh the model with all fine mesh. To improve calculation efficiency, regions with large deformation or strain concentration should be divided by using fine mesh. Figure 6 shows the finite element mesh of the 2D in weld joint.

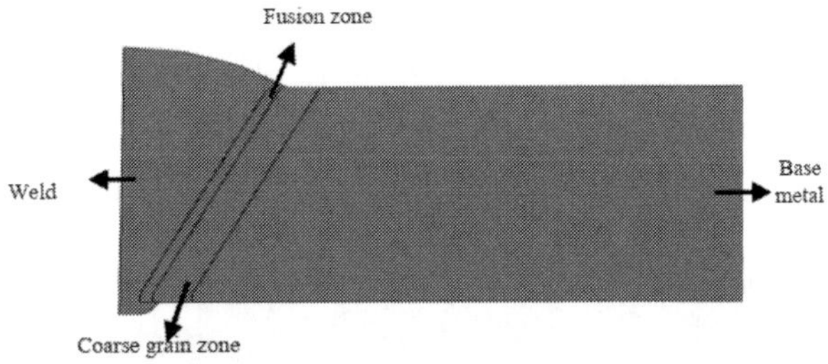

Figure 5. 2D dimension in weld joint

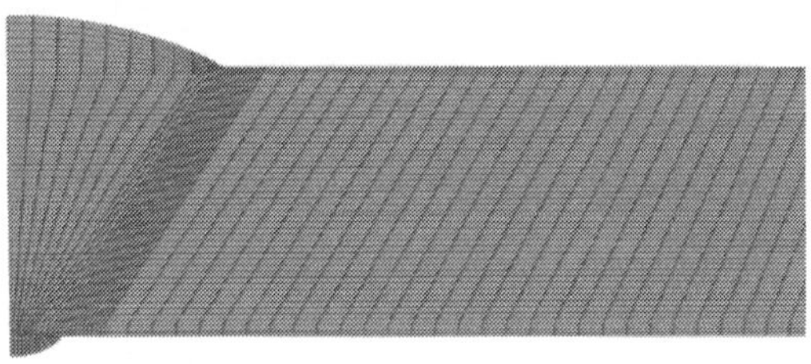

Figure 6. Mesh in weld joint

3.3. Boundary conditions
Two reference points are set on both sides of the steel pipe, which are set up with two surfaces to establish kinematic pairs and constraints. The axial tensile displacement is applied to the reference points on both sides of the model in first stage. Then the bending load of pipe is realized by extrusion with a rigid body, while internal pressure was loaded in the second stage. Figure 7 shows the boundary condition of the whole model.

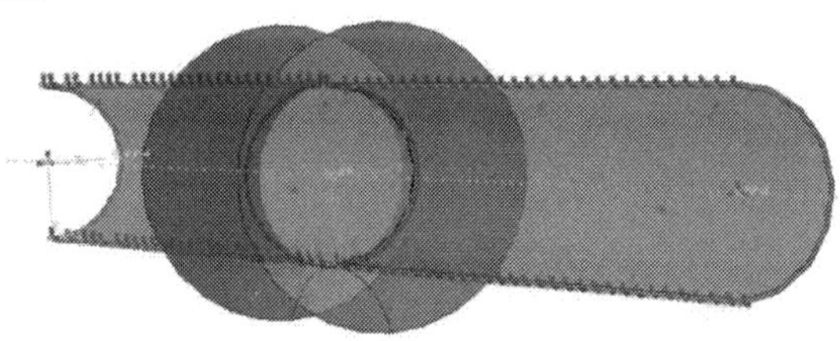

Figure 7. The whole simulation model with boundary condition

3.4. Finite element verification
As shown in Figure 8, the simulation results of bending test show that with the displacement of the reference points increasing, the force occurs rises rapidly until the displacement of 500mm, then slowly.

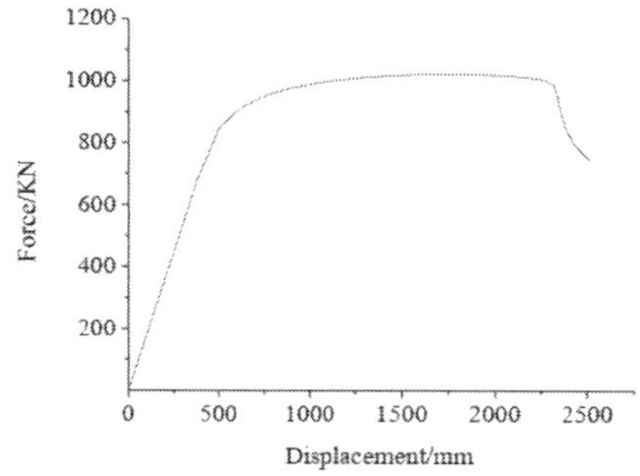

Figure 8. Simulated force-displacement curve of the tension bending test

When the load reaches the peak value, buckling phenomenon occurs, then the applied load decreases quickly. When the displacement reaches about 2276mm, the pipeline buckling occurs shown in the Figure 9.

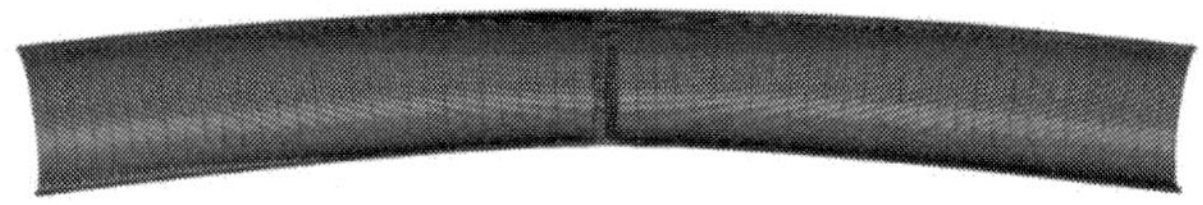

Figure 9. Buckling phenomenon in the pipe

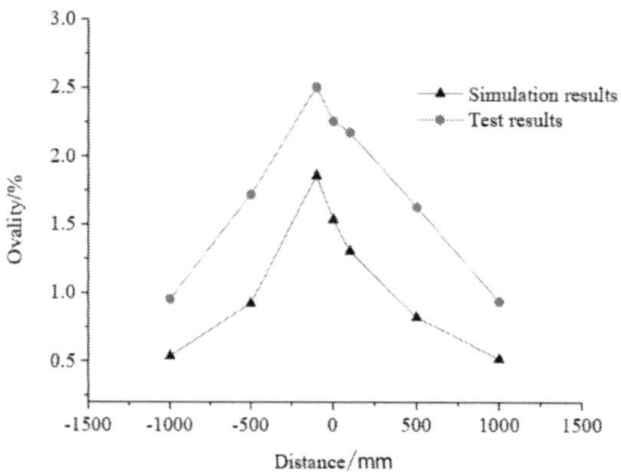

Figure 10. Simulation and measured results of bending test

With the bending angle of 5.5 degrees, the ovality was determined at the same position with the actual bending test. The simulation and test results are shown in Figure 10. The peak values of the ovality exits in the weld zone. The ovality decreases with increasing the distance from the weld zone, which is in coherent with the theory of buckling deformation. On one hand, for the simulation process, the initial ovality of the simulation is 0%. The bending process is carried out according to the ideal process and the simulated results are smaller. On the other hand, many factors affect the change of ellipticity in bending process such as residual stress and residual strain, which leads to large deformation and large ovality.

Taking into account the actual factors, the simulated results compared with the experimental ones are similar, which shows the validity and accuracy of the model.

4. Results and discussion

4.1. Evaluation of softening zone width

The dimensions of the pipeline are Φ1219mm × 16.3 mm, the groove angle is 24°, the align deviation value is 0, the weld overmatching strength of weld metal is 10%, softening rate of the coarse grain zone is 10%, the design coefficient is 0 and the internal pressure is 11.52 MPa. The deformation processes of weld pipe with softening zone width of 1mm, 2mm, 3mm, 4mm and 5mm were simulated. Figure 11 shows strain contour figure of weld pipe with softening zone width of 4mm. Under tension-bending load, the tensile and compressive strain of the five different weld pipes are similar. As shown in Figure 12, with the displacement of the reference points increasing, the force that the pipe bears rises rapidly and then slowly. When the displacement reaches about 180mm, the peak values of load presents and buckling phenomenon occurs, while the applied load decreases quickly. Force-displacement curves of the five weld pipes are almost coincident before buckling phenomenon occurs. In other words, their peak loads are basically the same. After pipeline buckling, the decreasing rates of the loads on the weld pipe with softening zone widths of 1mm, 2mm, 3mm and 4mm are equal to each other, while that of the weld pipe with softening zone width of 5mm is slower than the other pipes.

In conclusion, softening zone width has little influence on actual performance of the pipelines under tensile-bending load due to wide cap welding. The softening zone is covered with higher strength materials through wide cap welding. Therefore, it is better to use wide cap welding considering the premise of economic and technical conditions.

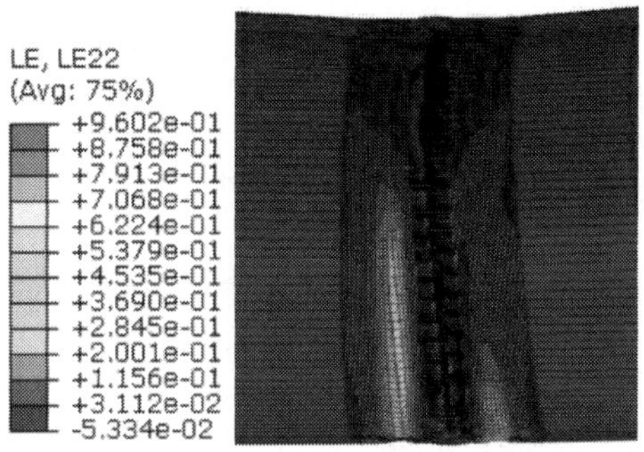

Figure 11. Strain contour figure of weld pipe with softening zone width of 4mm

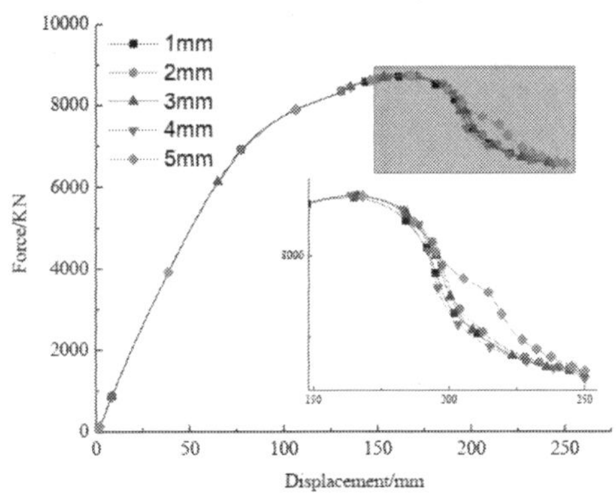

Figure 12. Force-displacement curves of weld pipes with different softening zone widths

4.2. Evaluation of softening rate

The dimensions of the pipeline are Φ1219mm × 16.3 mm, the groove angle is 24°, the align deviation value is 0, the weld overmatching strength of weld metal is 10%, softening zone width is 3mm, the design coefficient is 0 and the internal pressure is 11.52 MPa. The deformation processes of weld pipes whose softening rate are 5%, 10% and 15% were simulated.

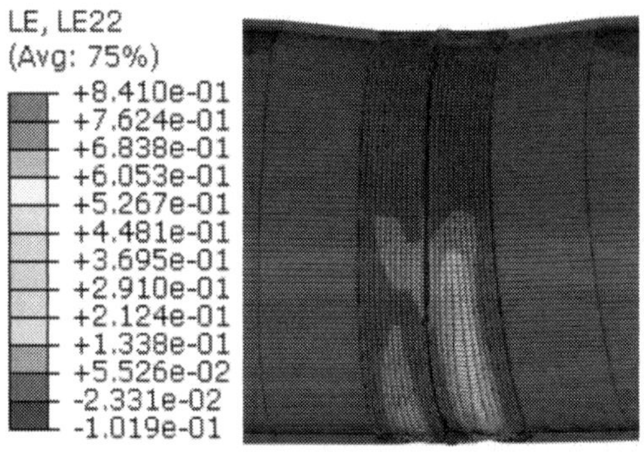

Figure 13. Strain contour figure of weld pipe with 10% softening rate

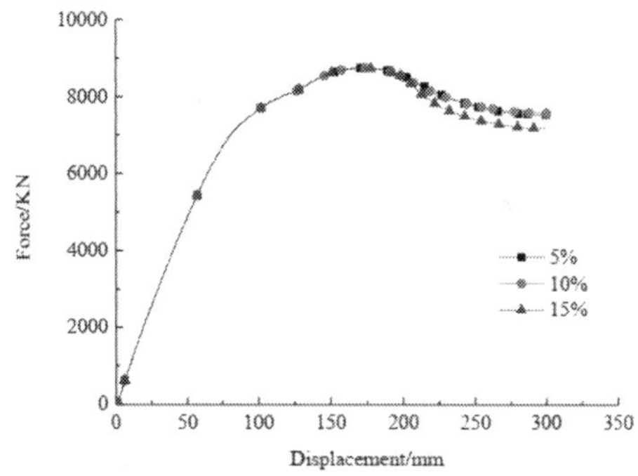

Figure 14. Force-displacement curves of different softening rates

Figure 13 shows strain contour figure of weld pipe with 10% softening rate. Under tension-bending load, the tensile and compressive strain of weld pipes with softening rates 5% and 10% are similar. The weld pipe whose softening rate is 15% has a higher value compared with another two in tensile and compressive strain. As shown in Figure 14, with the displacement of the reference points increasing, the force that the pipe bears rises rapidly and then slowly. When the displacement reaches about 190mm, the peak values of load presents and buckling phenomenon occurs while the applied load decreases quickly. Force-displacement curves of the three weld pipes are almost coincident before buckling phenomenon occurs. In other words, their peak loads are basically the same. The pipeline with softening rate 15% is faster and earlier than the other pipelines in load drop and in loss of performance. The load of the pipelines with softening rate 5% and 10% whose decreasing rates are approximate is higher than another one after buckling phenomenon occurs.
In conclusion, when the softening rates are 5%, 10% and 15%, the peak values of load are approached. However, structural instability of the pipeline with softening rate 15% is worse and more serious, and it has a greater strain value.

4.3. Evaluation of design coefficient

The dimensions of the pipeline are Φ1219mm × 16.3 mm, the groove angle is 24°, the align deviation value is 0, the weld overmatching strength of weld metal is 10%, softening zone width is 3mm. The deformation processes of weld pipe with design coefficients of 0 and 0.5 were simulated. When the design coefficient is 0, the pipeline internal pressure is 0, when the design coefficient is 0.5, the pipeline internal pressure is 11.52 MPa.

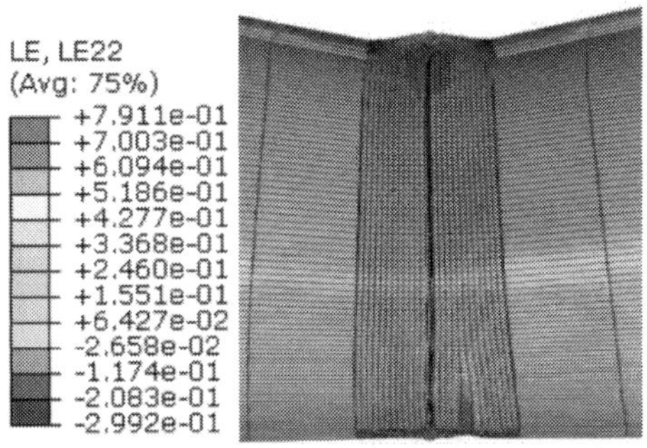

Figure 15. Strain contour figure of weld pipe with the design coefficient 0

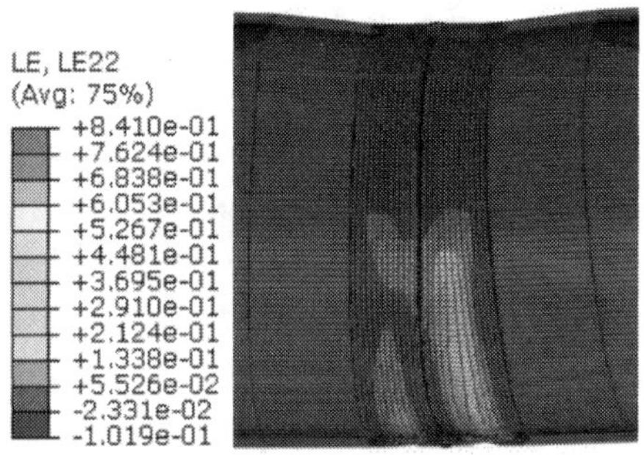

Figure 16. Strain contour figure of weld pipe with the design coefficient 0.5

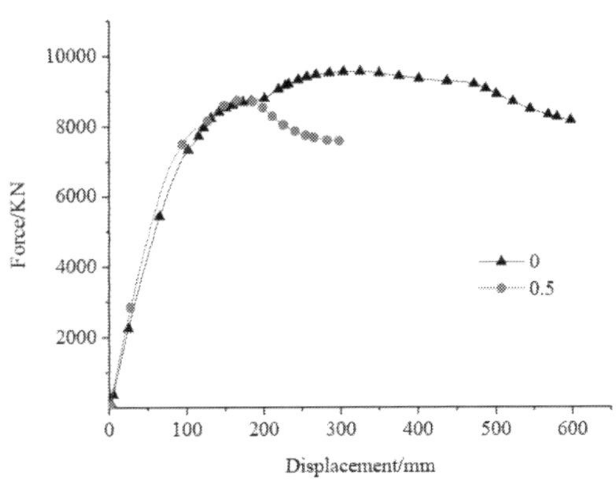

Figure 17. Force-displacement curves of weld pipes with different design coefficients

Figure 15 and Figure 16 are strain contour figures of weld pipes with design coefficient 0 and 0.5. Under tension load combined with bending load, the tensile strain of weld pipe with design coefficient 0.5 is bigger than the pipe of design coefficient 0 while the compressive strain is on the contrary. And large deformation was observed at compression side in Figure 15. As the Figure 17 shows, with the displacement of the reference points increasing, the force that the pipes bear rises rapidly, then slowly. When the displacement reaches about 190mm, the peak load of design coefficient 0.5 pipe's presents and buckling phenomenon happens with structural instability, then the applied load decreases quickly. However, the load of the pipe with design coefficient 0 increases until the displacement reaches about 300mm. Force-displacement curves of the two pipes are almost coincident before buckling phenomenon occurs. The pipeline with design coefficient 0.5 is faster and earlier than the another in load drop and in loss of performance. The load of the pipeline with design coefficient 0 is higher than the other after buckling phenomenon occurs.

In conclusion, the pipeline with design coefficient 0 can withstand greater deformation under tension load combined with bending load, and it can keep good performance.

5. Summary

(1) Taking into account the actual factors, simulated results were compared with experimental ones, which show the validity and accuracy of the model.
(2) The softening zone width has little influence on performance of the pipelines under the effect of tensile-bending load. It is better to use wide cap welding considering the premise of economic and service conditions.
(3) When the softening rates are 5%, 10% and 15%, the peak values of load are approached. However, structural instability of the pipeline with softening rate of 15% is faster and worse while it has a greater strain value.
(4) The pipe with design coefficient of 0 can withstand greater deformation under tension-bending load. So it can keep good service performance.

6. Acknowledgements

The authors are grateful for the financial support from the National Natural Science Foundation of China (No. 51205342 and No. 51275445) and Natural Science Foundation of Hebei Province of China (No. E2016203217).

7. Reference

[1] Dake Yi, Zhong Min Xiao, Sridhar Idapalapati, Shashi Bhushan Kumar. Fracture analysis of girth welded pipelines with 3D embedded cracks subjected to biaxial loading conditions[J]. *Engineering Fracture Mechanics,* 2012, **96**: 570 – 587.

[2] LI He-lin, LI Xiao, JI Ling-kang, CHEN Hong-da. Strain-based Design for Pipeline and Development of Pipe Steels with High Deformation Resistance [J]. *Welded Pipe and Tube,* 2007, **30** (5): 57-62.

[3] HU Meijuan, WANG Peng, HAN Xinli, JI Lingkang. Microstructure and properties of coarse grain region for high-strain pipeline X80 steel [J]. *Transactions of the china welding institution,* 2012, **33** (9): 5-13.

[4] You Zong-bin, LI Ye-zheng, LIU Yu. Thermal simulation for effect of welding peak temperature on properties of welded joint heat affected zone of X80 pipeline steel[J]. *Material for Mechanical Engineering.* 2016, Vol. **40** NO. 9 : 54-57, 61.

[5] SHI Lei, YANG Yong-he, XU Zhen, CHEN Xu. Mechanical properties of welded joint of domestic X80 pipeline steel pipes[J]. *Material for Mechanical Engineering.* 2015, **39** (7): 99-103, 116.

[6] Motohashi H, Hagiwara N. Effect of strength matching and strain hardening capacity on fracture performance of X80 line pipe girth welded joint subjected to uniaxial tensile loading[J]. *Journal of Offshore Mechanics and Arctic Engineering,* 2007, **129**(4) : 318-326.

[7] ZHENG Xiaofei, KANG Yonglin, MENG Deliang. Effect of cooling start temperature on microstructure and mechanical properties of X80 high deformability pipeline steel[J]. *Journal of Iron and Steel Research International,* 2011,**18** (10): 42-46.

[8] A. BASTOLA, J. WANG, H. SHITAMOTO, A. MIRZAEE-SISAN, M. HAMADA and N. HISAMUNE. Full- and small-scale tests on strain capacity of X80 seamless pipes[J]. *ScienceDirect,* 2016, **2**: 1894-1903.

[9] Fengping Yang, Chunyong Huo, Jinheng Luo, He Li, Yang Li. Crack propagation and arrest simulation of X90 gas pipe[J]. *International Journal of Pressure Vessels and Piping,* 2017, **149**: 120 - 131.

[10] Satoshi Igi, Takahiro Sakimoto, Shigeru Endo. Effect of Internal Pressure on Tensile Strain Capacity of X80 Pipeline[J]. *ScienceDirect,* 2011, **10**: 1451-1456.